M000231841

NEXT GENERATION IPTV SERVICES AND TECHNOLOGIES

BICENTENNIAL
1807
⊛WILEY
2007
BICENTENNIAL

THE WILEY BICENTENNIAL—KNOWLEDGE FOR GENERATIONS

*E*ach generation has its unique needs and aspirations. When Charles Wiley first opened his small printing shop in lower Manhattan in 1807, it was a generation of boundless potential searching for an identity. And we were there, helping to define a new American literary tradition. Over half a century later, in the midst of the Second Industrial Revolution, it was a generation focused on building the future. Once again, we were there, supplying the critical scientific, technical, and engineering knowledge that helped frame the world. Throughout the 20th Century, and into the new millennium, nations began to reach out beyond their own borders and a new international community was born. Wiley was there, expanding its operations around the world to enable a global exchange of ideas, opinions, and know-how.

For 200 years, Wiley has been an integral part of each generation's journey, enabling the flow of information and understanding necessary to meet their needs and fulfill their aspirations. Today, bold new technologies are changing the way we live and learn. Wiley will be there, providing you the must-have knowledge you need to imagine new worlds, new possibilities, and new opportunities.

Generations come and go, but you can always count on Wiley to provide you the knowledge you need, when and where you need it!

WILLIAM J. PESCE
PRESIDENT AND CHIEF EXECUTIVE OFFICER

PETER BOOTH WILEY
CHAIRMAN OF THE BOARD

NEXT GENERATION IPTV SERVICES AND TECHNOLOGIES

GERARD O'DRISCOLL

WILEY-INTERSCIENCE
A JOHN WILEY & SONS, INC., PUBLICATION

Published by John Wiley & Sons, Inc., Hoboken, New Jersey.

Published simultaneously in Canada.

For general information on our other products and services please contact our Customer Care Department within the U.S. at 877-762-2974, outside the U.S. at 317-572-3993 or fax 317-572-4002.

Wiley also publishes its books in a variety of electronic formats. Some content that appears in print, however, may not be available in electronic format.

Wiley Bicentennial Logo: Richard J. Pacifico

Library of Congress Cataloging-in-Publication Data:

O'Driscoll, Gerard.
    Next generation IPTV services and technologies / by Gerard O'Driscoll.
        p. cm.
    Includes index.
    ISBN 978-0-470-16372-6 (cloth)
1. Internet television. I. Title.
  TK5105.887.O37 2008
    621.388–dc22   2007029092

Printed in the United States of America.
10 9 8 7 6 5 4 3 2 1

This book is dedicated to my loving wife Olive and our three precious children; princess number 1 (Aoife AKA our Baby Fifes), princess number 2 (Ciara our little rascal), and of course the new boss in the house baby Ger (AKA Gerdie). Also a big dedication goes to my mother and father living in Dear Old Skibbereen, County Cork; my two young Celtic Cub brothers — Owen and Brian; Sarah Maddie, Ruairi, and baby Alice (sister-in-law, nieces and nephew); and finally, for my old drinking buds in *Electronic Production*!

CONTENTS

Preface   xi
    Organizational and Topical Coverage   xii
    Who Should Read This Book   xiii
    Acknowledgments   xiv
About the Author   xvii

1 IPTV: The Ultimate Viewing Experience   **1**

    1.1   Defining IPTV   2
    1.2   Differences between IPTV and Internet TV   3
    1.3   Overview of an IPTV Networking Infrastructure   4
    1.4   Key IPTV Applications and Services   6
    1.5   Growth Drivers for IPTV   10
    1.6   Market Data   12
    1.7   Industry Initiatives to Standardize IPTV   13
         Summary   18

2 IPTV Network Distribution Technologies   **20**

    2.1   "Last Mile" Broadband Distribution Network Types   21
    2.2   IPTV over a Fiber Access Network   21
    2.3   IPTV over an ADSL Network   26
    2.4   IPTV over Next Generation Cable TV Networks   32
    2.5   IPTV over Wireless Networks   48
    2.6   IPTV over the Internet   53
    2.7   IPTV Backbone Technologies   56

2.8   Network Factors Associated with Deploying IPTV   60
      Summary   62

3 IPTV Real-Time Encoding and Transportation   64

3.1   Introduction to Real-Time Encoding   64
3.2   Compression Methods   66
3.3   Packetizing and Encapsulating Video Content   81
      Summary   116

4 Broadcasting Linear Programming over IPTV   118

4.1   Underlying Video Components of an End-to-End
      IPTV System   119
4.2   Different Approaches to Streaming IPTV Content   124
4.3   Multicasting across an IPTV Network   129
4.4   IPTV Multicasting Networking Architecture   130
4.5   Multicasting IPTV Content across IPV6 Networks   155
4.6   Introduction to Channel Changing   158
4.7   Fundamentals of Channel Changing   163
4.8   Techniques for Speeding up IPTV Channel Changing Times 166
4.9   Discovering Channel Program Information   169
4.10 Time-Shifting Multicast IPTV   172
4.11 Channel-Changing Industry Initiatives   173
      Summary   173

5 IPTV Consumer Devices (IPTVCDs)   175

5.1   About Residential Gateways   175
5.2   RG Technology Architecture   177
5.3   RG Functionality   178
5.4   RG Industry Standards   184
5.5   Introduction to Digital Set-top Boxes   193
5.6   The Evolution of Digital Set-top Boxes   195
5.7   Categories of Digital Set-top Boxes   197
5.8   Major Technological Trends for Digital Set-Top Boxes   197
5.9   IP Set-top Boxes Defined   200
5.10 Types of IP Set-top Boxes   201
5.11 Other Emerging IPTV Consumer Devices   223
      Summary   224

6 IPTVCD Software Architecture   229

6.1   What Makes an IPTVCD Tick?   229
6.2   Interactive TV Middleware Standards   232
6.3   Proprietary Middleware Solutions   247
      Summary   247

7   IPTV Conditional Access and DRM Systems   249

    7.1   Introduction to IPTV Security   249
    7.2   Defining IPTV CA Security Systems   250
    7.3   CA Industry Initiatives   263
    7.4   Introduction to Next Generation DRM Solutions   265
    7.5   IPTV Intranet Protection   282
        Summary   283

8   Moving IPTV Around the House   285

    8.1   About Whole Home Media Networking (WHMN)   286
    8.2   WHMN Enabling Technologies   287
    8.3   Fast Ethernet and Gigabit Ethernet (GIGE)   288
    8.4   802.11n   294
    8.5   HomePlug AV   298
    8.6   UPA-DHS   305
    8.7   HomePNA^{TM} 3.1   307
    8.8   Multimedia Over Coax Alliance (MOCA^{TM})   311
    8.9   WHMN Middleware Software Standards   314
    8.10 QoS and WHMN Applications   328
        Summary   329

9   Video-on-Demand (VoD) over IP Delivery Networks   332

    9.1   History of Pay-Per-View   332
    9.2   Understanding PPV   333
    9.3   The Emergence of RF and IP Based VoD   334
    9.4   Types of IP-VoD Services   335
    9.5   Underlying Building Blocks of an End-to-End
        IP-VoD Infrastructure   340
    9.6   Integrating IP-VoD Applications with Other
        IP Based Services   358
    9.7   Protecting IP-VoD Content   364
        Summary   365

10 IP Based High Definition TV   367

    10.1 Overview of SDTV and HDTV Technologies   367
    10.2 HDTV over IP Defined   369
    10.3 An End-to-End IP HDTV System   370
        Summary   380

11 Interactive IPTV Applications   382

    11.1 The Evolution of iTV   382
    11.2 About iTV   387

11.3   Interactive IPTV Applications   387
11.4   IPTV Program Related Interaction Applications   417
11.5   Deploying Interactive IPTV Applications   419
11.6   Accessibility to IPTV Services   420
        Summary   420

12 IPTV Network Administration   423

12.1   An Introduction to IPTV Network Administration   424
12.2   Supporting the IPTV Networking Management
        System   424
12.3   Managing Installation, Service Problems,
        and Terminations   428
12.4   Network Testing and Monitoring   431
12.5   Managing Redundancy and Ensuring Service Availability   434
12.6   IP Address Space Management   436
12.7   Routine IT and Network Administrative Tasks   436
12.8   Managing IPTV QoS Requirements   437
12.9   Monitoring the IPTV Subscriber Experience   441
12.10   Remotely Managing in-Home Digital Consumer Devices   469
12.11   Scheduling and Managing Delivery of Software
        Updates to IPTVCDs   471
12.12   Troubleshooting IPTV Problems   473
12.13   IPTV and Business Continuity Planning   478
        Summary   481

Index   483

PREFACE

The highly anticipated IPTV industry sector has become a reality. Commercial deployments of IPTV services by telecommunication companies around the world continue to increase.

Not only has IPTV become a proven technology that allows telecom companies to deploy advanced services such as high quality multicast IPTV channels, IP based HDTV, and Whole Home Media Networking (WHMN) services but it also provides providers with new streams of revenue. Other advantages of evolving IPTV services include personalization and immediate access to a wide variety of on-demand digital content.

For wireline and wireless telcos that have already moved into the video services' sector, IPTV has the potential to generate additional revenue stream.

Although cable and satellite providers have already made significant investments into non-IP-based set-top boxes, networking infrastructure, and headend equipment, the migration to an IPTV platform is expected to accelerate over the next decade.

The deployment of IPTV services poses a host of unique operational challenges for telecoms, cable, and satellite TV providers. First and foremost, service providers have to make difficult decisions when choosing between the myriad of encoding, Digital Rights Management (DRM), set-top box, networking infrastructures, and security solutions.

Second, a commercially viable IPTV system needs to be streamlined and effectively supported on a day-to-day basis.

Finally, today's IPTV systems require technologies that deliver video content to end users in a manner that provides high quality of experience levels during the consumption of TV services.

These three primary challenges come against a backdrop where customer expectations are at an all-time high.

The mission of this book is to aid its readers in addressing these challenges and meeting the demands of designing, implementing, and supporting end-to-end IPTV systems.

Furthermore, the publication provides global sectors with a detailed technical analysis of deploying and managing end-to-end IPTV systems.

ORGANIZATIONAL AND TOPICAL COVERAGE

IPTV is a new method of delivering digital video and audio content across an IP broadband network. Chapter 1 defines IPTV and presents an overview of the networking infrastructure typically used by IP based video services. Growth drivers, 5-year market forecasts, and industry initiatives for the sector are also outlined.

A wide variety of network delivery technologies are available to provide IPTV services to end users. The second chapter focuses on the six types of broadband access networks commonly used to transport IPTV services and applications.

Encoding is one of the core functions associated with preparing video content for transmission across an IP network. Chapter 3 addresses readers who wish to gain an in-depth understanding of the various compression technologies used by IPTV systems. This chapter also details the communication protocols used by end-to-end IPTV networking systems.

In multicasting a number of users can receive the same video through a single stream with the help of routers and industry standard protocols. Chapter 5 covers the various logical and physical components required to deploy multicasting services across both IPv4 and IPv6 networks. The final section of this chapter deals with issues that affect the channel changing process and identifies various techniques, which may be used by service providers to speed up channel changing times.

The latest IPTV Consumer Devices (IPTVCDs) include a confluence of several new technologies, including, multicore processors, and hard disks, not to mention high capacity home networking interfaces. The fifth chapter in this book deals extensively with several contenders in the IPTVCD marketplace—residential gateways, IP set-top boxes, game consoles, and media servers.

The development of open interactive and IPTV standards has started to pave the way for the deployment of a whole range of next generation interactive IPTV based services. Many of the existing interactive TV standards are being extended to support the delivery of IPTV services. Chapter 6 overviews these standards—DVB's MHP, CableLabs' OCAP, the GEM specification, and ATSC's Digital ACAP.

Content security is a critical issue for the IPTV industry sector. From a service provider's perspective it is no longer acceptable to simply deploy a conditional access system to ensure that only authorized subscribers can access IPTV VoD and broadcast TV services. Content providers are now insisting that service providers incorporate advanced DRM systems into their end-to-end IPTV networking platforms, which will ensure that the content delivered to consumers is done so in a protected format. A DRM system is used by service providers to maintain control over the distribution IPTV content. Chapter 7 examines the hardware and software architectures required to implement both of these security systems.

Adding WHMN applications to a home network dramatically increases the demand for higher transmission rates. There are a number of interconnection technologies available that allow service providers to effectively implement WHMN services across their subscriber's in-home networks. Chapter 8 takes a detailed look at six of these competing technologies—GigE, 802.11N, HomePlugAV, UPS-DHA, HomePNA, and MoCA.

IP VoD is a core IPTV service that allows end users access a library of on-demand content. In addition to receiving immediate access to various on-demand titles, IP-VoD also allows end users to perform VCR functions on the video streams. IP-VoD is one of the most important services offered by IPTV service providers and is covered in Chapter 9.

Technology advancements in compression, backend servers, security, and IP set-top boxes combined with increased consumer demand are increasing the number of HDTV over IP broadband networks across the globe. Chapter 10 describes the building blocks that comprise an end-to-end IP HDTV system.

The two way capabilities of next generation IPTV networks allows for the deployment of a range of interactive TV applications. Chapter 11 describes 16 of the most popular interactive IPTV applications including caller ID for TVs, EPGs, and IPTV e-mail services.

To ensure that IPTV services compare favorably with existing pay TV providers and to ensure that quality of the viewing experience for end-users of the service is high, IPTV providers need to effectively manage their networking infrastructure. This is a major challenge, which needs to be met through the use of sophisticated network administration methodologies. Chapter 12 outlines a number of engineering and operational functions that are an essential part of delivering high quality video services to end users.

WHO SHOULD READ THIS BOOK

Job title	Role within the organization
System Integrators	There are a variety of companies offering turnkey IPTV system integration services ranging from the large telecommunications manufacturers firms such as Alcatel, Motorola and Lucent to the smaller software based companies that provide professional services to the smaller telco operators.
	Typical job functions for engineers working with these companies range from integrating encoders and conditional access systems at the headend to deploying video servers and IP video monitoring tools.

Job title	Role within the organization
Engineering directors and managers	These are people who work in telecommunication and cable TV companies who are responsible for overseeing the technical and operational aspects of deploying an IPTV networking infrastructure.
Technicians and engineering staff	The *Next Generation IPTV Services and Technologies* book will be a useful reference book for technicians who are involved in installing, maintaining, repairing, and troubleshooting the hardware and software functionality of IPTV products.
IPTV project managers	These are people who have responsibility for developing and launching IPTV and interactive television applications across broadband networks.
Software engineers	This group of engineers are typically involved in developing interactive TV applications for Film and TV companies that produce content for delivery over IP based networks.

Students and academics on postgraduate courses related to telecommunications, especially networking or IP protocols, will also find the *Next Generation IPTV Services and Technologies* book ideal for supplementary reading.

ACKNOWLEDGMENTS

I would like to take the opportunity to thank the many people that provided assistance and input into the creation of this book:

Paul Petralia, *Senior Editor at Wiley Interscience*
Shaheed Haque, *Director of Development at Microsoft TV EMEA*
Joel E. Welch, *Director Certification & Program Development at the SCTE*
Mark Rooney, *Head of IPTV at Pace Micro Technology plc*
Robert Gelphman, *Chair of MoCA Marketing Work Group*
Mike Schwartz, *Senior VP of communications for CableLabs*
Allen R. Gordon, *Senior Software Engineer at CableLabs*
Greg White, *Engineer at CableLabs*
Stuart Hoggan, *Engineer at CableLabs*
Richard Nesin, *President of the HomePNA alliance*
Brian Donnelly, *VP Marketing at Corinex*
Bruce Watkins, *President at Pulse~LINK, Inc.*
Prof. Dr.-Ing. Ulrich H. Reimers, *Chairman of the Technical Module of the DVB*
Dotan Rosenberg, *Product manager at Bitband*
Karthik Ranjan, *VP of InternetTV at Amino*
Karen Moore, *Executive Director at Coral Consortium*

Bill Foote Chair, *DVB-MUG Blu-ray Java Architect, Sun Microsystems*
Meredith Dundas, *Marketing Communications Manager at Espial IPTV*
Ekta Handa, *Project Manager for Thomson Digital and production team*
Danielle Lacourciere, *Associate Managing Editor at John Wiley & Sons, Inc*

REFERENCES

The Internet Engineering Task Force (IETF) Request for Comments Repository—www.ietf.org

Alliance for telecommunications Industry Solutions (ATIS) technical documents—www.atis.org

Internet Streaming Media Alliance technical documents—www.isma.tv

CableLabs standard specification documents—www.cablelabs.com

Juniper Networks white papers—www.juniper.net

WiMAX Forum technical documents—www.wimaxforum.org

Ixia product datasheets—www.ixiacom.com

Home Gateway Initiative technical requirements documents—www.homegatewayinitiative.org

HomePlug Powerline Alliance technology white papers—www.homeplug.org

Multimedia over Coax Alliance (MoCA™) presentations and white papers—www.mocalliance. org

Universal Powerline Association (UPA) technical documentation—www.upaplc.org

UPnP™ Forum standards and specifications—www.upnp.org

IneoQuest application notes—www.ineoquest.com

Cisco IPTV implementation guides and manuals—www.cisco.com

International Telecommunication Union Focus Group on IPTV output documents—www.itu.int/ITU-T/IPTV/

DSL forum technical reports—www.dslforum.org

Envivio product whitepapers—www.envivio.com

Espial product and technical documentation—www.espial.com

European Telecommunications Standards Institute (ETSI) technical specifications—www.etsi.org

ABOUT THE AUTHOR

Gerard O'Driscoll, is an accomplished international telecom-
munications expert, entrepreneur, and globally renowned
authority on emerging technologies. Over the past 15 years
Gerard has served in a variety of management, engineering,
and commercial positions. He has worked across the full
spectrum of technology equipment and services, including
broadband, digital TV headend systems, IP networks,
home networking, and enterprise IT systems.

O'Driscoll is a frequent commentator on industry trends in the cable, telecoms
and digital home industry sectors and has been quoted in a number of premier business publications. In addition, he speaks regularly at leading telecommunication
industry events. Gerard's other professional achievements include authoring of
books on topics ranging from set-top boxes to home networking technologies
and the publication of several market research reports related to the IPTV industry
section.

Gerard holds electronics and information technology qualifications from the
University of Limerick in Ireland. Author contact details are available at
www.tvmentors.com.

1

IPTV: THE ULTIMATE VIEWING EXPERIENCE

Digital Television, also known as Digital TV, is the most significant advancement in television technology since the medium was created over a century ago. Digital TV offers consumers more choice and makes the viewing experience more interactive. The analog system of broadcasting television has been in place for well over 60 years. During this period, viewers experienced the transition from black-and-white sets to color TV sets. The migration from black-and-white television to color television required viewers to purchase new TV sets, and broadcasters had to acquire new transmitters, pre, and post production equipment. Today, the industry is going through a profound transition, migrating from conventional TV to a new era of digital technology. Most TV operators have upgraded their existing networks and have deployed advanced digital platforms in an effort to migrate their subscribers away from traditional analog services to more sophisticated digital services. A new technology called Internet Protocol-based television (IPTV), has started to grab headlines across the world with stories about several large telecommunication, cable, satellite, terrestrial, and a slew of Internet start-ups delivering video over an IP based service. As the name suggests, IPTV describes a mechanism for transporting a stream of video content over a network that uses the IP networking protocol. The benefits of this mechanism of delivering TV signals vary from increased support for interactivity to faster channel changing times and improved interoperability with existing home networks. Before describing the various technologies that make up an end-to-end IPTV system, this chapter will start by defining IPTV. The growth drivers for the industry sector are then examined, and

Next Generation IPTV Services and Technologies, By Gerard O'Driscoll
Copyright © 2008 John Wiley & Sons, Inc.

the chapter concludes with a review of the main organizations developing standards for the industry.

1.1 DEFINING IPTV

There is a lot of buzz and excitement at the moment with regard to IPTV. The technology is growing in importance and is starting to have a disruptive effect on the business models of traditional pay TV network operators.

But what does the IPTV acronym mean and how will it affect TV viewing? For a start, IPTV, also called Internet Protocol Television, Telco TV, or broadband TV, is about securely delivering high quality broadcast television and/or on-demand video and audio content over a broadband network. IPTV is generally a term that is applied to the delivery of traditional TV channels, movies, and video-on-demand content over a private network. From an end user's perspective, IPTV looks and operates just like a standard pay TV service. The official definition approved by the International Telecommunication Union focus group on IPTV (ITU-T FG IPTV) is as follows:

> IPTV is defined as multimedia services such as television/video/audio/text/graphics/ data delivered over IP based networks managed to provide the required level of quality of service and experience, security, interactivity and reliability.

From a service provider's perspective, IPTV encompasses the acquisition, processing, and secure delivery of video content over an IP based networking infrastructure. The type of service providers involved in deploying IPTV services range from cable and satellite TV carriers to the large telephone companies and private network operators in different parts of the world.

IPTV has a number of features:

- *Support for interactive TV*—The two-way capabilities of IPTV systems allow service providers to deliver a whole raft of interactive TV applications. The types of services delivered via an IPTV service can include standard live TV, high definition TV (HDTV), interactive games, and high speed Internet browsing.
- *Time shifting*—IPTV in combination with a digital video recorder permits the time shifting of programming content — a mechanism for recording and storing IPTV content for later viewing.
- *Personalization*—An end-to-end IPTV system supports bidirectional communications and allows end users personalize their TV viewing habits by allowing them to decide what they want to watch and when they want to watch it.
- *Low bandwidth requirements*—Instead of delivering every channel to every end user, IPTV technologies allows service providers to only stream the channel that the end user has requested. This attractive feature allows network operators to conserve bandwidth on their networks.

- *Accessible on multiple devices*—Viewing of IPTV content is not limited to televisions. Consumers often use their PCs and mobile devices to access IPTV services.

1.2   DIFFERENCES BETWEEN IPTV AND INTERNET TV

IPTV is sometimes confused with the delivery of Internet TV. Although both environments rely on the same core base of technologies, their approaches in delivering IP based video differ in the following ways.

1.2.1   Different Platforms

As the name suggests Internet TV leverages the public Internet to deliver video content to end users. IPTV, on the contrary, uses secure dedicated private networks to deliver video content to consumers. These private networks are managed and operated by the provider of the IPTV service.

1.2.2   Geographical Reach

Networks owned and controlled by the telecom operators are not accessible to Internet users and are located in fixed geographical areas. The Internet, on the contrary, has no geographical limitations where television services can be accessed from any part of the globe.

1.2.3   Ownership of the Networking Infrastructure

When video is sent over the public Internet, some of the Internet Protocol packets used to carry the video may get delayed or completely lost as they traverse the various networks that make up the public Internet. As a result, the providers of video over the Internet content cannot guarantee a TV viewing experience that compares with a traditional terrestrial, cable, or satellite TV viewing experience. In fact, video streamed over the Internet can sometimes appear jerky on the TV screen and the resolution of the picture is quite low. The video content is generally delivered to end users in a "best effort" fashion.

In comparison to this experience, IPTV is delivered over a networking infrastructure, which is typically owned by the service provider. Owning the networking infrastructure allows telecom operators to engineer their systems to support the end-to-end delivery of high quality video.

1.2.4   Access Mechanism

A digital set-top box is generally used to access and decode the video content delivered via an IPTV system whereas a PC is nearly always used to access Internet

TV services. The type of software used on the PC will depend on the type of Internet TV content. For instance, downloading to own content from an Internet TV portal site sometimes requires the installation of a dedicated media player to view the material. A robust digital rights management (DRM) system is also required to support this access mechanism.

1.2.5   Costs

A significant percentage of video content delivered over the public Internet is available to consumers free of charge. This is however changing as an increasing number of media companies are starting to introduce fee based Internet TV services. The costing structure applied to IPTV services is similar to the monthly subscription model adopted by traditional pay TV providers. Over time, many analysts expect Internet TV and IPTV to converge into a central entertainment service that will ultimately become a mainstream application.

1.2.6   Content Generation Methodologies

A sizeable portion of video content generated by Internet TV providers is user-generated and niche channels, whereas IPTV providers generally stick with distributing traditional television shows and movies, which are typically provided by the large and established media companies.

1.3   OVERVIEW OF AN IPTV NETWORKING INFRASTRUCTURE

Figure 1.1 shows the typical high level functional requirements of an end-to-end IPTV system.

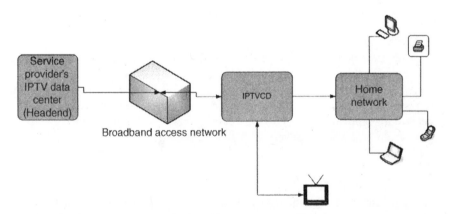

FIGURE 1.1   Simplified block diagram of an end-to-end IPTV system

1.3.1 IPTV Data Center

Also known as the "headend," the IPTV data center receives content from a variety of sources including local video, content aggregators, content producers, cable, terrestrial, and satellite channels. Once received, a number of different hardware components ranging from encoders and video servers to IP routers and dedicated security hardware are used to prepare the video content for delivery over an IP based network. Additionally, a subscriber management system is required to manage IPTV subscriber profiles and payments. Note that the physical location of the IPTV data center will be dictated by the networking infrastructure used by the service provider.

1.3.2 Broadband Delivery Network

The delivery of IPTV services requires a one-to-one connection. In the case of a large IPTV deployment, the number of one-to-one connections increases significantly and the demands in terms of bandwidth requirements on the networking infrastructure can be quite large. Advancements in network technologies over the past couple of years now allow telecom providers to meet this demand for large amounts of bandwidth networks. Hybrid fiber and coaxial based cable TV infrastructures and fiber based telecommunication networks are particularly suited to the delivery of IPTV content.

1.3.3 IPTVCDs

IPTV consumer devices (IPTVCDs) are key components in allowing people to access IPTV services. The IPTVCD connects to the broadband network and is responsible for decoding and processing the incoming IP based video stream. IPTVCDs support advanced technologies that minimize or completely eliminate the effect of network problems when processing IPTV content. As broadband starts to become a mainstream service, the functionality of IPTVCDs continues to change and increase in sophistication. The most popular types of IPTVCDs (residential gateways, IP set-top boxes, game consoles, and media servers) are detailed in Chapter 5.

1.3.4 A Home Network

A home network connects a number of digital devices within a small geographical area. It improves communication and allows the sharing of expensive digital resources among members of a family. The purpose of a home network is to provide access to information, such as voice, audio, data, and entertainment, between different digital devices all around the house. With home networking, consumers can save money and time because peripherals such as printers and scanners, as well as broadband Internet connections, can be easily shared. The home networking market is fragmented into a range of different technologies, which will be covered in Chapter 8.

1.4   KEY IPTV APPLICATIONS AND SERVICES

The two key IPTV applications typically deployed by service providers are broadcast digital TV and Video on demand (VoD).

1.4.1   Broadcast Digital TV

Before going into the world of ones and zeros it is important to take a perspective of where television has come from over the past number of years. The history of television started in 1884 when a German student, Paul Gottlieb, patented the first mechanical television system. This system worked by illuminating an image via a lens and a rotating disk (Nipkow disk). Square apertures (small openings) were cut out of the disk, which traced out lines of the image until the full image had been scanned. The more apertures there were, the more lines were traced and hence the greater the detail.

In 1923, Vladimir Kosma Zworykin replaced the Nipkow disk with an electronic component. This allowed the image to be split into many more lines, which allowed a higher level of detail without increasing the number of scans per second. Images could also be stored between electronic scans. This electronic system was patented in 1925 and was named the *Iconoscope*.

J.L. Baird demonstrated the first color (mechanical) television in 1928. The first mechanical television used a Nipkow disk with three spirals, one for each primary color (red, green, and blue). At the time, very few people had television sets and the viewing experience was less than impressive. The small audience of viewers was watching a blurry picture on a 2- or 3-in. screen.

In 1935, the first electronic television system was demonstrated by a company called Electric Musical Industries (EMI). By late 1939, sixteen companies were making or planning to make electronic television sets in the United States.

In 1941, the National Television System Committee (NTSC) developed a set of guidelines for the transmission of electronic television. The Federal Communications Commission (FCC) adopted the new guidelines and TV broadcasts began in the United States. Television benefited from World War II, in that much of the work done on radar was transferred directly to television set design. One area that was improved greatly was the cathode ray tube.

The 1950s were an exciting time period and heralded the golden age of television. The era of black-and-white television commenced in 1956 and prices of TV sets eventually dropped. Toward the end of the decade, U.S. manufacturers were experimenting with a range of different features and designs.

The 1960s began with the Japanese adoption of the NTSC standards. Toward end of the 1960s, Europe introduced two new television transmission standards:

(1) Systeme Electronique Couleur Avec Memoire (SECAM) is a television broadcast standard in France, the Middle East, and parts of Eastern Europe.

(2) Phase Alternating Line (PAL) is the dominant television standard in Europe.

The first color televisions with integrated digital signal processing technologies were marketed in 1983. At a meeting hosted in 1993, the Moving Picture Experts Group (MPEG) completed a definition of MPEG-2 Video, MPEG-2 Audio, and MPEG-2 Systems. Also in 1993, the European Digital Video Broadcasting (DVB) project was born. In 1996, the FCC established digital television transmission standards in the United States by adopting the Advanced Television Systems Committee (ATSC) digital standard. As of 1999, many communication mediums have transitioned to digital technology. In recent years, a number of countries have started to launch standard definition and high definition TV services and are acting as the primary driving force behind a new type of television systems—liquid crystal display (LCD) panels and plasma display panels (PDPs). A summary of significant historical TV developments is shown in Table 1.1 and illustrated in Fig. 1.2.

1.4.1.1 DTV Formatting Standards The standard for broadcasting analog television in most of North America is NTSC. The standard for video in other parts of the world are PAL and SECAM. NTSC, PAL, and SECAM standards will all be replaced over the next 10 years with a new suite of standards associated with digital television. Making digital television a reality requires the cooperation of a variety of industries and companies, along with the development of many new standards. A wide variety of international organizations have contributed to the standardization of digital TV over the past couple of years. Most organizations create formal standards by using specific processes: organizing ideas, discussing the approach, developing draft standards, voting on all or certain aspects of the standards, and then formally releasing the completed standard to the general public.

TABLE 1.1 TV Development History

Year	Historical Event
1884	Paul Gottlieb, patented the first mechanical television system.
1923	Vladimir Kosma Zworykin replaced the Nipkow disc with an electronic component.
1925	The first TV electronic system was patented.
1935	The first electronic television system was demonstrated by EMI.
1941	The NTSC developed a set of guidelines for the transmission of electronic television.
1956	The era of black and white television commenced.
1993	The European DVB project was founded.
1996	The FCC established digital television trans mission standards in the United States.
1999	Implementation of digital TV systems across the globe.

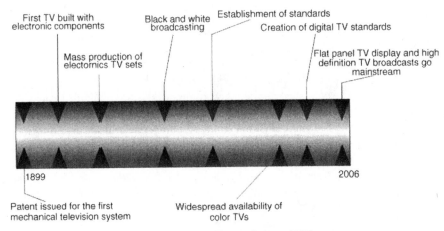

FIGURE 1.2   The evolution of TV

Some of the best-known international organizations that contribute to the standardization of digital television include:

- ATSC
- DVB
- Association of Radio Industries and Businesses (ARIB)

ATSC   The ATSC is an organization that was formed to establish a set of technical standards for broadcasting television signals in the United States. ATSC digital TV standards cover a number of different key broadcasting techniques including the delivery of high definition, standard definition, and satellite direct-to-home signals to homes across the United States. The ATSC was formed in 1982 by the member organizations of the Joint Committee on Intersociety Coordination (JCIC): the Electronic Industries Association (EIA), the Institute of Electrical and Electronic Engineers (IEEE), the National Association of Broadcasters (NAB), the National Cable and Telecommunications Association (NCTA), and the Society of Motion Picture and Television Engineers (SMPTE). Currently, there are approximately 200 members representing the broadcast, broadcast equipment, motion picture, consumer electronics, computer, cable, satellite, and semiconductor industries. ATSC has been formally adopted in the United States where an aggressive implementation of digital TV has already begun. Additionally, Canada, South Korea, Taiwan, and Argentina have agreed to use the formats and transmission methods recommended by the group. For more information on the various standards and specifications produced by this organization visit www.atsc.org.

DVB   The DVB project was conceived in 1991 and was formally inaugurated in 1993 with approximately 80 members. Today DVB is a consortium of around 300 companies in the fields of broadcasting, manufacturing, network operation, and regulatory matters that have come together to establish common international standards for the move from analog to digital broadcasting. The work of the DVB project has resulted in a comprehensive list of standards and specifications that describe solutions for implementing digital television in a variety of different environments. The DVB standards cover all aspects of digital television from transmission through interfacing, security and interactivity for digital video, audio, and data.

Because DVB standards are open, all the manufacturers making compliant systems are able to guarantee that their digital TV equipment will work with other manufacturers' equipment. To date, there are numerous broadcast services around the world using DVB standards. There are hundreds of manufacturers offering DVB compliant equipment, which are already in use around the world. DVB has its greatest success in Europe; however, the standard has its implementations in North and South America, China, Africa, Asia, and Australia. For more information on the various standards and specifications produced by this organization visit www.dvb.org.

ARIB As per the organization's Web site, ARIB conducts studies and research and development, establishes standards, provides consultation services for radio spectrum coordination, cooperates with other overseas organizations, and provides frequency change support services for the smooth introduction of digital terrestrial television broadcasting. The organization has produced a number of standards that are particularly relevant to the digital TV sector, including the video coding, audio coding, and multiplexing specifications for digital broadcasting (ARIB STD-B32). For more information on the various standards and specifications produced by this organization visit http://www.arib.or.jp/english/.

1.4.1.2 Benefits of Digital TV Transmissions

When compared to analog technology the broadcasting of television in computer data format provides digital TV viewers and service providers with a number of benefits.

Improved Viewing Experience The viewing experience is improved through cinema quality pictures, CD quality sound, hundreds of new channels, the power to switch camera angles, and improved access to a range of exciting new entertainment services, additionally, any of the picture flaws that are present in analog systems are absent in the new digital environment.

Improved Coverage Both analog and digital signals get weaker with distance. However, while the picture on an analog TV system slowly gets worse for viewers that live long distances away from the broadcaster, a picture on a digital system will stay perfect until the signal becomes too weak to receive.

Increased Capacity and New Service Offerings By using digital technologies to transmit television, service providers can carry more information than is currently possible with analog systems. With digital TV, a movie is compressed to occupy just a tiny percentage of the bandwidth normally required by analog systems to broadcast the same movie. The remaining bandwidth can then be filled with programming or data services such as

- Video on demand (VoD)
- E-mail and Internet services
- Interactive education
- Interactive TV commerce

Increased Access Flexibility   Traditionally, it was only possible to view broadcast quality analog content on a TV set. With the introduction of digital technologies, video is accessible on a whole range of devices ranging from mobile phones to standard PCs.

Note that eventually, all analog systems will be replaced with digital TV. The transition from analog to digital will be gradual to allow service providers to upgrade their transmission networks and for manufacturers to mass produce digital products for the buying public. In development for more than a decade, the digital TV system that has evolved today is the direct result of work by scientists, technologists, broadcasters, manufacturers, and a number of international standard bodies. Till a couple of years ago it was only practical to use radio frequency (RF) based signal technologies to deliver digital TV to consumers. Recent advancements in compression and broadband technologies are however changing this situation, and many service providers have started to use IP based networks to deliver broadcast digital TV services to their customers.

1.4.2   Video on Demand (VoD)

In addition to allowing telecommunication companies to deliver linear TV channels to their subscribers, IPTV provides access to a wide range of downloadable and VoD based content. In contrast to traditional TV services where video programs are broadcasted according to a preset schedule, VoD provides IPTV end users with the ability to select, download, and view content at their convenience. The content delivered through an IPTV VoD application typically includes a library of on-demand movie titles and a selection of stored programming content.

Facilitating access for VoD is a pretty major challenge for all telecommunication companies. For a start, broadband subscribers that regularly access on-demand content consume huge amounts of bandwidth. On top of this the server architecture required to stream video content to multiple subscribers is quite large.

Chapter 9 provides a more detailed insight into the various VoD types supported by next generation IPTV systems.

1.5   GROWTH DRIVERS FOR IPTV

A confluence of forces has brought us to this point.

1.5.1 The Digitization of Television

Most satellite, terrestrial, and cable TV providers have started to switch their delivery platforms from analog over to digital. In addition, most if not all of the video production studios are using digital technologies to record and store content. These factors have negated the need to support legacy analog technologies and encouraged the adoption of IP based video content.

1.5.2 Enhancements in Compression Technologies

The delivery of video content over an IP network is nothing new, with a number of Internet streaming video sites in operation for a number of years, at this stage. Traditionally, the quality of the material streamed over the Internet was poor due to limited bandwidth capacities. Increasing numbers of broadband subscribers combined with improvements in compression techniques for digital video content has in recent years changed the whole dynamic of sending TV content over IP connections.

1.5.3 Business and Commercial Drivers

Increased competition combined with declining revenue streams is forcing many telecommunication companies to start the process of offering IPTV services to their subscribers. These new IPTV services typically extend the current broadband, and telephony offerings to form a product bundle called a triple play. For both fixed and wireless telecommunication companies, the triple-play bundle of IP based products is identified as being a key part of growing their businesses in the years ahead.

1.5.4 Growth in Broadband Use

The pervasiveness of the Internet has brought the need for high speed, always on Internet, access to the home. This need is being satisfied through broadband access technologies such as digital subscriber line (DSL), cable, fiber, and fixed wireless networks. The adoption of broadband Internet access by many households in turn has become a very powerful motivation for consumers to start subscribing to IPTV services.

1.5.5 Emergence of Integrated Digital Homes

People's homes and lifestyles are evolving and undergoing a number of positive changes. Many of these changes are underpinned by a range of new technologies that are helping to make life easier in addition to keeping consumers entertained. Digital entertainment devices such as gaming consoles, multiroom audio systems, digital set-top boxes, and flat screen televisions are quite common. In addition, the dramatic reduction in the costs of PCs is increasing the number of households that own multiple PCs. All of these technologies have finally spawned the emergence of a number of households that can be classified as "digital homes." The increase in these types of homes has started to drive demand for whole home media networking (WHMN) services such as IPTV.

1.5.6 A Wide Range of Companies Are Deploying IPTV

In addition to allowing traditional telephone companies to add video services to their product portfolios, IPTV also allows satellite and terrestrial companies to provide their customer bases with IP based pay TV services.

1.5.7 The Migration of Standard Definition (SD) Television to High Definition TV (HDTV)

HDTV has finally arrived and is here to stay. Demand from consumers is exploding, and improved adoption of digital networking technologies is enticing multiservice network operators to start offering HDTV channels to their program lineup. Additionally, the delivery of HDTV over IP broadband networks is now a required option for telecommunication providers.

The simultaneous combination of all of these drivers has made IPTV a practical reality that is both commercially and technologically successful.

1.6 MARKET DATA

IPTV is projected to be an extremely fast-growing sector over the next 5–10 years. Detailed market research data are set out in Table 1.2 and graphically illustrated in Fig. 1.3.

1.6.1 Messages in the Data

The following points should be noted from Table 1.2 and Fig. 1.3 and an analysis of related data:

- The Compound Annual Growth Rate (CAGR) for the global IPTV market-place is projected to be 63.31% over this period.

TABLE 1.2   Global IPTV Forecast (Millions of Subscribers), 2005–2010

	2005A	2006A	2007E	2008E	2009E	2010E	5 Year CAGR
Units (millions)	3.22	4.84	10.12	16.4	26.8	37.4	63.31%

Courtesy: TVMentors (leading market research provider to the global IPTV industry).

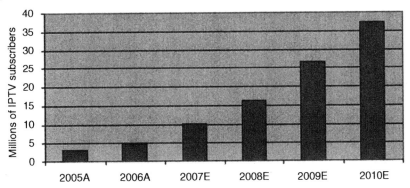

FIGURE 1.3   Worldwide IPTV forecast subscriber forecast, 2005–2010

- At the end of 2006, TVMentors estimated that approximately 4.8 million households around the world subscribed to an IPTV service. TVMentors forecasts that the number of households around the world subscribing to IPTV services offered by network carriers will reach 37.4 million in 2010.
- Over the long term, the number of households adopting IPTV will grow at a steady pace. In fact, TVMentors believes that IPTV services will start to become a mainstream product in 2009.

Note that TVMentors publishes reports and databases, periodically and methodologically, to forecast the number of global IPTV subscribers on an ongoing basis. Readers who are interested in this level of information are encouraged to visit www.tvmentors.com for further details.

1.7   INDUSTRY INITIATIVES TO STANDARDIZE IPTV

Similar to the cable and satellite pay TV sectors, the IPTV industry sector also requires a set of standards that will promote competition, lower costs for subscribers, minimize confusion in the market, and improve the delivery of compelling IPTV services. Standardizing IPTV is not an easy task because there are a whole range of components and systems from different vendors involved with building an end-to-end IPTV system. However, as with any emerging technology, a number of standard bodies and industry consortiums have got involved in standardizing IPTV.

1.7.1   DSL Forum

The DSL Forum is a nonprofit corporation organized to create guidelines for DSL network system development and deployment. The organization has created a

TABLE 1.3   Parts of MPEG-E Standard

Part Number	Part (API) Description
ISO/IEC 23004-1	Architecture
ISO/IEC 23004-2	Multimedia API
ISO/IEC 23004-3	Component model
ISO/IEC 23004-4	Resource and quality management
ISO/IEC 23004-5	Component download
ISO/IEC 23004-6	Fault management
ISO/IEC 23004-7	System integrity management
ISO/IEC 23004-8	Reference software

number of recommendations, which are particularly relevant to the IPTV industry sector. For more information visit www.dslforum.org.

1.7.2   Moving Pictures Experts Group

The Moving Picture Experts Group is a working group of ISO/IEC in charge of the development of international standards for compression, decompression, processing, and coded representation of moving pictures, audio, and their combination. The MPEG group is progressing a number of specifications that are relevant to IPTV. In addition to the various video coding standards, the group has also developed the multimedia middleware ISO/IEC 23004 (MPEG-E M3W) standard. MPEG-E comprises a number of application program interfaces (APIs), which are defined in eight separate parts (see Table 1.3).

For more information visit www.chiariglione.org/mpeg/.

1.7.3   European Telecommunications Standards Institute (ETSI)

ETSI formed a group called Telecoms & Internet converged Services & Protocols for Advanced Networks (TISPAN) in 2003 to develop specifications for next generation wireless and fixed networking infrastructures. TISPAN in turn has structured itself into groups that work to deliver specifications on topics that are particularly important to the IPTV industry sector ranging from home networks and security to network management and addressing. For more information on the various TISPAN specifications visit www.etsi.org/tispan/.

1.7.4   Open IPTV Forum

At the time of writing a group of network operators, network infrastructure equipment providers, and consumer electronic suppliers formed an IPTV standardization consortium called the Open IPTV Forum. The aim of this group is to work with existing standardization bodies to define end-to-end specifications for delivering IPTV services across a variety of different networking architectures. For more information visit www.openiptvforum.org.

1.7.5 Broadband Services Forum (BSF)

According to the organization's Web site "BSF is an international industry resource that provides a forum for dialogue and development, along with the tools and information to address the fundamental business and technology issues vital to the growth and health of the broadband industry." This consortium of companies has a particular focus with regard to IPTV and is promoting the industry through its participation in various industry conferences and trade shows. For more information visit www.broadbandservicesforum.org.

1.7.6 WirelessHD Consortium

The WirelessHD Consortium is a group of technology and consumer electronics companies that were formed in 2006. At the time of writing the group had started to work on a wireless digital interface specification that sends uncompressed IP- and RF-based high definition (HD) TV to HD display panels. When completed the technology is to be incorporated into a range of audio–video equipment types including IP set-top boxes and HD flat panel displays.

1.7.7 State Administration of Radio, Film, and Television (SARFT)

The Chinese state run organization SARFT in conjunction with the Ministry of Information is responsible in China for issuing standards related to the deployment of IPTV technologies in the country.

1.7.8 ITU-T FG IPTV

In the spring of 2006, the ITU established a focus group on IPTV, known as the IPTV FG, to coordinate and promote the development of global IPTV standards. The group is concentrating its energies in five key areas:

- Architecture
- DRM
- Quality of Service (QoS) metrics
- Metadata
- Interoperability and test

For more information visit http://www.itu.int/ITU-T/IPTV/index.phtml.

1.7.9 The Alliance for Telecommunications Industry Standard (ATIS)

ATIS is a telecom industry organization that includes more than 350 companies including the major service providers. To further the standardization work for the IPTV industry sector, ATIS has launched a subgroup called the IPTV Interoperability Forum (IIF). According to the group's Web site, the primary remit of the IIF is to produce an overall reference architecture for deploying IPTV services, which focuses on four major areas, infrastructure equipment, content security, interoperability testing, and quality of service. The company recently published a number of guidelines in the areas of IPTV digital rights management and architecture:

- *ATIS-0800001:* This document defines the interoperability specifications associated with implementing IPTV DRM systems. The organization plans to use this document as a basis for creating an IPTV DRM/security interoperability specification in the future.
- *ATIS-0800002:* This document provides guidelines to content and service providers on the architecture required to deliver IPTV services.
- *ATIS-0800003:* Published in 2006, this document sets out a roadmap consisting of a number of phases for standardizing the architecture of IPTV systems.
- *ATIS-0800004:* This document defines a framework for monitoring QoS metrics for various types of IPTV services.
- *ATIS-0800005:* This technical document covers the topic of packet loss across IPTV networking infrastructures. In addition to identifying the various causes of packet loss the document also provides readers with a set of recommendations with regard to reducing the impact of packet losses in a live IPTV networking environment.

The organization has agreed to share these documents with other IPTV standard organizations such as the ITU-T FG IPTV to ensure interoperability between the various technologies. The organization also has plans to establish a certification process for IPTV hardware and software vendors in the future. For more information visit www.atis.org.

1.7.10   The Internet Protocol Detail Record Organization (IPDR)

IPDR.org is an industry consortium of service providers, and equipment suppliers exclusively focused on developing and driving the adoption of next generation IP service usage exchange standards worldwide. This organization has taken on the responsibility of defining interoperability standards for IPTV billing, network management, and back-office systems. For more information visit www.ipdr.org.

1.7.11   Internet Streaming Media Alliance (ISMA)

Founded in the year 2000, ISMA is a nonprofit industry alliance of companies, and since its inception, it has received wide industry support. Its mission is to facilitate and promote the adoption of an open architecture for streaming audio and video over IP networks. The organization has developed a number of specifications ranging from improving the channel changing times for IPTV systems to synchronizing graphics and data with streaming video content. All of its specifications produced to date make extensive use of open Internet standards that have been produced by the IETF. For more information visit www.isma.tv.

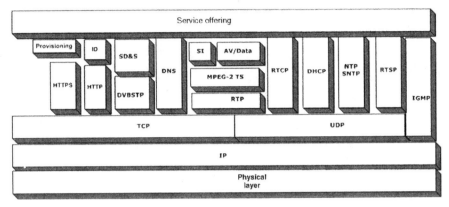

FIGURE 1.4   DVB-IPI protocol framework

1.7.12   DVB-IPI

To develop standards for the transmission of digital TV services over IP broadband networks the DVB organization has formed a group called the DVB Technical Module Ad Hoc Group on IP Infrastructure (DVB-IPI). The goal of the IPI group is to specify technologies that allow consumers to purchase a DVB-IP set-top box in any shop, connect it to a broadband network, switch it on and, without further ado start to receive DVB services over IP based networks. As shown in Fig. 1.4, the

TABLE 1.4   Five IPTV Principles Put Forward by the CEA and a Number of U.S. Telecom Operators

Principle name	Description
Nationwide compatibility	This principle aspires to defining a set of a nationwide (United States initially) common protocols that allow consumer electronics (CE) manufacturers to manufacture devices, which will interoperate with all home networks that run IPTV services.
Open standards	The establishment of a forum that drives the adoption of open standards for the sector.
Reasonable licensing terms	This principle hopes to introduce reasonable and non-discriminatory licensing terms that allow CE manufacturers and video service providers to include improved features in their IP based products.
Reasonable testing and certification procedures	As the name suggests the group is planning to establish a testing and certification process for products that support IPTV services.
Reasonable terms of service for consumers	This fifth and final principle aims to provide consumers a choice when deciding on a digital device to access their IPTV services.

DVB-IPI uses a number of existing and mature technologies to build a framework that supports the overall goal of DVB-IPI.

Note that all of the technologies and protocols, shown in Fig. 1.4, are explained later in the book.

1.7.13   IPTV "Principles" Initiative

The Consumer Electronics Association (CEA) in conjunction with a number of U.S. telecom operators have defined a set of five principles, designed to ensure the availability of digital devices that connect to networks that run IPTV services. The five principles are shown in Table 1.4.

Owing to the broad diversity of IPTV architectures around the world and the diversity of technologies associated with IPTV, the implementations of the above standards will typically vary from network to network.

SUMMARY

Digital TV technology offers fundamental improvements over analog services. For instance, where an analog signal degrades with distance, the digital signal will remain constant and perfect as long as it can be received. The advent of digital TV benefits the general public because of crystal clear pictures, CD quality sound and access to a range of new entertainment services. Depending on geographical location, analog television systems are based on either NTSC, PAL or SECAM standards. There are two main global digital TV standards, namely, DVB and ATSC. By using digital technologies to transmit television, service providers can carry more information than is possible with analog systems.

IPTV is a new method of transporting digital TV content over a network and is seen as part of the larger triple-play bundle that is typically on offer from network operators worldwide. IPTV is a term that describes a system that enables the delivery of real-time television programs, movies, and other types of interactive video content over an IP based network.

Consumers often do not realize that behind the simple end-user IPTV environment is a series of powerful components that seamlessly work together to make the delivery of TV over broadband networks possible. These components or subsystems include the processing of video, security, and the delivery platform. An end-to-end IP video network infrastructure can include some or all of the following elements:

- The IPTV data center that is responsible for processing and preparing content for distribution across a broadband network.
- An IPTV distribution network consisting of a mix of technologies that carry IPTV content from the data center to end users.
- IP digital set-top boxes or residential gateways that are installed at the subscribers home and provide connectivity between the TV and the IP based access network.

- A home network enabling the distribution of data, voice, and video between different consumer devices.

A recipe of increased broadband adoption combined with advancements in compression technologies and the need for telecommunication companies to offer video services to their customers is helping to grow the size of the global IPTV marketplace. A number of organizations are involved in developing technological standards and products to encourage consumers to adopt IPTV services, including SARFT, ITU, ATIS, DVB, IPDR, ISMA, and the CEA.

2

IPTV NETWORK DISTRIBUTION TECHNOLOGIES

It looks like IPTV is well on its way to becoming a popular means for delivering digital TV services to consumers. Owing to the nature of IPTV, a high speed distribution networking platform is required to underpin the delivery of IP based content. The purpose of this network is to move bits of data back and forth between the IPTV consumer device and the service provider's IPTV data center. It needs to do this in a manner that does not affect the quality of the video stream delivered to the IPTV subscriber, and it is up to each IPTV provider to decide on the type and sophistication of the network architecture required to support their IPTV services. An IPTV network architecture consists of two parts the "last mile" broadband distribution and the centralized or core backbone. A wide variety of networks, including cable systems, copper telephone, wireless, and satellite networks may be used to deliver advanced IPTV network services over the last mile section of the network. The delivery of video over all of these different types of networks comes with its own set of challenges. The majority of this chapter focuses on describing these key technology platforms. Once the key delivery platforms are covered, the chapter concludes with an analysis of the two primary core networking technologies and a brief overview of network factors that affect the deployment of IPTV services.

Next Generation IPTV Services and Technologies, By Gerard O'Driscoll
Copyright © 2008 John Wiley & Sons, Inc.

2.1 "LAST MILE" BROADBAND DISTRIBUTION NETWORK TYPES

One of the primary challenges faced by IPTV service providers is providing enough bandwidth capacity in the network segment that lies between the core backbone and the end-users' home. A number of terms are used to describe this segment ranging from local loop and last mile to edge and broadband access network. There are six different types of broadband access networks that are scalable enough to meet the bandwidth requirements of IPTV:

- Through a network built with fiber
- Via an DSL network
- Via a cable TV network
- Via a satellite based network
- Via a fixed wireless broadband connection
- Via the Internet

Different service providers operate each system. The following sections give a technical overview of these platforms when used in an IPTV end-to-end networking infrastructure.

2.2 IPTV OVER A FIBER ACCESS NETWORK

Increasing demand for bandwidth combined with lower operating costs and immunity to electromagnetic interference are some of the factors that are driving deployments of optical fiber based access networks. Networks that utilize fiber have been used by multiple service operators (MSOs) to build networks for decades. Owing to a recent reduction in equipment and deployment costs over the last couple of years, interest in using fiber based networks to deliver emerging IP based services such as IPTV has risen dramatically. Additionally, fiber links provide end consumers with a dedicated connection, which is well suited to the delivery of IPTV content. Bringing fiber technologies and higher bandwidth capabilities nearer to the user can be implemented using one of the following network architectures:

Fiber to the regional office (FTTRO)—This refers to the installation of fiber from the IPTV data center to the nearest regional office owned by the telecommunication or cable company. Existing copper wiring is then used to carry signals from the regional office to the IPTV end user.

Fiber to the neighborhood (FTTN)—Also known as fiber to the node, FTTN entails installing fiber from the IPTV data center to a neighborhood splitter. This node is generally located less than 5000 ft away from the subscriber. Another mechanism such as digital subscriber line (DSL) over copper wire is then utilized to make the final link to the customer. The deployment of FTTN allows end users to receive a complete bundle of pay services, including IP based TV, high definition TV, and video on demand (VoD).

Fiber to the curb (FTTC)—An FTTC networking infrastructure involves the installation of optical fiber to within a thousand feet of a home or a business. A coaxial cable or copper wire is typically used to establish the connection from the optical cable terminated in a cabinet located at a "curb" on a street to the residential gateway located in the IPTV subscriber's premises. This configuration is typically installed during the construction of a new housing development.

Fiber to the home (FTTH)—With fiber to the home, the entire route from the IPTV data center to within the home is connected by optical fiber. FTTH based optical networks are capable of delivering very high volumes of data to end users of the system. This architecture is also quite popular for new construction sites, since the trenches are cut and the cost of laying fiber is relatively similar to installing copper cables. FTTH is a full-duplex communication system and supports the interactive nature of IPTV services.

Fiber to the apartment (FTTA)—The deployment of an FTTA network entails the installation of a number of fiber cables between a central gateway hub, typically located in the basement of an apartment block and each individual apartment.

The delivery of these architectures is typically enabled through two different variants PON and AON.

2.2.1 PON Networks

Passive optical network (PON) refers to a point-to-multipoint networking topology that makes extensive use of fiber optic cabling and optical components. They use lightwaves of different colors to carry data across the network and require no electrical components between the IPTV data center and the destination point.

The network architecture used to build a PON based FTTx network will typically comply with international standards. The G.983 specification from the International Telecommunications Union (ITU) is the most popular option at the time of writing this book.

A G.983 compliant PON typically consists of an optical line termination (OLT), which is located at the IPTV data center and a number of optical network terminals (ONTs), which are installed at the end users premises. Note that ONTs may also be installed at different neighborhoods where the optical fiber terminates. In these situations, high speed copper data transfer technologies such as DSL are used to carry the IPTV signals into the end-users' household.

The OLT uses components such as fiber cable and optical splitters to route network traffic to the ONTs.

- *Fiber cable*—The OLT and the various ONTs are interconnected by fiber optic cabling. With few transmission losses, low interference, and high bandwidth potential, optical fiber is an almost ideal transmission medium. The core of the fiber optic cable is made of glass and carries data in the form of lightwave signals. The diameter of the fiber cable is relatively small and is designed to allow network engineers splice the cable at various locations along the physical route. The purity of today's glass fiber, combined with improved system electronics in the cable, permits the transmission of high speed services over long distances. In fact, the G.983 standard allows the PON to carry digitized light signals up to a maximum distance of 20 km without amplification.

- *Optical splitters*—The optical splitters are used to split a single optical signal into multiple signals. It achieves this function while not altering the state of the signal; in other words, it does not convert it to electrical pulses. Optical splitters are also used to merge multiple optical signals back into a single optical signal. These splitters allow up to 32 households to share the FTTx network bandwidth and are typically housed in accessible mechanical closures.

Fiber cable and optical splitters are "passive" optical components. The use of passive components to guide the lightwaves through the network eliminates the need for remote powering, which cuts down on operational and maintenance costs.

The main purpose of the ONT is to provide IPTV subscribers with an interface to the PON. It receives traffic in optical format, examines the address contained within the network packets, and converts it into electrical signals. The ONT can be located inside or outside the residence, and is typically powered from a local source, and includes bypass circuitry that allows the phone to operate normally in the event of a power failure. The majority of ONTs will include an Ethernet interface for data traffic, an RJ-11 connection for connecting into the home phone system, and a coaxial interface to provide connectivity to the TV. The ONT is also responsible for converting data into optical signals for transmission over the PON.

Figure 2.1 shows how the basic PON networking infrastructure could be built to support the delivery of IPTV and high speed Internet services to six different households.

As shown, a single piece of optical fiber is run from the backend office to an optical splitter, which is typically located in close proximity to the subscriber's house. The bandwidth on this fiber is typically shared and is capable of supporting high bandwidth capacities ranging from 622 Mbps all the way up to several gigabytes of data per second.

In addition to the physical components of a PON Fig. 2.1 also illustrates the transmission of three different light wavelengths (channels) over the network. The first wavelength is used to carry high speed Internet traffic. The second wavelength is allocated to carry IP video services and the third wavelength may be used to carry interactive traffic from the subscriber's home network back to the service provider's backend equipment. As shown specialized filters called wavelength division

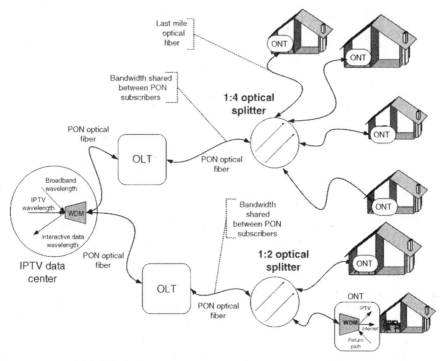

FIGURE 2.1   IPTV FTTH network using PON technologies

multiplexers (WDMs) are installed at the data center and inside the OLT that allow
a PON to support the transmission of multiple parallel channels or wavelengths on
the one piece of fiber. Thus, creating a number of virtual fiber channels over a single
fiber pair. Under WDM, the capacity of the network is increased by assigning
signals that originate from optical sources to specific wavelengths on the optical
transmission spectrum.

There are various flavors of PON technologies including BPON, EPON, and
GPON that support both traditional RF based TV services and IPTV. Details of
each of these technologies are provided in the following sections.

2.2.1.1 BPON   Broadband PON or BPON is based on the ITU-T G.983
specification. This type of FTTx networking topology supports data rates up to
622 Mbps downstream and up to 155 Mbps on the upstream. Thus, it is an
asymmetrical transmission method. In other words, the downstream communication speeds are higher than the upstream speeds. This is because downstream data
flows in a point-to-point fashion between the OLT and each ONT whereas on the
upstream each ONT is given a time slot for transmitting data. Assigning time slots
reduces the possibility of traffic collisions between ONTs on the network; however,

it reduces the overall data rates of the upstream communication channel. Note that BPONs can also be configured to support symmetrical data traffic.

BPON uses Asynchronous Transfer Mode (ATM) as the bearer protocol. ATM based networks are quite popular for delivering high speed data, voice, and video applications. ATM is a cell relay technology capable of very high speeds. It divides all information to be transferred into blocks called cells. These cells are fixed in size—each has a header with 5 bytes and an information field containing 48 bytes of data. The information field of an ATM cell carries the IPTV content, whereas the header contains information relevant to the functioning of the ATM protocol.

ATM is classified as a connection orientated protocol. In other words, a connection between the receiver and the transmitter is established prior to sending the IP video data over the network. The ability to reserve bandwidth to time sensitive applications is another feature of ATM networks. This is particularly a useful feature for delivering IPTV services. The allocation of specific channels to different services helps to remove interference.

2.2.1.2 EPON   Ethernet PON or EPON is an optical based access technology that was developed by a subgroup of the IEEE called the Ethernet in the first mile (EFM) group and adopted as a standard in 2004. As the name suggests this flavor of PON uses Ethernet as its transport mechanism. The rates supported depend on the distances between the OLT and ONTs. Note EPONs only support Ethernet network traffic.

2.2.1.3 GPON   Gigabit PON or GPON is an optical access system, which is based on the ITU-T G.984 specification. GPON is basically an upgrade to the BPON specification and includes support for

- higher transmission rates—downstream rates of 2.5 Gbits/s and upstream rates of 1.5 Gbits/s. This rate is achievable at distances up to 20 km.
- protocols such as Ethernet, ATM, and SONET are supported.
- enhanced security features.

The multiprotocol support provided by GPONs allows network operators to continue providing customers with traditional telecommunication services, while also having the facility to introduce new services such as IPTV onto their networking infrastructure. Table 2.1 summarizes the characteristics of the various PON technologies used to carry IPTV signals.

In addition to their extensive use by FTTx networks, PONs can also serve as a backbone infrastructure for hybrid fiber coaxial and wireless broadcast networks.

With regard to the future development of PON technology the Full Service Access Network (FSAN) consortium in conjunction with the IEEE continue to develop next generation PON technologies. At the time of writing exploratory work

TABLE 2.1 PON Technologies Comparison: BPON, EPON, and GPON

	ITU-T Specification	Data Rates	Transmission Protocol
BPON	G.983	622 Mbps downstream and 155 Mbps upstream	Primarily ATM however IP over Ethernet will also operate on the network
GPON	G.984	2.5 Gbps downstream and 1.5 Gbps upstream	Ethernet and SONET
EPON	P802.3ah	1.25 Gbps downstream and 1.25 Gbps upstream	Gigabit Ethernet

has commenced on two possible successors to GPON—wavelength division multiplexing—passive optical network (WDM-PON) and 10G-PON.

2.2.2 AON Networks

Active optical networks (AON) makes use of electrical components between the IPTV end user and the data center. In particular, the AON networking architecture utilizes Ethernet switches that reside between the IPTV data center and the endpoint of the fiber network.

2.3 IPTV OVER AN ADSL NETWORK

In the last couple of years a number of telephone companies (telcos) in different parts of the world have entered the IPTV market. Their entrance into this market has been driven by efforts to counteract the threat to their revenue streams from cable television operators and wireless broadband providers who have started to offer a variety of Internet access and telephone services. In response to this challenge, the telcos are taking advantage of their DSL based networking infrastructure to start rolling out next generation television services to their subscribers. Note that DSL is a technology that enables telecommunication providers to deliver high bandwidth services over existing copper telephone lines. It transforms the existing telephone cabling infrastructure between a local telephone exchange and a customer's telephone socket into a high speed digital line. This capability allows telephone companies to use their existing networks to provide multiple high speed Internet data services to their subscriber bases.

Bandwidth is a key issue in the delivery of next generation IPTV services. This is particularly true for the local DSL loop. Many of the existing DSL based broadband networks are built around legacy DSL standards, which are simply not capable of meeting the growing demand to support high speed video services. Most of these networks are restricted to delivering one IP video stream to each household. In some cases it is impossible to send a standard quality TV signal over these DSL access networks. The boost in performance required for IPTV can however be achieved through the deployment of DSL technologies such as ADSL, ADSL2+, and VDSL. An overview of the features of these technologies and how they work is provided in the following sections.

2.3.1 ADSL

Asymmetric Digital Subscriber Line (ADSL) is currently the most popular flavor of DSL deployed on telecommunication networks across the globe. It has made considerable inroads into the residential market, where it competes with cable modems for customers who are looking for high speed broadband connections.

ADSL is a point-to-point connection technology. This feature enables telecommunication providers to deliver high bandwidth services such as IP video over existing copper telephone lines. It is called "asymmetric" because the transmission of information from the central data center or regional office to an IPTVCD is quicker than information traveling from the IPTVCD to the central data center. Also the point-to-point characteristic of ADSL eliminates the variation in bandwidth capacities of a shared networking environment.

By using specialized techniques, ADSL typically allows a downstream rate of 8 Mbps and an upstream rate of 1.5 Mbps. Therefore, one ADSL connection is sufficient to simultaneously support two MPEG-2 standard definition broadcast TV channels and a high speed Internet connection.

ADSL's major drawback is that the availability is contingent upon distance from the service provider's central office (CO). ADSL is a distance sensitive technology so subscribers near the central data center receive a better quality service than those further away. A basic ADSL service is limited to homes within 18,000 ft from the nearest telephone exchange or regional office.

From a technical perspective, telephone lines were originally designed to support the transmission of low frequency voice traffic. Traffic that is sent over a telephone line at high frequencies normally experiences distortion and interference. Apportioning the bandwidth of a telephone line helps to minimize interference and increase data rates. The frequency allocation of an ADSL circuit reserves the lower 4 kHz for the existing telephone service, while upstream and downstream data channels occupies 26 kHz to 1.1 MHz frequency range (Fig. 2.2).

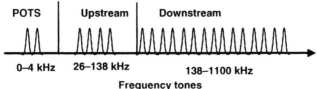

FIGURE 2.2   ADSL frequency allocations

ADSL equipment provides a digital connection over the PSTN network; however, the signal that is transmitted across this connection is modulated as an analog signal. ADSL circuits must use analog signaling because the local loop network is not capable of carrying signals coded in digital format. Thus, the modem at the IPTV data center is responsible for converting digital data into analog signals that are acceptable for transmission. The residential modem, which connects to the IPTVCD then converts the analogue signals that are received over the ADSL circuit back into appropriate digital signals.

The two primary techniques used to modulate digital IPTV data into an analog signal for ADSL transmission are carrierless amplitude and phase (CAP) and discrete multitone (DMT).

CAP—CAP was the original approach used to modulate digital data into an analog signal for ADSL transmission. While the name specifies that the modulation is "carrierless" an actual single carrier is used to transmit the data over the telephone network. CAP is closely related to the quadrature amplitude modulation (QAM) scheme. QAM is a very well understood and widely used technique, common in satellite and cable TV applications for many years.

DMT—Discrete multitone is now the preferred modulation alternative over CAP and is used in by modern DSL technologies. It separates the DSL signal frequency range into a number of small subchannels or frequency tones. During transmission, each one of these subchannels carries a portion of the total data rate. By dividing the transmission bandwidth into a collection of subchannels, DMT is able to adapt to the distinct characteristics of each telephone line and maximize the data transmission rate. DMT is closely related to Orthogonal Frequency Division Multiplexing (OFDM) or C-OFDM (Coded OFDM). COFDM is used by Europe's DVB television standards.

2.3.1.1 ADSL Equipment

2.3.1.1 ADSL Equipment   ADSL is a technology that can expand the useable bandwidth of existing copper telephone lines. The equipment used in the installation of a full rate ADSL service is shown in Fig. 2.3, and consists of the following:

An ADSL modem—In the subscriber's home there is an ADSL transceiver or modem. The modem connects usually through an USB or Ethernet connection, from the home network or PC to the DSL line. Many modems nowadays also incorporate a router, which supports high speed Internet access and data services.

A POTS splitter—Users getting connected to the Internet with an ADSL broadband connection use a device called a POTS splitter to separate the data signals from the voice signals. The splitter divides the incoming signal into

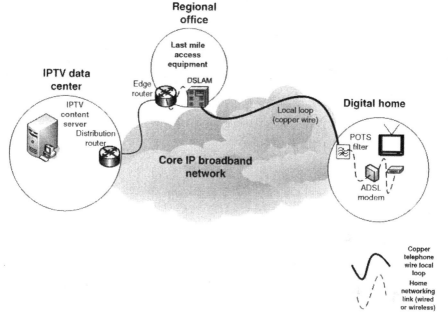

FIGURE 2.3   IP over ADSL equipment architecture

low frequencies to send to a telephone and high frequencies for data to the home network.

A DSLAM—DSLAM stands for Digital Subscriber Line Access Multiplexer. At the IPTV service provider's regional office, the DSLAM receives the subscriber's connections over copper cable, aggregates them, and connects back to the central IPTV data center through a high speed fiber based network backbone. For IPTV deployments it is typical that the DSLAM supports multicast transmission. This negates the need to replicate channels for each request originating from an IPTV viewer. The DSLAM has overall responsibility for distributing the IPTV content over the "last mile" to the IPTV subscribers. DSLAMs fall into two broad categories: Layer 2 and IP-aware.

(1) *Layer 2 DSLAMs* operate at level 2 of the OSI (Open Systems Interconnection) communications model and perform functions such as switching traffic between Ethernet and ATM, passing network traffic upstream, and preventing interference between IPTV subscribers. The switching between ATM virtual circuits and upstream Ethernet packets is facilitated through the used of a bridging mechanism.

(2) *IP-aware DSLAMs* include limited support for level 3 IP networking protocols. Advanced functions supported by this category of DSLAMs

include replication of broadcast TV channels and executing channel changing instructions.

ADSL technology is ideal for a range of different interactive services; however, it is not an optimal solution for delivering IPTV content due to the following reasons:

Data rates—An ADSL maxes out at around 8 Mbps, which supports two SD channels and some Internet traffic; however, it will not be able to meet the needs of IPTV providers who plan to deliver high definition programming to their subscribers.

Interactivity—Because the technology is ADSL the upload data rate is lower than the download rate. This limitation means that ADSL is unsuitable for applications, such as peer-to-peer services that require as much upstream bandwidth as download bandwidth.

Therefore, network service providers are starting to deploy more advanced DSL technologies that overcome these limitations.

2.3.2   ADSL2

The ADSL2 family of standards was created to address the increased demand for capacity to support bandwidth intensive applications such as IPTV. There are three different members of the ADSL2 family:

ADSL2—The initial version of ADSL2 was approved by the ITU in 2003 and included a number of enhancements to the original ADSL standard, namely, higher downstream data rates and longer reach from the CO to the subscriber's modem.

ADSL2+—Soon after the standardization of ADSL2, a further flavor of DSL was adopted by ITU known as ADSL2+. This standard builds on ADSL2 and allows network service providers to offer speeds up to 20 Mbps to subscribers that live within a distance of approximately 5000 ft (1.5 km) away from the CO. ADSL2+ operates within the signal bandwidth of 138 kHz to 2.208 MHz.

ADSL-Reach Extended—The deployment of ADSL2+ is of no use to subscribers who live over 5000 ft (1.5 km) away from the CO. Therefore, a technology called ADSL-Reach Extended, also known as RE-ADSL2 was standardized in 2003 to allow IPTV service providers to increase the range of their offerings to subscribers that are up to 19,000 ft (6 km) away from their nearest CO. It shows good performance in terms of reach and speed over long copper lines.

2.3.3 VDSL

VDSL (Very high speed Digital Subscriber Lines) is based on the same underlying technology as ADSL2+. It is the newest and most sophisticated DSL technology at the time of writing and was developed to overcome the shortcomings of previous versions of ADSL access technologies. It eliminates last mile bottlenecks and supports huge data rate capacities that allow service providers to comfortably offer IPTV subscribers a whole range of services including video-on-demand and multichannel high definition TV broadcasting. VDSL was also designed to support the transmission of ATM and IP based traffic over the copper loop that is very useful for providers who want to migrate their legacy ATM networks over to an IP based infrastructure. There are different members of the VDSL family:

VDSL1—This flavor of DSL was ratified in 2004. It operates at an upper limit of 55 Mbps on the downstream and 15 Mbps over the upstream channel. It does however have a very short range and is typically installed inside multiple dwelling units (MDUs).

VDSL2—is an enhancement to VDSL1 and is defined in the ITU-T Recommendation G.993.2. It can be subclassified into VDSL2 (Long Reach) and VDSL2 (Short Reach)

VDSL2 (Long Reach)—Owing to the fact that DSL is highly dependent upon the length of the local loop, a version of VDSL was created to deliver VDSL services to as many customers as possible, while enjoying the high speed capacities of the VDSL broadband access platform. In fact, VDSL with enhanced long reach capabilities can provide IPTV subscribers with 30 Mbps broadband access at distances of between 4000 and 5000 ft (1.2−1.5 km) away from the CO. These long reach capabilities are enabled through the use of relatively high power levels when transmitting the data. A frequency spectrum of 30 MHz is used to achieve the high throughput as opposed to the 12 MHz frequency range used by VDSL1. Advanced error correction mechanisms also help to improve the reliability of end-to-end VDSL2 connections.

VDSL2 (Short Reach)—Based on the DMT modulation scheme, the technology uses 4096 tones that are separated by 4 and 8 KHz frequency bands. The VDSL2 standard uses channel bonding techniques to allow it to operate at over 12 times the speed of the original ADSL standard, namely, a massive 100 Mbps on the downstream path over relatively short distances—approximately 1200 ft (350 m). Although the data rates do not typically reach 100 Mbps on the upstream path, the rates do exceed the upstream data rates of ADSL2+ based networks. These performance levels make the assumption that no interference exists on the copper cable and the quality of the cable is quite good. The ability to provide IPTV subscribers with 100 Mbps access service enable operators to start offering a wide variety of advanced interactive services to their customers.

New VDSL2 features such as advanced quality of service (QoS), dual latency (the ability to segregate delay sensitive data such as IPTV multicast traffic), and improved coding techniques are all particularly suited to the delivery of triple-play applications. One final benefit of VDSL that solidifies its position as the ultimate DSL technology is its backward compatibility and interoperability with previous versions of ADSL networks. This enables IPTV providers to smoothly and efficiently migrate to VDSL based next generation networks.

There are two main approaches used by IPTV service providers when integrating VDSL2 into their existing networking infrastructure. The first approach is to add new VDSL2 equipment to the regional office and allow the DSLAM run in parallel to the existing ADSL and ADSL2 DSLAM systems. The various flavors of DSL will subsequently continue in operation. The second approach is to locate the VDSL2 equipment closer to the IPTV subscriber. Possible locations for the new equipment include street cabinets and underground chambers for new housing estates. Table 2.2 provides a feature comparison between the various flavors of DSL technologies used to carry IPTV signals.

The main advantage of DSL for IPTV systems is the fact that it utilizes the existing wires that already run into most houses around the world. On the negative side, all DSL systems have to make a trade-off between distance and bandwidth capacity. In other words the DSL access speed reduces as the distance between the IPTV subscriber and the CO increases.

2.4   IPTV OVER NEXT GENERATION CABLE TV NETWORKS

Cable TV operators have made significant investments in the recent past to upgrade their networks to support advanced communication services such as IPTV. To understand the delivery of IPTV content over a cable TV network and to put the technology in context, it is first necessary at least on a high level, to review the basic principles of hybrid fiber networks and traditional digital TV transmission technologies.

2.4.1   Overview of HFC Technologies

If a cable television network is available in a particular area, then consumers access IPTV from a network based on hybrid fiber/coax (HFC) technology. HFC technology refers to any network configuration of fiber-optic and coaxial cable that may be used to redistribute a variety of digital TV services. Most cable television companies are already using it. Networks built using HFC technology have many characteristics that make it ideal for handling the next generation of communication services:

- HFC networks are capable of simultaneously transmitting analog and digital services. This is extremely important for network operators who are rolling out digital and IPTV to their subscribers on a phased basis.

TABLE 2.2 Feature Comparison Between Different Types of DSL Technology Systems

DSL flavor	Max Downstream Bandwidth Capacity (Mbit/s)	Max Upstream Bandwidth Capacity (Mbit/s)	Max Distance (Approximate Values and Dependent on Quality of Copper)	Types of Services Supported
ADSL	8	1	18,000 ft (5.5 km)	One standard definition (SD) MPEG-2 compressed video channel, high speed Internet access, and VoIP services.
ADSL2	12	1	18,000 ft (5.5 km)	Two SD MPEG-2 compressed video channels or one high definition (HD) channel, high speed Internet access, and VoIP services.
ADSL2+	25	1	5000 ft (1.5 km)	Five SD MPEG-2 compressed video channels or two HD MPEG-4 compressed channels, high speed Internet access, and VoIP services.
ADSL-Reach Extended	25	1	20,000 ft (6 km)	Five SD MPEG-2 compressed video channels or two HD MPEG-4 compressed channels, high speed Internet access, and VoIP services.
VDSL1	55	15	Designed for multiple dwelling units (MDUs) and has a range of several hundred metres	Twelve SD MPEG-2 compressed video channels or five HD MPEG-4 compressed, high speed Internet access, and VoIP services.
VDSL2 (Long Reach)	30	30	Four and five thousand feet (1.2–1.5 km)	Seven SD MPEG-2 compressed video channels or three HD MPEG-4 compressed, high speed Internet access, and VoIP services.
VDSL2 (Short Reach)	100	100	Twelve hundred feet (350 m)	Twenty five SD MPEG-2 compressed video channels or 10 HD MPEG-4 compressed, high speed Internet access, and VoIP services.

- HFC meets the expandable capacity and reliability requirements of an IPTV system. The expandable capacity feature of HFC based systems allows network operators to add services incrementally without major changes to the overall network infrastructure.
- HFC is essentially a "pay as you go" architecture that matches infrastructure investment with new revenue streams, operational savings, and reliability enhancements.
- The physical characteristics of coaxial and fiber cables support a network operating at several gigabits per second.

The topology of an end-to-end HFC network is illustrated in Fig. 2.4.

From the diagram we can see that the HFC network architecture consists of a fiber based backbone connected via an optical node to a coaxial network. The optical node acts as an interface that connects upstream and downstream signals that traverse the fiber optic network with the coaxial cabling. The coaxial portion of the HFC network uses tree-and-branch topology and is used to connect cable TV subscribers via specialized devices called taps to the HFC network. The digital TV signal is transmitted from the headend in a star-like fashion to the fiber nodes. The fiber nodes, in turn, distribute the signals over coaxial cable, amplifiers, and taps throughout the customer serving area.

2.4.2 Deploying IPTV Over a Cable TV Network

The debate within the cable TV industry to start carrying video traffic across an IP based architecture continues as of this writing. Threats to their core pay TV

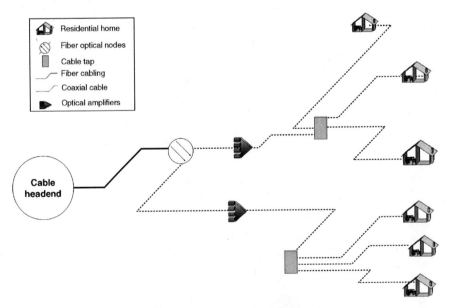

FIGURE 2.4 End-to-end HFC network

business from telecommunication operators combined with bandwidth efficiencies associated with IP delivery mechanisms are two of the key factors driving cable operators toward a more IP centric model of delivering video content to end users.

Switching a radio frequency based network over to an IP based switched digital video (SDV) environment however requires the installation of a number of new pieces of equipment ranging from routers to IP set-top boxes and high speed networking switches. Some of the advantages associated with deploying an SDV environment include:

- A large amount of network bandwidth is freed up due to the fact that the operator is only required to transmit a single TV channel to a subscribers IP set-top box. This contrasts quite sharply from traditional systems where all channels on the operator's lineup are broadcasted across the network and unused channels still occupy bandwidth.
- Spare bandwidth allows cable operators to deliver advanced IPTV content and services to their subscribers.
- Cable operators can accurately measure and monitor what video content is watched by each of its subscribers. This is an important feature for operators who want to generate revenue through advertising.

From a technical perspective a typical cable IPTV system constitutes a mix of IP and RF based hardware devices that are used to carry the video signal across the networking infrastructure. Figure 2.5 shows an example of a cable IPTV architecture that constitutes a mix of IP and RF technologies.

As shown the network consists of a variety of hardware devices, including

- *GigE routers or switches*—Gigabit Ethernet (GigE) is emerging as the transport protocol of choice for connecting IP networking components. GigE is typically used by higher capacity applications, such as VoD. The GigE router aggregates the IPTV traffic and provides interconnectivity to the core access network.
- *Optical transport network*—The core network provides the network path between the video servers in the headend and the modulators at the edge of the network. Synchronous optical network (SONET), ATM, and dense wavelength division multiplexing (DWDM) are examples of technologies that may be used in the core network.
- *Edge modulators*—The modulators located at the regional offices receive the IPTV content from the core network, convert the video content from IP based packets to RF, and distribute it via an HFC network to the set-top boxes.

This is an example of a large scale cable IPTV deployment and makes use of a hierarchical architecture through the establishment of regional distribution headends.

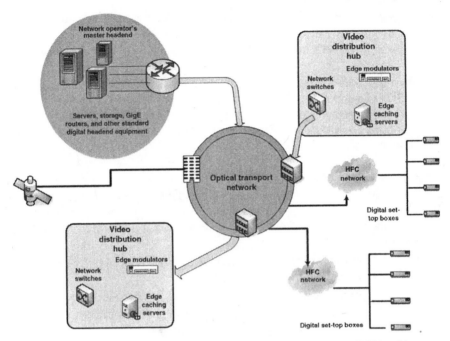

FIGURE 2.5   Sample deployment of a mixed IP and RF based cable IPTV architecture

2.4.3   IPTV via DOCSIS

Data over Cable Service Interface Specification (DOCSIS®) and its European sister specification, EuroDOCSIS, were originally designed to carry high speed Internet traffic over wide area networks. The specifications have evolved over the years and the latest version of DOCSIS provides enough capacity to support the delivery of IPTV services across HFC networks. Both DOCSIS and EuroDOCSIS are discussed in the following sections.

2.4.3.1 Understanding DOCSIS   Two-way cable TV networks deployed in the United States that support high speed broadband services are likely to use a specification called DOCSIS. CableLabs, a research and development consortium of cable television system operators representing the Americas, developed this technology. The specification defines the protocols and modulation formats used for delivering IP broadband services over a cable TV network.

The first revision of the technology, known as DOCSIS 1.0, was approved as a standard by the ITU in 1998 and there has been a family of DOCSIS specifications issued over the past 8 years. An overview of the technical features supported by each generation of DOCSIS technology is presented in Table 2.3.

As shown in Table 2.3, the DOCSIS versions through to DOCSIS 2.0 support a data rate of approximately 40 Mbps on the downstream. This bandwidth capacity is typically used to service a number of high speed Internet access subscribers. If this bandwidth was to be utilized for delivering IPTV multicast TV channels, then 10 or possibly 15 IPTV streams could be simultaneously provided on any individual downstream channel. This calculation is based on streaming 10–15 standard definition IPTV channels that each occupy between 2.5 and 4 Mbps of bandwidth. Once the channels are configured, cable operators use a networking technique called multicasting (described in Chapter 3) to assign multiple users to each IPTV stream.

The latest release of the specification, DOCSIS 3.0, enhances cable TV operator's capacity for delivering IP-VoD and multicast IPTV services to their customers. DOCSIS 3.0 is a next generation wideband technology that allows cable operators to bond multiple channels together to form high speed IP pipes capable of operating at hundreds of megabits per second. In North America 6 MHz channels are bonded together whereas in Europe the channels bonded together have a capacity of 8 MHz. Note that the bonding feature applies to both upstream and downstream channels. Other features of DOCSIS 3.0 include its in-built support for next generation IPv6 management addresses. Advanced support for improved network security, IP multicasting and associated QoS mechanisms are all major features of DOCSIS 3.0.

2.4.3.2 *Inside the DOCSIS 3.0 Technical Specification*   Before discussing the DOCSIS technical specification, it is first helpful to gain a high level understanding of how an end-to-end DOCSIS 3.0 system operates. A highly simplified graphical view of an end-to-end DOCSIS 3.0 system is shown in Fig. 2.6.

As illustrated the cable modem communicates via a two-way HFC network to a device located in the headend called a Cable Modem Termination System (CMTS). DOCSIS defines two variants of CMTSs—an integrated CMTS and a modular CMTS. An integrated CMTS consists of a single unit with RF interfaces and upstream network interface(s). A modular CMTS (M-CMTS) implements the upstream RF interfaces and the network interface(s); however, the downstream traffic is processed separately via cable modulators.

Figure 2.6 also includes a connection from the CMTS to network management system (NMS) and provisioning systems. The provisioning system comprises of a number of servers that provide different types of functionality including dynamic host configuration protocol (DHCP) IP address allocation, cable modem configuration parameters, and providing time services. The NMS provides monitoring and management services to the end-to-end networking architecture. A connection to the Internet is also provided to facilitate subscriber's access to high speed Internet access service. The DOCSIS 3.0 technology itself is a significant standard, which is broken down into four specifications.

(1) *CM-SP-PHYv3.0*—This specification deals with physical layer aspects of the technology.

TABLE 2.3 DOCSIS Technical Characteristics

	DOCSIS 1.0	DOCSIS 1.1	DOCSIS 2.0	DOCSIS 3.0
Maximum downstream broadband capacity (Mbps)	40 and 55	40 and 55	40 and 55	160
Downstream frequency range (MHz)	50–750	50–750	88–870	88–1002
Maximum upstream broadband capacity (Mbps)	10	10	30	120 and above
Upstream frequency range (MHz)	5–42	5–42	5–42	Support for 5–42 MHz is mandatory. Manufacturers also have an option of using the 5–85 MHz frequency range
Modulation scheme	QPSK and 16 QAM	QPSK and 16 QAM	QPSK, 8 QAM, 16 QAM, 32 QAM, 64 QAM, and 128 QAM	QPSK, 8 QAM, 16 QAM, 32 QAM, 64 QAM, and 128 QAM
Upstream channel width	200 kHz, 400 kHz, 800 kHz, 1.6 MHz, 3.2 MHz	200 kHz, 400 kHz, 800 kHz, 1.6 MHz, 3.2 MHz	200 kHz, 400 kHz, 800 kHz, 1.6 MHz, 3.2 MHz, 6.4 MHz	1.6,3.2 and 6.4 MHz. Note that support for the 0.2, 0.4 and 0.8 MHz channels are optional for this version
Support for version 6 of the IP addressing system	No	No	No	Yes

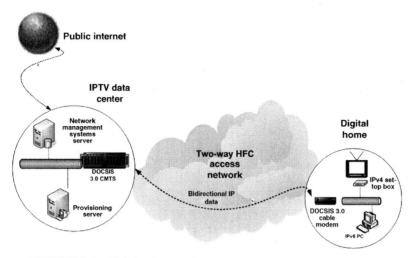

FIGURE 2.6   High level view of DOCSIS 3.0 networking infrastructure

(2) *CM-SP-MULPIv3.0*—This specification includes implementation details for the Media Access Control (MAC) and upper layer protocols used in an end-to-end DOCSIS 3.0 system.

(3) *CM-SP-OSSIv3.0*—This specification defines the requirements for configuring and managing features introduced in DOCSIS 3.0.

(4) *CM-SP-SECv3.0*—This final specification provides the necessary details required to secure DOCSIS 3.0 end-to-end systems.

The following sections give a brief overview of the technologies covered in these DOCSIS 3.0 documents when used in the context of an IPTV networking environment.

Examining the DOCSIS 3.0 PHY Layer   The PHY layer specification defined by DOCSIS 3.0 is covered in the CM-SP-PHYv3.0 document, which is available for download at the CableLabs Web site. The document itself provides technology implementation details for two network types:

• Systems deployed in North America, which use 6 MHz channeling to deliver multiprogram TV signals.

• Systems deployed in Europe, which use 8 MHz channeling to deliver multiprogram TV signals.

All descriptions described in this book apply to both technology types. DOCSIS 3.0 cable modems and CMTSs operate in the 50–1002 MHz frequency band for downstream transfer of services such as IPTV. For upstream communications, the

5–42 MHz or 85 MHz frequency band is used. Operating DOCSIS 3.0 frequencies varies between implementations. The physical media portion of the document defines the electrical characteristics and signal processing operations used between cable modems and CMTS(s) installed at the data center. With regard to gaining access to the physical network DOCSIS 3.0 defines two primary methods: time division multiple access (TDMA) and SCDMA (synchronous code division multiple access). The modulation of IPTV data onto these channels is achieved through the use of a wide variety of modulation schemes. The modulation scheme is dependent on the access method and the direction of the network traffic, namely, upstream or downstream. These schemes deal with the hostilities that are often present on a HFC network when used to deliver real-time data traffic such as IPTV.

Examining the DOCSIS 3.0 MAC Layer   DOCSIS 3.0 introduces a number of features that build upon what was present in previous versions of DOCSIS. These enhancements are detailed in the CM-SP-MULPIv3.0 document. These features are explored in the following subsections.

ADVANCED FRAME STRUCTURE TO SUPPORT NEW FEATURES   DOCSIS 3.0 has defined new ways of utilizing the MAC frame in order to support some of the new features incorporated into the technology. A frame is the basic unit of transfer used for communication at the MAC layer between the CMTS and cable modems. The length of the frame is variable and consists of two parts — a header and the video payload. The structure of a DOCSIS 3.0 packet header is illustrated in Fig. 2.7 and explained in Table 2.4.

Note that a DOCSIS 3.0 packet also includes overhead details that are used by the upper layers of the protocol stack.

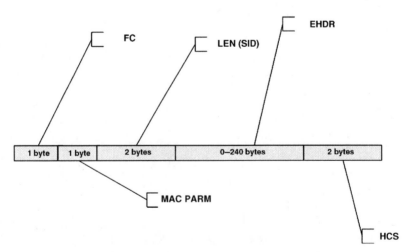

FIGURE 2.7   DOCSIS 3.0 MAC layer frame structure

TABLE 2.4   DOCSIS 3.0 MAC Layer Field Descriptions

Field Name	Size (Byte)	Description of Functionality
FC	1	Short for frame control, this field identifies the header type and identifies whether an extended header (EHDR) is present or not.
MAC_PARM	1	When an EHDR sequence field is present, this provides the EHDR length.
LEN (SID)	2	This defines the length of the frame.
EHDR	0–240	The extended MAC header is used to support a number of functions, including the tagging of Multicast IPTV packets and the sequencing of out-of-order packets.
HCS	2	HCS stands for header check sequence and is used to identify transmission errors that affect the header.

CHANNEL BONDING   The ability of DOCSIS 3.0 devices to use a technique called channel bonding to increase the data throughput over a HFC network allows cable operators to implement IPTV services. Under this mechanism, multiple smaller channels are bonded together to create a larger logical channel with high bandwidth capabilities. In addition to providing greater throughput when compared to a single channel, this mechanism also reduces the congestion delays associated with sending packets over a single channel. DOCSIS cable modems include multiple tuners, which are used to access the various channels available as part of the group of bonded channels.

The mechanism operates by concurrently distributing IPTV packets across a number of channels, which gets delivered to a cable modem. The distribution of packets across multiple channels can cause difficulties for the cable modem as some of the channels suffer from jitter and latency. If this is the case the packets will arrive out of order. To mitigate against such problems, DOCSIS 3.0 introduces a method of tagging or marking each packet with a particular sequence number, which is used by the cable modem to reassemble the original IPTV stream before forwarding onwards to an IP set-top box. The packet sequence number is 2 bytes in length and in the case of downstream traffic is included in the DS-EHDR field of the MAC header.

Overall management and allocation of bonded channels is the responsibility of the CMTS. The channel bonding mechanism can be used for transmitting data both upstream and downstream across a cable TV network.

LOAD BALANCING   DOCSIS 3.0 builds on the load balancing mechanisms used by previous versions of the technology. Load balancing is a critical feature of large cable TV networks that support hundreds and in some cases thousands of cable modems. The load balancing section of the DOCSIS 3.0 specification makes provisions to cope with the large volumes of traffic that are generated by multimedia applications such as IPTV.

MULTICASTING DOCSIS 3.0 allows cable operators to use IP multicasting as a method of delivering IPTV sessions and services to their subscribers. The use of multicasting is particularly attractive to cable TV providers because it reduces the amount of network bandwidth required to deliver a multichannel pay TV service.

To facilitate the use of multicasting DOCSIS 3.0 features an identifier called a multicast downstream service ID (MDSID). The MDSID allows cable modems that have been authorized to view a particular IPTV multicast stream to efficiently identify packets associated with the stream. It is 20 bits in length and is added by the CMTS to the packet header. Applying a MDSID to various packets allows cable modems to process only multicast and IP-VoD streams that are intended for transmission onwards to an IPTVCD connected to its home networking port. In addition to tagging network traffic and bonding multiple channels together DOCSIS 3.0 also accommodates support for multicast specific communications models and protocols such as

- Source Specific Multicast (SSM)
- Version 3 of the Internet Group Membership Protocol (IGMP)
- Multicast Listener Discovery (MLD) versions 1 and 2

Chapter 4 presents further details on these technologies.

INITIALIZATION AND PROVISIONING OF DOCSIS 3.0 MODEMS under DOCSIS 3.0, the cable modem is authorized by the CMTS for use on the network, and configures itself according to parameters that are passed to it from a provisioning server. This sequence of events occurs automatically without user involvement. The process of initializing a DOCSIS 3.0 modem on a HFC network is divided into the following phases:

(1) Once the cable modem is powered up, it scans the network for both a downstream and upstream channel that supports the transfer of data.
(2) After locating the channels is complete, the cable modem informs the headend of its presence on the broadband network by sending a variety of parameters back to the CMTS.
(3) Once the cable modem has been assigned to the network topology, it needs to be authenticated. This typically requires the cable modem to send a digital certificate, which is installed during manufacturing, to the CMTS for validation. Once validated a series of keys are generated and used to encrypt the data transferred across the network.
(4) After the modem is connected to the cable network, it must invoke DHCP mechanisms to obtain an IP address, which is part of the operators

authorized address space. Note that DOCSIS 3.0 allows IPTV network administrators to configure cable modems for both IPv4 and IPv6 address formats.

(5) Once the modem receives an IP address, it downloads a configuration file that has parameters the cable modem needs to configure itself. Time of day information is also downloaded to the modem.

(6) Once the modem has been configured and authorized, the CMTS authorizes resources at the MAC layer and the cable modem can use the network like any standard Ethernet network device.

SUPPORT FOR 61 MANAGEMENT MESSAGE TYPES   The DOCSIS 3.0 specification has defined 61 messages for exchanging management data between cable modems and a CMTS. The specification also provides a provision to extend this number up to 255. Management messages that are particularly relevant to IPTV implementations include the dynamic bonding change request, response, and acknowledge commands. A detailed description of all 61 management messages types is available in the specification on CableLabs Web site.

Examining the DOCSIS 3.0 Operations Support System Interface (OSSI)   The CM-SP-OSSIv3.0 document provides details on the various OSSI technologies. DOCSIS 3.0 OSSI specification defines the requirements to support fault, configuration, performance, security, and accounting management functions.

FAULT MANAGEMENT   In addition to the various mechanisms used by previous versions of DOCSIS to identify, monitor, correct, and record faults, DOCSIS 3.0 also provides for some new features:

- A diagnostic log that records operational problems such as repeated cable modem initialization and registration details.
- Enhancements that identify and resolve issues related to new functionality integrated into DOCSIS 3.0.

The management of faults in a DOCSIS 3.0 environment requires the use of protocols such as the simple network management protocol (SNMP) in combination with a number of different types of event logging and recording mechanisms. Note that Chapter 12 on IPTV network administration provides a description of the SNMP protocol.

CONFIGURATION MANAGEMENT   As the name suggests this function ensures that configuration parameters are kept current and are suitable for the smooth operation of the network. One of the key differences to this management function when compared to previous versions is an updated configuration file structure. The parameters within the file itself have been adjusted to support DOCSIS 3.0 technologies such as IPv6, multicasting, and channel bonding. CableLabs has adopted existing standardized protocols to support the execution of configuration management within a DOCSIS networking infrastructure.

PERFORMANCE MANAGEMENT   The purpose of this function is to use either IPDR or SNMP protocols to gather event and error data at the PHY, MAC, and IP layers of the DOCSIS protocol stack. This information is used to help engineers determine the overall health of the networking infrastructure.

SECURITY MANAGEMENT   This function is responsible for protecting various types of data in particular access to SNMP traffic. This helps to strengthen the security aspect of DOCSIS by minimizing the risk of hackers using SNMP as a mechanism to damage a cable providers networking infrastructure.

ACCOUNTING MANAGEMENT   This function is executed by the CMTS and its core responsibility is to gather network usage statistics for individuals and businesses who have subscribed to one of the available services, such as high speed Internet access. Once collected this usage information is passed to the billing system. The mechanism for delivering subscriber usage billing records is facilitated through a schema that includes the use of a streaming protocol developed by the IPDR organization. This networking protocol has been designed to efficiently transport high volumes of XML based usage billing records over a Transmission Control Protocol (TCP) network connection. To reduce the size of these records, CableLabs has additionally specified the use of a compression mechanism called IPDR/XDR. As the name suggests this encoding format was developed by the IPDR organization and uses the External Data Representation Standard (XDR) coding format defined in RFC 1832.

Examining DOCSIS 3.0 Security Mechanisms   The security mechanisms required to support communication between different devices across a DOCSIS 3.0 enabled infrastructure is described in CM-SP-SECv3.0. These various mechanisms help to ensure that subscriber's privacy is not compromised and preventing theft of services such as IPTV. The core portion of the DOCSIS 3.0 security system is based on the Baseline Privacy Plus (BPI+) scheme. The architecture of this scheme is logically portioned into two protocols.

AN ENCAPSULATION PROTOCOL   This protocol encrypts the IPTV packets as they traverse the cable TV network. In addition to encrypting packets that are destined for IPTV and triple-play users, the BPI+ encapsulation protocol is also used to encrypt other types of protocol information, which is used in the provisioning of cable modems, namely DHCP, Trivial File Transfer Protocol (TFTP), and various types of management messages that are transported via the MAC layer. Note that DOCSIS 3.0 provides stronger network traffic encryption compared to its predecessors through its support for the 128-bit Advanced Encryption Standard (AES) algorithm.

A KEY MANAGEMENT PROTOCOL   The protocol designed by CableLabs to secure the distribution of keying data between the CMTS and cable modems is called

Baseline Privacy Key Management (BPKM). Standard technologies such as digital certificates and public-key encryption algorithms are used by BPKM to secure key communications across the HFC network. Note that MAC management messages discussed previously are used to transport the BPKM protocol information.

DOCSIS 3.0 uses BPI+ to secure initialization of a cable modem onto the network. The BPI+ security process commences when the modem identifies a communications channel. At this stage the modem sends an authentication information message to the CMTS. The details contained in this message are described in Table 2.5.

Once the authentication information message arrives at the CMTS, it is verified and the CMTS responds back to the cable modem using an authorization reply message. This message includes identification details and an encrypted key.

The BPI+ also uses a technique called source address verification to eliminate IP spoofing by in-home networking devices. To enforce this security policy DOCSIS 3.0 specifies that any network packets that originate from a device whose IP source address has not been assigned by the IPTV service provider is discarded.

2.4.3.3 EuroDOCSIS The European cable industry has developed its own standards for high speed data transfer across a cable TV network. For the most part the technical details follow the DOCSIS system very closely. The primary difference between the two standards is the difference in channel widths. European cable

TABLE 2.5   Structure of BPI + Authentication Information Message

Item of Information	Purpose
Identifiers	The identifiers include the cable modems hardware address and details about the manufacturer.
Public key	This security component is incorporated into the device during manufacturing.
Digital certificate	The use of a digital certificate allows the modem to be authenticated by the CMTS and restricts network access to authorized devices. Digital certificates are supplied to "trusted vendors" who incorporate it into DOCSIS modems during manufacturing.
Cryptographic algorithm descriptor	This provides the CMTS with information on the authentication and data encryption algorithms supported by the cable modem.
Security association identifier (SAID)	This item of information identifies the security information, which is shared between the CMTS and one or more of its client cable modems.

set-top boxes use 8 MHz tuners whereas set-top boxes deployed in American homes
are optimized to access channels that use 6 MHz of radio spectrum. The wider
channel width allows cable operators to deploy broadband services with
approximately 33% more bandwidth than is typically configured in a DOCSIS-
compliant system.

It is also important to note that for HFC networks, the MPEG-2 standard is the
most ubiquitous mechanism of compressing video content. Therefore, additional
hardware needs to be installed at the headend to facilitate the transportation of
MPEG-2 encoded transport streams across an IP cable broadband network.

2.4.4   IPTV Over a Satellite Based Network

IP is also emerging as a preferred method of distributing video content via satellite
links. Satellite links can provide higher bandwidth than terrestrial transmission
networks and are starting to get used for IP based triple-play services that comprise
of digital video content, VoIP, and high speed Internet access.

Many of the satellite network providers have started to use their satellite based
networking platforms to deliver IP video content to cable and telecommunication
headends and IPTV data centers. The networking infrastructure used to support this
mechanism of IPTV distribution is shown in Fig. 2.8.

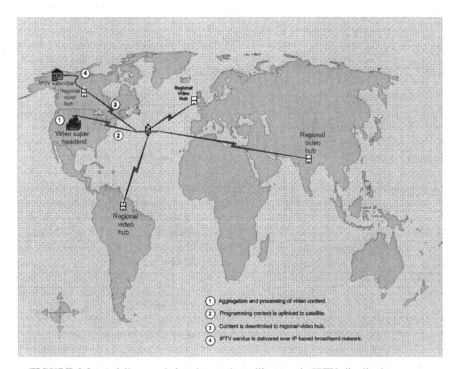

FIGURE 2.8   A fully rounded end-to-end satellite centric IPTV distribution system

As shown the original content is received, aggregated, encoded in MPEG-2, MPEG-4, or Windows Media format and encrypted at the satellite operator's video operations center. Once processed at the operations center, the content is uplinked to a satellite and relayed back down to various video hubs. These video hubs are generally operated by cable or telecommunication companies and use their own existing networking infrastructure to deliver IPTV services to residential subscribers.

With regard to delivering IPTV services directly to consumers a number of options are currently available.

2.4.4.1 Deploy Hybrid Satellite IP Set-top Boxes The deployment of hybrid satellite IP set-top boxes allows consumers to access traditional satellite services via the satellite link and IPTV services via a standard broadband connection.

2.4.4.2 Utilize Standard IP Set-top Boxes This model involves the delivery of programming channels using standard satellite transmission techniques to large housing developments, converting the channels to IP, and streaming to IP set-top boxes.

2.4.4.3 Provide Subscribers with Set-top Boxes That Include a Hard Disk The large bandwidth capacity requirements of delivering on-demand content makes IP-VoD based applications for most satellite systems impractical. However, some satellite service providers have started to circumvent the limitations of satellite based delivery systems by deploying set-top boxes that incorporate a hard disk. Once installed at the subscribers home, the IP-VoD content is automatically downloaded to the hard disk. This allows IPTV end users to view the content at their own convenience.

2.4.4.4 Use Satellite Broadband Modems Broadband modems may also be used to deliver satellite IPTV services. These modems generally comply with one of three international standards:

(1) *IP over satellite (IPoS)*—This standard has been ratified by U.S. based Telecommunication Industry Association (TIA) and ETSI. It utilizes a technology called DVB-S2 (please refer to Chapter 5 for specific details on DVB-S2) and supports data throughputs of up to 120 Mbps.

(2) *DVB return channel via satellite (DVB-RCS)*—DVB-RCS was developed by European based DVB. It defines a forward-link data transmission rate of 40 Mbps and a return channel capacity of approximately 2 Mbps. DVB-RCS is officially defined in ETSI EN 301 790.

(3) *DOCSIS over satellite*—This mechanism for transmitting IPTV content over a satellite link is based on an adapted version of the DOCSIS standard. The main difference between cable DOCSIS and satellite DOCSIS is the use of the Quadratare Phase Shift Keying (QPSK) modulation scheme instead of the QAM, which has been designed for HFC networks. Early versions of the protocol provide support for speeds of 1.5 Mbps, whereas newer versions

can achieve much higher data rates. These newer rates mean that satellite operators can start to leverage IP based technologies to deliver video content across their networks.

2.5 IPTV OVER WIRELESS NETWORKS

New broadband wireless networks provide telecom operators with another alternative distribution platform to deliver IPTV services to households. Various options are available and are described in the following sections.

2.5.1 Fixed WiMAX

Demands from consumers and the telecommunication sector to use WiMAX as a platform to carry IPTV content is growing at a steady pace. WiMAX (Worldwide Interoperability for Microwave Access) is a high capacity IP broadband wireless technology that is considered by the industry to be a close "relative" of the Wi-Fi family of wireless standards. It defines a number of services that conform to the IEEE 802.16 technical standard. The WiMAX Forum, an industry association, is responsible for developing WiMAX specifications, promoting the technology, and

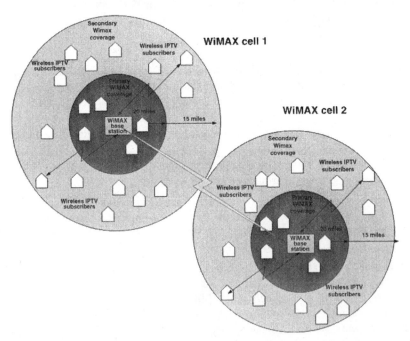

FIGURE 2.9 Simplified block diagram of a WiMAX System carrying IPTV traffic

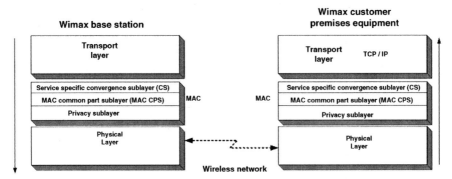

FIGURE 2.10   WiMAX communications model

managing the overall certification of WiMAX products. The organization boasts a membership of over 370 companies at the time of writing this book. Figure 2.9 shows a simplified diagram of two WiMAX broadcast cells connected together and delivering video content to a number of IPTV end users.

The technical characteristics of fixed version WiMAX follow.

2.5.1.1 Operating Frequencies   WiMAX will operate within licensed and unlicensed frequency bands. These bands have been allocated by various communication regulation bodies around the world. The licensed bands are the preferred operating frequency option for real-time applications such as IPTV because there is less chance of interference occurring. Fixed WiMAX operates in frequencies of 3400–3600 MHz.

2.5.1.2 Physical and MAC Layer Protocols   As depicted in Fig. 2.10, the 802.16 communications model defines three layers: physical, MAC, and transport.

WiMAX Physical Layer Properties   Under the WiMAX standard, equipment manufacturers have a choice between three different PHY options when building products:

(1) The *Single Carrier* physical layer option is intended for straightforward line of sight applications.
(2) The *Orthogonal Frequency Division Multiplexing (OFDM)* option is the most popular physical layer choice for most WiMAX equipment manufacturers because of its ability to deal with the issue of multipath propagation. This feature means that OFDM based WiMAX technologies are particularly suited to the delivery of IPTV services.
(3) *Orthogonal Frequency Division Multiple Access (OFDMA)* is the most sophisticated option and is capable of separating user connections on the upstream frequency channels.

WiMAX MAC Layer Properties   The MAC layer is subdivided primarily into three sublayers:

(1) *Service specific convergence sublayer (CS)*—The main purpose of this sublayer is to interface with the higher layers in the WiMAX communication model.

(2) *MAC common part sublayer (MAC CPS)*—This sublayer takes care of core MAC functionalities such as security, management of connections, and access to the physical network.

(3) *Privacy sublayer*—As suggested by the name this manages authentication of IPTV subscribers and encryption of the video content.

WiMAX Transport Layer Properties   Standard TCP/IP is generally used at the network and transport layers to ensure delivery of IPTV services.

2.5.1.3   Transmission Ranges   Geographic topologies combined with other factors such as equipment specifications and weather conditions can all have an impact on the distance between a IPTV consumer device and a WiMAX base station. WiMAX has a theoretical maximum speed of approximately 60 Mbps within a coverage area of 6–10 km. This varies between implementations and equipment vendors. Assuming that the contention ratios are correctly planned, these data throughput levels will comfortably allow subscribers within the WiMAX coverage area to access IPTV services. It is important to note that WiMAX can support both line of sight (LoS) and nonline of sight (NLOS).

2.5.2   Mobile WiMAX

IEEE 802.16 cannot be used to provide broadband services in a mobile environment. For this purpose, an amendment to the IEEE Standard 802.16 standard was developed called IEEE 802.16e. Also known as Mobile WiMAX, IEEE 802.16e was approved in 2005 and certified products were released to the market in 2006. It operates in several licensed spectrum bands: 2.5, 3.3, and 3.4–3.8 GHz. Mobile WiMAX incorporates a number of key features that are necessary to transport IPTV services and applications:

- The technology supports peak data speeds of around 32–46 Mbps. These types of speeds if deployed correctly allow delivery of compressed IP based high definition content to mobile handsets.
- It utilizes technologies such as OFDMA and optimized handoffs to allow IPTV viewers access multicast broadcast TV channels in geographical areas that are susceptible to the affects of multipath transmission paths.
- It integrates with the IP multimedia subsystem (IMS), which simplifies interworking between IPTV applications and other IP based services such as

high speed Internet access and VoIP. Note that IMS is an emerging technology architecture that allows network operators to accelerate and simplify the deployment of IP based services.

• Mobile WiMAX provides support for advanced quality of services (QoS) mechanisms, which are beneficial to real-time applications such as IPTV.

In addition to the above characteristics, it is important to note that at the time of writing the WiMAX forum continues its work on expanding the multicasting capabilities of Mobile WiMAX. This is expected to further enhance the ability of Mobile WiMAX to meet the stringent requirements associated with broadcasting live IP based TV channels to mobile devices.

2.5.3   Wireless Municipal Mesh Networks

Municipal Mesh Networks, also known as muni networks are another wireless platform that promises to support the delivery of IPTV services to end users. A number of these network types have been deployed in various cities and towns around the world.

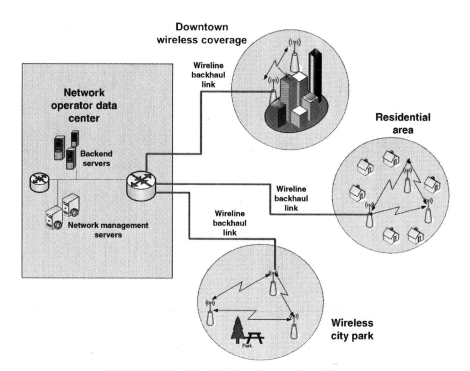

FIGURE 2.11   Municipal mesh network architecture

Municipal networks operate in an outdoor environment in either the unlicensed 2.4 or the 5 GHz spectrum range. Wi-Fi, also known as the 802.1x family of wireless products has been the technology of choice for building mesh networks, because most if not all notebooks and handhelds manufactured nowadays come with in-built Wi-Fi interfaces. Constructing Wi-Fi networks in an outdoor environment requires the use of a number of access points interconnected to each other and to a wired connection that provides backhaul to the broadband service provider (Fig. 2.11).

2.5.3.1 Mesh Wi-Fi Access Points (APs)   APs used by municipal networks cover a much greater area compared to conventional indoor APs. They are generally attached or mounted onto fixed physical structures that provide a good line of sight and easy access to power. Suitable locations for mounting outdoor APs include light posts, tall buildings, and communication towers. Interconnecting outdoor APs back to a central point using some type of physical cable is cost prohibitive for most deployments. Therefore, all APs in a municipal wireless networking architecture dynamically connect wirelessly with each other and a gateway AP in a cluster type configuration. The gateway aggregates the 802.11x signals within the cluster and interfaces via an Ethernet port with the broadband backhaul link. The number of APs in a cluster varies between implementations. Each AP in the cluster is ruggedized and uses a mesh routing protocol to interconnect with other APs and back to the backhaul point of presence. The main responsibility of the routing protocol is to provide the most efficient route for IP packets through the mesh from each AP to and from the backhaul link. It does this by continuously monitoring the wireless network and identifying wireless paths that provide the greatest bandwidth throughput capabilities. Mesh APs come in two variants single and multiple radio configurations:

- *Single Wi-Fi radio-mesh APs* use one channel to support access from various client devices in addition to carrying interconnectivity traffic to and from the mesh network. To minimize constraints and improve performance, these types of APs are typically configured into clusters that operate at different frequencies to their neighboring clusters.
- *Dual Wi-Fi radio-mesh APs* use separate channels for carrying mesh traffic and providing access to client Wi-Fi devices in the 2.4 GHz frequency band. The use of two separate channels that operate in different frequency bands offer improved performance levels and reduced latency levels that make dual channel APs more suitable to carrying time-sensitive applications such as IPTV.

2.5.3.2 Wired Backhaul Connectivity   A wired backhaul is required to provide connectivity to the IP data center and onwards to the public Internet. A technology called virtual LANs (VLANs) is often used to segment the different types of traffic that traverse a wireless municipal network. Note that VLAN technology will be described in greater detail later in the book.

At the time of writing the average downstream data rate of a municipal wireless network is approximately 1 Mbps, which is more than adequate for public Internet access applications. However, IPTV has more demanding throughput needs, therefore, deployments of video centric applications over these types of networks are typically confined to specialized functions such as streaming IPTV content from Wi-Fi cameras.

2.5.4   3G Networking Technologies

Mobile networks based on 3G technologies such as EV-DO and HSDPA are also capable of delivering a range of mobile IPTV applications.

2.5.4.1   EV-DO   Evolution-Data Optimized (EV-DO) is a wireless radio broadband data standard that boosts maximum data rates up to 4.9 Mbits/s.

2.5.4.2   HSDPA   High-Speed Downlink Packet Access (HSDPA) supports rates up to approximately 14 Mbps on the downstream path, with higher speeds planned in the future.

Although not an ideal platform for delivering IPTV services, EV-DO and HSDPA do provide network operators with the ability to deliver IPTV services to consumers who live in areas that are poorly served by DSL and cable broadband systems.

2.6   IPTV OVER THE INTERNET

Since the invention of television, a number of different distribution technologies have been developed to deliver signals to consumers around the world. Until recently, there were five primary networking platforms used to distribute TV content, namely, wireless off-air, satellite, DSL, fiber and cable TV networks. In more recent times a new platform has emerged that also allows consumers to view broadcast and on-demand video content — the Internet.

Improved broadband speeds combined with advances in compression technologies and greater viewing choices are some of the reasons why consumers have started to increasingly turn to the Internet for video entertainment over the last couple of years. IPTV over the Internet is available in a number of formats.

2.6.1   Streamed Internet TV Channels

The delivery of TV channels over the Internet is a popular IPTV application and involves the streaming of video content from a server to a client device, which is capable of processing and displaying the video content. The type of device used to view Internet TV channels is typically a PC or a media center PC. Streamed Internet TV channels can however also be accessed via a mobile phone or an IP set-top box. The content available via streamed Internet TV channels is delivered in real time, and the viewing experience mimics the traditional approach to TV viewing, namely, a channel is chosen and viewing commences immediately. The technical process of streaming an Internet TV channel usually starts at the streaming server where the video content is broken into multiple IP packets, compressed, and then transmitted or streamed across the Internet to the client PC. The PC includes software, typically a browser, which decompresses the video content and generates a live video feed. The time period between selecting a TV channel and the commencement of viewing is generally quite short and depends on the connection speed that is available between the client and the server. A highly simplified graphical view of a networking architecture used to deliver a single streamed TV channel over the Internet is presented in (Fig. 2.12).

In all Internet TV channel deployments a streaming server is required. In addition to playing out video content upon request from an IPTV subscriber, a standard streaming server will also support the following functions:

- Storing and retrieving source video content.
- Controlling the rate at which the IP video packets are delivered to the client viewing device.
- Executing forward and back commands as requested by the Internet TV viewer.

A single streaming server works fine for delivering a small number of Internet TV channels to a limited number of users. To support the delivery of multiple channels to thousands and possibly hundreds of thousands of IPTV subscribers, a number of streaming servers need to be deployed in different parts of the network.

From a security perspective the streaming of video content is quite secure because the material is not stored on the client's access device. Thus, unauthorized

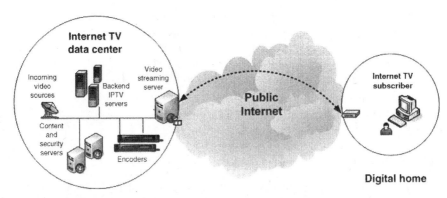

FIGURE 2.12 Internet TV channels network infrastructure

copying of content is prohibited. Other benefits of this flavor of IPTV include its ability to operate effectively over low bandwidth connections and viewers have the ability to start watching content at any point in the IPTV stream.

What separates the delivery of Internet TV with the other delivery mechanisms discussed in this chapter is the fact that Internet portal sites do not own or control the underlying infrastructure used to stream IP video services to Internet users. The networking infrastructure is generally owned by either cable TV providers or telecommunication companies.

2.6.2 Internet Downloads

As the name suggests this flavor of IPTV allows consumers to download and watch on-demand content. Most Internet download services are subscription based or pay-per-download and can include a mix of local news and weather, movies, local films and music, an entertainment guide, and classified advertising. A number of the major online Internet based portal sites have recently started to offer their own libraries of downloadable IPTV programming content to Internet users. In most cases standard or media center PCs are used to view Internet downloads; however, some companies have started to provide broadband enabled IP set-top boxes for

TABLE 2.6 Technical Characteristics of an Internet Download IPTV Service

Characteristic	Description
Networking protocols	Standard file transfer protocol (FTP) and hypertext transfer protocol (HTTP) standards are generally used to transfer the IPTV content from the server to the client device. The use of these protocols minimizes the possibility that the IPTV content will be blocked by firewalls.
Server technology	Standard Web server software is generally used to service requests for on-demand video content.
Network speed	The time taken to download an Internet movie depends on the speed of the broadband connection and quality of the video content. Standard definition movies and programs download relatively quickly in comparison to HD based video content. Although broadband is the preferred connection type it is possible to also use slower dial-up links to access Internet download services.
Network errors	Standard TCP error correction techniques are used to correct problems during the downloading process.
Storage requirements	Both server(s) and client(s) require advanced storage capabilities to support the processing of large IPTV files. Some Internet download applications allow IPTV subscribers to burn a copy of the downloaded video content to a DVD and play in a DVD player.

consumers who do not want to watch videos on their PCs. The technical characteristics of an end-to-end IPTV based Internet download service are shown in Table 2.6.

2.6.3 Peer-to-Peer (P2P) Video Sharing

A peer-to-peer video sharing application allows users to watch, share, and create online video content. Using a peer-to-peer video sharing application is quite straightforward and typically involves the download and installation of specialized software. Once the software is operational on the PC, the user simply clicks on a link to download a particular video file. Once the download process is initiated, the P2P video sharing application software establishes connections and starts to retrieve the requested video content from a variety of different sources. Once the video file is downloaded to the local hard disk, viewing can commence.

2.7 IPTV BACKBONE TECHNOLOGIES

The backbone or core of an IPTV networking infrastructure is required to carry large volumes of video content at high speed between the IPTV data center and the last mile broadband distribution network. There are several different types of backbone transmission standards that provide multipath and link protection capabilities that are necessary to ensure high reliability capabilities. Each standard has a number of specific features including data transfer speed and scalability. Three of the major backbone transmission technologies used in IPTV network infrastructures are ATM over SONET/SDH, IP over MPLS and metro ethernet. As

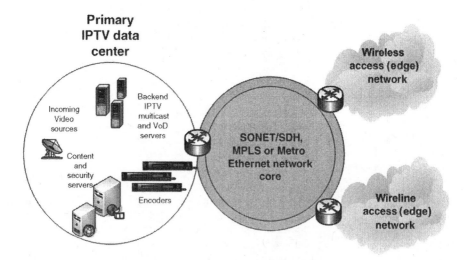

FIGURE 2.13 IPTV core-networking infrastructure.

illustrated in Fig. 2.13 these core networking technologies provide connectivity between the centralized IPTV data center and various types of access networks.

2.7.1   ATM and SONET/SDH

As described, ATM can support demanding applications such as IPTV that require high bandwidth and low transmission delays. ATM will operate over different network media including coaxial and twisted pair cables, however, it runs at its optimum speed over fiber optic cable. A physical layer called Synchronous Optical Network (SONET) is typically used by a number of telecommunication carriers to transport ATM cells over the backbone network.

SONET is a protocol that provides high speed transmission using fiber optic media. The term Synchronous Digital Hierarchy (SDH) refers to the optical technology outside the United States. The SONET signal rate is measured by optical carrier (OC) standards. Table 2.7 illustrates the available transmission rates (called OC levels).

SONET uses time division multiplexing (TDM) to send multiple data streams simultaneously. With TDM, the SONET network allocates bandwidth to a certain portion of time on a specific frequency. The preassigned time slots are active regardless of whether there is data to transmit.

In the context of an IPTV networking environment the SONET equipment receives a number of bit streams and combines into a single stream, which then is sent out onto the fiber network using a light emitting device. The rates of the combined input rates will equal the rate outputted from the SONET device. For example, four input streams carrying IPTV traffic at 1 Gbps will be combined by the SONET device and a 4 Gbps stream is forwarded onto the fiber network.

2.7.2   IP and MPLS

A number of larger telecommunication companies have started to deploy the Internet Protocol in their core networks. Although IP was never originally designed

TABLE 2.7   SONET Optical Carrier Standards

OC level	Signal Transmission Rate
OC-1 (base rate)	51.84 Mbps
OC-3	155.52 Mbps
OC-12	622.08 Mbps
OC-24	1.244 Gbps
OC-48	2.488 Gbps
OC-192	10 Gbps
OC-256	13.271 Gbps
OC-768	40 Gbps

with features such as QoS and traffic segregation capabilities the protocol works quite well in these environments when combined with a technology called Multiprotocol Label Switching (MPLS). An MPLS enabled network supports the efficient delivery of various video traffic types over a common networking platform.

An MPLS platform is designed and built using advanced Label Switch Routers (LSRs). These LSRs are responsible for establishing connection-oriented paths to specific destinations on the IPTV network. These virtual paths are called Label Switched Paths (LSPs) and are configured with enough resources to ensure the smooth transition of IPTV traffic through an MPLS network. The use of LSPs simplifies and speeds up the routing of packets through the network because deep packet inspection only occurs at the ingress to the network and is not required at each router hop.

The other main function of LSRs is to identify network traffic types. This is achieved by adding a MPLS header onto the beginning of each IPTV packet. The key elements of an MPLS header are explained in Table 2.8.

As illustrated in Fig. 2.14 the header is added at the ingress LSR and removed by the egress LSR as it leaves the MPLS core network.

While the IPTV traffic traverses across MPLS enabled routers a number of local tables called Label Information Bases (LIBs) are consulted to determine details about the next hop along the route. In addition to examining the table, a new label is applied to the packet and forwarded onward to the appropriate router output port. Other added benefits of MPLS networks include their support for high levels of resilience when a failure occurs.

2.7.3 Metro Ethernet

Another technology, which may be deployed in the core network, is Metro Ethernet. An alliance of leading service providers, equipment vendors, and other

TABLE 2.8 MPLS Header Format

Field Name	Field Length (Bits)	Description of Functionality
Label	20	Contains specific next hop routing details that are specific to each IPTV packet.
Experimental bits	3	Reserved for other uses. For instance the MPLS-Diffserv mechanism described in Chapter 12 uses this field.
Stacking bit	1	A header can contain one or more labels. Once the stacking bit is set to one, the LSR will identify the last label in the packet.
Time to live (TTL)	8	This value is copied from the TTL field in the IP header.

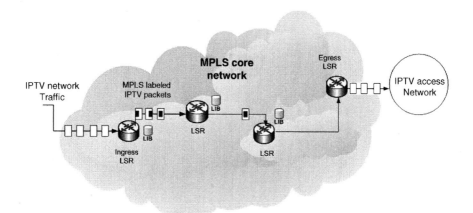

FIGURE 2.14   MPLS core network topology

prominent networking companies promote the technology through a consortium called the Metro Ethernet Forum (MEF). The MEF is responsible for establishing specifications for integrating Ethernet technologies into high capacity backbone and core networks. In addition to developing specifications the MEF also certifies Ethernet equipment for use in service provider's core networking infrastructures. The key technical and operational characteristics of Metro Ethernet based core networks include:

- It meets the various requirements that are typical of a core networking technology, namely, resilience, high performance, and scalability.
- Some of the modern Metro Ethernet networking components can operate at speeds up to 100 Gbps across long geographical distances. This provides service providers with an ideal platform for efficiently delivering new value-added services such as IPTV to geographically dispersed regional offices.
- It implements a sophisticated recovery mechanism in the event of a network link failure thereby, ensuring that services such as IPTV are unaffected by the outage.
- Metro Ethernet technologies support the use of connection orientated virtual circuits that allow IPTV service providers to guarantee the delivery of high quality video content within the network core. These dedicated links are called Ethernet Virtual Connections (EVCs). Figure 2.15 shows how four EVCs are used to provide connectivity between the IPTV data center and a number of regional offices.

In addition to the above characteristics, the low delay and packet loss features of Metro Ethernet make it an ideal core networking technology for carrying IPTV services.

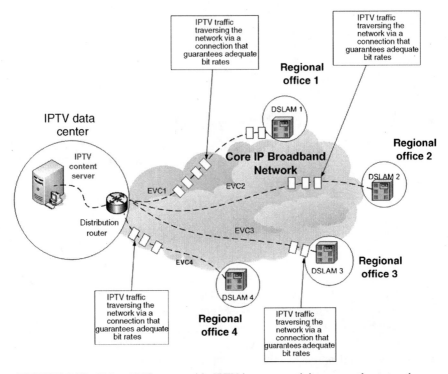

FIGURE 2.15   Using EVCs to provide IPTV interconnectivity across the network core

2.8   NETWORK FACTORS ASSOCIATED WITH DEPLOYING IPTV

A number of network factors need to be considered before commercially launching IPTV services.

2.8.1   Network Dimensioning

To support the transport of video, IPTV distribution networks need to feature high bandwidth carrying capacities. The amount of bandwidth required to carry IPTV services is generally a multiple of the bandwidth required to support Voice over IP (VoIP) and Internet access services. The total bandwidth required to implement IPTV services depends on a couple of factors:

The number of IPTV multicast channels on offer—As noted a single copy of each channel is sent from the IPTV data center on to the distribution network. Once the channel is streamed onto the networking infrastructure the multicast process handles the copying of channels and routing to individual IPTV subscribers. Consider an example of a service provider who is offering its 10,000 subscribers a package of 100 standard definition (SD) IP broadcast TV channels. If we assume that the provider is using H.264 to compress the channels, this generally translates to a bandwidth requirement of at least 2 Mbps for each broadcast channel. In the scenario where at least one subscriber is accessing each channel at a particular instance in time then the Next Generation Network (NGN) core distribution network will require 200 Mbps of bandwidth capacity. This is the bandwidth requirement from the IPTV data center to each of the regional offices. At these offices techniques such as IGMP snooping can be used to reduce the bandwidth required over the local access section of the network. Note that H.264 and IGMP will be described in greater detail later in the book.

Inclusion of IP-VoD services—The dimensioning of the network is further complicated by the addition of IP-VoD applications. These types of applications use the unicast transport mechanism to provide communications between the IPTV consumer devices and the on-demand video server. This mode of operation consumes a large amount of bandwidth and the network needs to accommodate this level of network traffic. Consider the same network of 10,000 end users and assume that at a particular instance in time there is 5% of the subscriber base accessing IP-VoD titles. By assuming again that the H.264 compression standard is used, this translates to a peak usage on the network of 1 Gbps ($10,000 \times 5\% \times 2$ Mbps). This is a significant requirement for the core network.

2.8.2 Reliability

The IP networking infrastructure needs to be reliable in the event of device failures. There should be no single point of failure that could interrupt the delivery of IPTV services, both multicast or unicast applications. Redundant links should be used wherever possible.

2.8.3 Fast Responsiveness

The network needs to support minimum response times associated with channel zapping (refers to changing from one channel to another during a TV viewing experience).

2.8.4 Predictable Performance

The nature of video bit rate streams is variable due to the differing scene complexities, which are delivered to an IPTV access device on a frame-by-frame basis. Therefore, it is difficult to predict the exact requirements of a video transmission until the service is operating in real-time. IPTV operators have to bear this in mind and assign appropriate network resources to cope with variable bit rate streams.

2.8.5   Level of QoS

Owing to the fact that most IPTV services operate over a private IP broadband network, it is advisable to implement a QoS policy when delivering video content to paying subscribers. A QoS system preserves a video signal and lessens the probability of impairments as it gets transmitted over long distances. It allows operators to provide services that require strict performance guarantees such as IP-VoD and IP Multicast. It comprises of a number of network techniques and supporting protocols that guarantee IPTV subscribers a specific level of viewing quality. The subject of IPTV QoS is covered in greater detail later in the book.

SUMMARY

Today, telecom operators have a plethora of network technologies to choose from when deciding to deploy IPTV services. Table 2.9 summarizes the characteristics of these various networking media used to carry IPTV signals.

The virtually unlimited bandwidth capabilities of fiber based networks make it an ideal platform for carrying IPTV services.

Technologies such as ADSL2 and VDSL provide networks with high bandwidth speeds. They are fully scalable communication solutions and are suitable for delivering advanced IPTV applications to end users.

DOCSIS 3.0 is a next generation broadband networking platform that allows cable operators to offer advanced IPTV services to their subscribers.

The satellite industry has also started to augment their broadcast TV services with new IP video products that are aimed at the telecommunication and cable TV industry sectors. The networking architecture used to support these new IPTV based products includes a video headend for processing, satellites for distribution,

TABLE 2.9   IPTV Network Distribution Technologies

Networking Technology	Capacity Ranges	Modulation Techniques
FTTx	Several gigabits per second	Dependent on the MAC layer technology deployed to carry the IPTV services
DSL	Between 1 and 100 Mbps	CAP and DMT
HFC	Over 100 Mbps	QAM
Satellite	Between 1 and 10 Mbps	QPSK
Fixed wireless	Between 1 and 10 Mbps	Varies according to the type of networking platform
Internet	Dependent on the capacity of the network links used by the Internet to carry the video content	Open schemes are typically used

video hubs for regional aggregation, and broadband access networks for delivering the IPTV signals to subscribers.

The delivery of video over wireless networks is another approach for delivering IPTV services. Emerging wireless technologies such as Wi-Fi, WiMax, HSDPA, and EV-DO are seen as platforms that will improve adoption of mobile and IPTV anywhere services in the coming years.

The ability to efficiently distribute TV over the public Internet is fundamentally transforming the global digital TV industry. Guaranteeing a quality of service is one of the biggest issues that service providers have to grapple with when delivering IPTV over the public Internet. Delays, network congestion, and corruption of IP packets are only some of the conditions that Internet TV channel producers and large video service portal sites have to deal with on a daily basis.

IPTV has some very strict requirements when it comes to networking infrastructure, namely, adequate bandwidth capacities, predictable performance levels, support for a QoS system, and in-built redundancy.

3

IPTV REAL-TIME ENCODING AND TRANSPORTATION

With the convergence of television, telephony, and Internet access, various compression technologies are playing a pivotal role in the growth of new products and services for carriers and equipment vendors alike. In this chapter, we will explain the capabilities of the three most powerful compression technologies used by IPTV systems, namely, MPEG-2, H.264/AVC, and VC-1. Furthermore, the chapter provides a detailed explanation of a networking framework called the IPTV communication model (IPTVCM). IPTVCM is a conceptual environment that consists of seven and potentially eight layers that are stacked on top of each other, each specifying particular network functions. This chapter describes each of these layers and explains the protocols required to support the responsibilities assigned to each layer in the model.

3.1 INTRODUCTION TO REAL-TIME ENCODING

Before discussing encoding, it is first helpful to understand the two processes that occur prior to the video reaching the encoding stage. The first process is relatively straightforward and involves the use of a camera to take and capture the video content. Once captured, which is typically in analog format, the video content needs to pass through another process called digitization to convert a continuous analog signal into a series of digital bits. A specialized piece of hardware called an analog-to-digital (A/D) convertor is used to execute the conversion process.

Next Generation IPTV Services and Technologies, By Gerard O'Driscoll
Copyright © 2008 John Wiley & Sons, Inc.

Sampling and quantization techniques are applied during the processing of the signal. As the name suggests *sampling* refers to the number of samples taken from the incoming analog signal. The sample rate is typically measured on a per-second basis. *Quantization* is the second part of the conversion process and involves the assignment of a number of bits to each sample taken from the signal. Once in digital format the uncompressed video bit stream is ready for encoding.

The encoding of digital bit streams requires the use of specialized hardware devices called encoders. The encoding process at an IPTV data center is quite involved and typically occurs in three separate steps:

(1) A video feed is received from a particular source. The format of this feed can vary quite dramatically and ranges from low quality analog signals to high quality digital streams.

(2) Once received, the encoder applies a particular compression scheme to the content. There are a couple of popular compression schemes used by encoders, which are covered in this chapter.

(3) Once compressed, the video is prepared for transmission. Preparation effectively means that the content is inserted into data packets. There are a couple of approaches involved with packetizing and encapsulating video content.

The encoding of video content for an IPTV networking environment has numerous advantages and disadvantages. The primary advantages of the technology include the following:

● A large reduction in the amount of hard disk space required to store video files.

● Processing uncompressed video requires a significant amount of computational power. Applying compression techniques to a piece of video content cuts down on the number of processor instructions required for processing and rendering on a screen.

● Owing to the fact that the file size of a compressed file is smaller then the original, the amount of time required to send it over the network is reduced.

● Relatively low capacity broadband connections may be used to deliver IPTV content. For instance, an encoded stream of a SDTV channel will take up approximately 1.5 Mbps, whereas a HDTV channel will occupy approximately 8 Mbps. When compared to older compression standards that require 3.5 Mbps for SDTV and between 20 and 25 Mbps for HDTV, the benefits of modern compression techniques are clear.

There are however some downsides to using compression technologies when delivering IPTV services:

● The process of compressing and decompressing a signal introduces delays.

- Owing to the fact that some information is discarded during the compression process, the overall quality of the image will be less than an uncompressed signal.
- The transcoding of incoming signals and changing from one compression format to another can affect the quality of the signal.

So there are trade offs with regard to encoding signals for transport over an IP network.

3.2   COMPRESSION METHODS

Compression allows IPTV service providers to broadcast several high quality video and audio channels over an IP broadband network. It achieves this objective by taking advantage of deficiencies in the human and aural systems and exploiting this fact through the use of mathematical algorithms. For instance, the human eye cannot detect all image patterns. Therefore, compression reduces the size of the original signal by removing these sections of the image. The level of compression applied to video content is called "compression ratio" and is measured as a numerical representation. For instance, a compression ratio of 100 : 1 means that the size of the original content has reduced by a factor of a 100. Note that as a rule of thumb increasing the compression ratios will often decrease the quality of the resulting video signal. Compression methods fall into two broad categories: lossless and lossy.

A lossless compression method allows a client IPTVCD to perfectly recreate the original image on a screen. Thus, no loss of image quality has been experienced during the compression and transfer of content. This is a rare occurrence on IPTV networks because virtually all compression techniques introduce a certain amount of loss during the encoding process. As a result, lossless compression algorithms are mainly used for encoding still images and not live video.

Many of the compression methods used in delivering IPTV services fall into the lossy compression category. During the execution of a lossy compression method some video image information is destroyed. Thus, the IPTVCD decoder is unable to fully recreate the original image that was outputted from the digitization process. However, modern day lossy compression algorithms are engineered to ensure that only limited amounts of data are destroyed during the encoding process.

The most popular and dominant lossy compression methods used by commercial IPTV providers are MPEG and VC-1 technologies. More details are provided on each one of these formats in the following sections.

3.2.1   MPEG Compression

MPEG technology is a compression standard, which is widely used by satellite, cable, and terrestrial TV systems. MPEG is an acronym for Moving Pictures

TABLE 3.1 Summary of MPEG Formats

MPEG Format	Description
MPEG-1	The MPEG-1 file format was originally developed in 1988 and was primarily used to compress video data at bit rates of 1.5 Mbps. MPEG-1 content is used for such services as DAB (Digital Audio Broadcasting). MPEG-1 is also the basis of the MP3 standard, which is widely used for music on the Internet.
MPEG-2	MPEG-2 builds on the powerful compression capabilities of the MPEG-1 standard. MPEG-2 is widely used in the delivery of broadcast-quality television and storing video content on DVDs. A number of international television standards are based on this compression format.
MPEG-4 (Part 2)	MPEG-4, whose formal ISO/IEC designation is ISO/IEC 14496, was finalized in October 1998 and became an international standard in 2000. Part 2 of the standard is divided into a number of profiles that address the requirements of various video applications ranging from mobile phones to surveillance cameras.
MPEG-4 Part 10	MPEG-4 Part 10 also called H.264/AVC is designed to deliver broadcast and DVD-quality video at minimum data rates.

Experts Group and represents an industry association that was formed to help develop compression techniques suitable for video transmission. The group was originally created by the International Organization Standardization (ISO) with the International Engineering Consortium (IEC) to work on developing video and audio compression standards. Since its foundation, the group has produced a family of major compression standards—MPEG-1, MPEG-2, MPEG-4 (Parts 2 and 10), MPEG-7, and MPEG-21. Table 3.1 summarizes these MPEG formats.

In addition to the above, the MPEG working group has also produced specifications that define standards for describing A/V content, delivery, and consumption—MPEG-7 and MPEG-21. MPEG-4 integrates closely with MPEG-7 and MPEG-21. Of the various types shown in the table above, MPEG-2 and MPEG-4 Part 10 (H.264) are the most widely deployed by IPTV service providers.

3.2.1.1 MPEG-2 MPEG-2 has been a hugely successful technology and is the dominant transport and compression standard for digital TV across a wide variety of network media. MPEG-2 coding technologies can be broadly classified into two categories: video and audio.

Video Compression Video in its most basic form is a sequence of images, which are displayed in a sequential order. The technical term for one of these video images is "a frame." These frames are identifiable within a bit stream by a header. The human eye in general can comfortably watch TV at around 25 frames per second (fps). There is a not a huge point in broadcasting a higher rate because TV viewers will not know the difference. It is possible to reduce the size of the original video through a process called compression. Devices called video encoders are used to compress video content contained in each of the frames, while maintaining a high level of picture quality. The first stage of compression involves a process called subsampling, which basically involves reducing the size of each frame. Reducing the frame size eliminates a number of bits, which decreases the bandwidth required to carry the signal. This process is not however without its drawbacks. For instance, frame size reduction can often produce aspect ratio problems when displayed on a low resolution television set.

The second stage of compressing a video signal is the division of a picture frame into 8×8 pixel blocks, the smallest coding unit in the MPEG algorithm. Blocks can be one of three types: grayscale luminance (Y), red chrominance (C_r), or blue chrominance (C_b). The chrominance types bear information about the different picture colors while luminance carries the black-and-white part of the picture.

Once this is complete, MPEG performs a mathematical function called Discrete Cosine Transform (DCT) to each video block independently. The principle behind DCT involves separating the video block into parts of differing importance. Important parts of the block are retained for further processing while the remaining parts are discarded. This approach ensures that the human eye does not notice the removal of less important parts of the video block while limiting the overall bit rate.

The next step in the MPEG compression process is called quantization. The quantization of digital video data is the reduction in the number of bits that are required to represent the various blocks contained in the picture frame. The level of quantization applied to a video signal is important. If the level is high, then higher compression rates are achieved because a good number of pixels have been removed. The drawbacks of high quantization levels will however become evident at the IPTV end-users screen due to degradation in image quality. Note that video blockiness is an example of high quantization levels.

Once all the blocks in the frame are compressed, MPEG breaks the picture frame into new sets of blocks called macroblocks. Each macroblock consists of a 16×16 pixel array of luminance and chrominance blocks. If there is a difference between the last and current picture frames, the MPEG compression equipment moves those new blocks of video to a new location on the current frame. This eliminates the need to broadcast a complete new picture frame that substantially reduces the bandwidth requirements. There are two ways of doing this:

Spatial compression refers to the bit reduction achieved on the pixels available in a single frame. This is achievable because pixels that are located side by side in a frame may often have similar values. So rather than encoding each individual pixel, the spatial redundancy technique encodes the difference between neighboring pixels. The amount of bits required to represent the difference between neighboring pixels is in most cases less than the amount of data required to compress each pixel individually.

Temporal compression refers to the bit reduction between successive frames. In some video productions there is information duplicated between consecutive frames. For example, if a wall is displayed, the information available in 30 images of that wall will not change over 1 s. So, rather than encoding the image 30 times during the second, temporal compression only sends information on the predictions of motions between frame images. In the case of the wall, the motion prediction is set to zero.

The method used to compress a particular video frame will vary quite considerably. For instance a complicated picture frame may have a low spatial redundancy factor because only a very small percentage of the pixels will be replicated. The nature of video content is that it is highly variable, which means that the bit rates associated with encoding video can also vary quite dramatically. Large variations in bit rate are not easily accommodated by IP networks, so many encoders include a buffering function that helps to control and manage the overall rate at which the bits are outputted to the next stage of the video processing system.

The next step in the MPEG compression process is coding of macroblocks into slices. A slice represents horizontal strips of pictures from left to right. Several slices can exist across the screen width and is the fundamental unit of synchronization for variable length and differential coding. A number of slices are combined to make a picture. Each slice is coded independently from other slices that helps to limit errors to a specific slice.

The MPEG standard specifically defines three types of pictures:

(1) *Intra-frames (I-frames)*—An I-frame is encoded as a single image, with no reference to any past or future frames. The encoding scheme used is similar to JPEG compression. They are self-contained and used as a foundation to build other types of frames.

(2) *Forward predicted frames (P-frames)*—A "P" frame is a predicted frame and is based on past "I" frames. It is not actually an encoded image and contains motion information that allows the IPTVCD to rebuild the frame. P-frames require less bandwidth than I-frames, which is particurlaly important for IPTV based networks.

(3) *Bi-directional predicted frames (B-frames)*—A "B" frame is a bidirectional frame made up from information from both I-frames and P-frames. The encoding for B-frames is similar to P-frames, except that motion vectors may refer to areas in the future reference frames. B-frames occupy less space than I-frames or P-frames. So, relating back to an IPTV environment a stream that contains a high density of B-frames will require less bandwidth compared to a digital stream built with a high density of I and P frames. Even though B-frames help to minimize the bandwidth requirements of MPEG video streams, there is one main drawback—time delays. The time delays are created because the IPTVCD will have to wait to examine two reference frames before it is capable of processing the B-frame details.

These three types of pictures are combined to form a sequence of frames called a group of pictures (GOP). Each GOP begins with an I-frame and has a number of P- and B-frames spaced throughout the video stream. A typical MPEG GOP has the following structure:

[I B B B P B B B P B B B P B B B P]

A GOP must begin with an I-frame. Although the size of GOPs varies, the average for a GOP in an IPTV environment is between 12 and 15 frames. A regular GOP structure can be described with two parameters: N, which is the number of pictures in the GOP, and M, which is the spacing between the frames. GOPs fall into two broad categories: open and closed. With closed GOPs the final B-frame does not require an I-frame from the next GOP for decoding, whereas open GOPs do require the I-frame contained in the next GOP. The GOPs are then combined together to produce a video sequence. Each video sequence begins with a start code, which is followed with a sequence header and is ended with a unique code.

It is also worth noting that the order that frames are transmitted onto the broadband network are typically different to the order of frames contained in the uncompressed bit-stream fed into the encoder. This is because the decoder chipset in the IPTVCD needs to first process I- and P-frames before reconstructing related B-frames. The overall relationship between the sequence of pictures, picture, slices, macroblocks, blocks, and pixels is illustrated in Fig. 3.1.

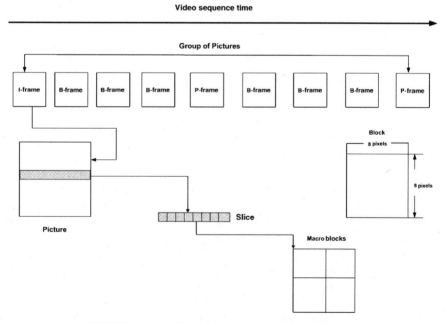

FIGURE 3.1 MPEG video stream hierarchical structure

MPEG-2 PROFILES AND LEVELS   MPEG-2 uses what are called profiles and levels for its encoding and decoding.

A profile defines a subset of the overall specification. In other words, it defines the complexity of the encoding process. MPEG-2 is divided into the following profiles.

- *Simple*—No B-frames are used in an MPEG system that is configured as a simple profile. The absence of B-frames means that picture reordering, which normally takes 120 ms, is not required.
- *Main profile*—This profile uses all three frame types I, P, and B. This profile is commonly used for deploying multicast IPTV services over a broadband network.
- *SNR*—The signal-to-noise (SNR) profile adds support for enhancement layers of DCT coefficient refinement, using the "SNR ratio scalability" tool. This type of profile is ideally suited to transmitting basic quality pictures with enhanced data services.
- *Spatial*—The spatial profile uses the "spatial scalability" tool to improve picture characteristics.
- *High profile*—This is intended for high definition TV applications in 4:2:0 and 4:2:2 formats.

To complement these variations in picture quality, MPEG has also subdivided profiles into levels. These different levels are used to define constraints on picture size, frame rate, bit rate, and buffer size for each of the defined profiles. Different profile and level combinations have already been established for different products and for different applications of digital compression.

For example, Main Profile at Main Level (MP@ML) refers to a resolution of 720×576, video at 30 frames/s, and bit rates up to 15 Mbps. This profile is widely used for digital television broadcasting and video-on-demand services. Higher quality images, such as those expected from high definition TV (HDTV), require additional bandwidth. Consequently, a main profile at high level (MP@HL) configuration that supports up to 80 Mbps and 1920×1152 pixel resolution is required.

VARIATIONS IN MPEG-2 BIT STREAM RATES   The bit rates outputted from an encoder fall into two broad categories, namely, constant bit rate (CBR) and variable bit rate (VBR).

- As the name suggests *CBR* MPEG-2 streams operate at a constant bandwidth data throughput regardless of the complexity of the video content. This bit rate option is ideal for DSL networks that typically provide fixed bandwidth connections. The generation of a stream in CBR mode is achieved by adjusting the encoder's quantization parameters. The inability of CBR streams to adjust according to the complexity of the video picture is a drawback that needs to be taken into account by IPTV service providers.

● The frames in a *VBR* MPEG-2 stream are encoded using different bit rates. So a complex frame requires a large number of bits, whereas less complex frames use a smaller amount of bits to represent the image. The cable industry sector often uses VBR in combination with a process called statistical multiplexing to "fit" multiple digital TV services in the same frequency range that was once occupied by a single analog channel. Statistical multiplexing operates on the principle that the bit rates between multiple channels running over the same connection at a particular instance in time will fluctuate and differ. Thus, a channel running a series of frames depicting a high action sports scene requires a large amount of bits to carry this detail over the network. By using statistical multiplexing this VBR encoded channel will borrow the extra bandwidth capacity required for these frames from another channel in the shared RF channel that is broadcasting a simple scene with low capacity frame types at that particular instance in time. Once the high action scene is finished, the VBR rate of that stream will drop to normal or below average rates and any excess bits available on this stream may be used by other streams within the shared RF channel. Thus, the peaks of one program are matched to the valleys of other programs. Note the overall bit rate for the various streams in the channel remain within a constant bit rate. Therefore, statistical multiplexing is a particularly desirable technique for transporting video content.

Note that many of the network architectures implemented to carry IPTV services use a switched-video mode of operation. Under this architecture each stream operates on a stand-alone basis and is not combined into a shared channel with other streams. Therefore, the CBR technique is the predominant mode of operation for carrying video services over DSL based networks at the time of writing.

Although MPEG-2 has served the cable and satellite industries well for the past decade, it has shortcomings when deployed on networks that have limited bandwidth capacities. A telephone network is a typical example of a system that was not optimized to carry MPEG-2 video content streams. Therefore, new advanced compression schemes with better capabilities have been developed in recent years for the purpose of delivering video content over bandwidth constrained networks. The most popular of these standards are MPEG-4 Part 10 (H.264) and VC-1. The following sections give a brief overview of these compression technologies when used in an IPTV infrastructure.

Audio Compression   The MPEG audio compression topic is covered in Chapter 5.

3.2.1.2 MPEG-4   The MPEG-4 standard (ISO/IEC 14496) is the successor to MPEG-2. In addition to compression MPEG-4 defines a complete ecosystem that features support for processing a whole range of multimedia formats. The MPEG-4 specification consists of a number of interrelated parts that can be implemented together or separately. Table 3.2 provides a brief explanation of each part.

TABLE 3.2 Parts Overview of the MPEG-4 Specification

Part Number	Official Title	Brief Description
1	ISO/IEC 14496-1 (systems)	This part contains details on various H.264 toolsets.
2	ISO/IEC 14496-2 (visual)	This part includes the DCT compression algorithm and descriptions of the various profiles and levels supported by the technology.
3	ISO/IEC 14496-3 (audio)	This part includes details on coding speech and general audio. Areas such as text to speech are also covered in this part of the standard.
4	ISO/IEC 14496-4 (conformance)	This part includes details on verification tests that are performed on encoders and decoders to ensure they conform with the standard.
5	ISO/IEC 14496-5 (reference software)	This part includes some sample reference software, which is used by IPTV vendors to commence developments of a H.264 compliant device.
6	ISO/IEC 14496-6 (delivery multimedia integration framework)	This part defines the interface between the upper layers of the IPTVCM and the transport layer.
7	ISO/IEC 14496-7 (optimized software for MPEG-4 tools)	This part is an extension to Part 5 in the sense that it provides a more optimal approach to implementing the MPEG-4 technology.
8	ISO/IEC 14496-8 (carriage on IP framework)	As the name indicates this part defines the requirements for transporting MPEG-4 content over an IP network.
9	ISO/IEC TR 14496-9 (reference hardware)	This part includes descriptions of the hardware video coding tools.
10	ISO/IEC 14496-10 (advanced video coding)	Also known as MPEG-4 Part 10, H.264 or AVC this part defines a high performance digital encoding technology that produces good quality video signals that are economical on bandwidth usage.
11	ISO/IEC 14496-11 (scene description and application engine)	This part defines a mechanism that specifies interactive audio–visual scenes called binary format for screens (BIFS).
12	ISO/IEC 14496-12: (ISO base media file format)	This part defines a file format for storing time-based audio and video content.

TABLE 3.2   (*Continued*)

Part Number	Official Title	Brief Description
13	ISO/IEC 14496-13 (intellectual property management and protection (IPMP) extensions)	This part defines syntax and semantics for MPEG-4 IPMP systems.
14	ISO/IEC 14496-14 (MPEG-4 file format)	This part describes the MPEG-4 file format.
15	ISO/IEC 14496-15 (AVC file format)	This part defines the storage file format of files compressed using AVC technology.
16	ISO/IEC 14496-16 (animation framework extension)	This part defines extensions relevant to the animation of multimedia objects.
17	ISO/IEC 14496-17 (timed text subtitle format)	This part describes a mechanism of coding text as part of a multimedia presentation at very low bit rate.
18	ISO/IEC 14496-18 (font compression and streaming (for opentype fonts)	The choice of font is an extremely important aspect associated with creating multimedia content. This part defines how to effectively embed fonts into MPEG-4 encoded content.
19	ISO/IEC 14496-19 (synthesized texture stream)	This part describes the assimilation of synthesized texture with MPEG-4 multimedia content.
20	ISO/IEC 14496-20 (lightweight scene representation)	This part defines some guidelines for representing and delivering advanced multimedia content to resource-constrained devices.
21	ISO/IEC 14496-21 (MPEG-J graphical framework eXtension)	This part defines the use of the MPEG standard Java application environment and some standard Java application program interfaces (APIs) to create applications for mobile devices.
22	ISO/IEC 14496-22 (open font format specification based on opentype)	This part outlines details of support for the OpenType specification of digital fonts.

The deployment of MPEG-4 requires specialized equipment in the IPTV data center and the introduction of new decoding technology into the IPTVCD. Till recent years one of the main drawbacks of MPEG-4 has been the fact that additional processing power and memory is required to decode MPEG-4 based video content. Advancement in IPTVCD and semiconductor technology over the last couple of years has meant however that MPEG-4 is now becoming the compression technology of choice for IPTV data center systems and IPTVCDs.

MPEG-4 Part 10 AVC/H.264 Overview   The proliferation of next generation networks is driving demand for advanced video services such as HDTV and video-on-demand applications. Bandwidth requirements for these types of services are enormous. For instance, a single HD channel can require the equivalent bandwidth of six SD channels. So to meet current and future bandwidth demands a powerful and sophisticated standard called MPEG-4 Part 10 AVC also known as H.264 was introduced in 2002. In this book we refer this compression technology as H.264/AVC. The main benefits of H.264/AVC are as follows:

- *Good performance:* It is a relatively new audio/video technology with better compression capabilities than previous standards. Thus, it allows the delivery of high quality video services over networks with limited bandwidth capacities.
- *Low bandwidth requirements:* The video quality of H.264/AVC is quite similar to MPEG-2; however, it requires less bandwidth to transport the same quality of signal. This feature makes H.264/AVC particularly suitable for IPTV systems.
- *Interoperable with existing video processing infrastructure:* H.264/AVC allows operators to utilize their existing MPEG-2 and IP based networking infrastructures.
- *Support for HDTV:* When deployed optimally the compression standard can double or even triple the carrying capacity of existing networks. Therefore, telecommunication operators can use this standard to deploy DVD quality HD video content over their existing IP access networks.
- *Selected by a wide range of organizations:* Owing to the fact that H.264/AVC is an open international standard it has received wide industry support. Some of the organizations recommending the use of H.264/AVC in their specifications and standards include
  DVD Forum
  Blue-ray Disk Association
  DVB
  ATSC
  DMB
  IETF
  ISMA

The fact that H.264/AVC is an open standard is one of the main reasons why it has received such wide industry support.

- *Reduced storage space:* H.264/AVC reduces the space required by servers to store video content.
- *Support for multiple applications:* MPEG-4 compression technology is used by a wide variety of multimedia applications. Each of these platforms has

their own unique requirements. For instance, an IPTV multicast application requires the rendering of an image on a standard TV display, whereas a mobile entertainment application produces images that are rendered on either a mobile phone or a portable media player. To support the myriad of MPEG-4 applications a number of profiles and levels are included in the MPEG-4 specification that allows manufacturers to develop products that are specific to their target market. Thus, products with the same profile and level will interoperate with each other. Characteristics that are unique to individual profiles and levels include bit rates and image sizes.

- *Transport independent:* H.264/AVC compressed content can be transmitted across a wide range of protocols including ATM, RTP, UDP, TCP, and MPEG-2 transport streams.

- *Adapts easily to poor quality networks:* In-built error concealment and recovery mechanisms allow H.264/AVC to operate across poor quality networks.

- *Used in a wide variety of applications:* H.264/AVC is a highly flexible compression technology and is used by a number of different markets ranging from IPTV and video conferencing to mobile entertainment and portable gaming.

H.264/AVC TECHNICAL ARCHITECTURE   At a basic level, the H.264/AVC coding engine is quite similar to previous MPEG compression standards and comprises of a number of underlying encoding modules and technical characteristics. We explore these modules and techniques in the following subsections.

Intraprediction and Coding   By using the intraprediction and coding mechanism, H.264/AVC is able to exploit the spatial redundancies that are part of a video picture. It operates similarly to MPEG-2 in the sense that the transformation is applied to each frame. The similarity between both technologies however stops at this point. In MPEG-2, the DCT transform is applied to each macroblock in the picture frame, whereas H.264/AVC applies the transform to adjacent macroblocks. Note the DCT transform used by H.264/AVC is quite similar to the one used by MPEG-2 but is not the same. This approach is based on the assumption that the difference between adjacent macroblocks is minimal. Thus, the H.264 compression engine uses the surrounding macroblocks, which are already encoded as references in order to predict a specific macroblock of interest. Predicting the macroblock reduces the amount of bits compared to directly transforming the macroblock.

Interprediction and Coding   The interprediction and coding component of H.264/ AVC relies on motion estimation to exploit various temporal redundancies that exist between video frames in a sequence. Similar to MPEG-2, H.264/AVC also uses this method to increase the efficiency of encoding video sequences.

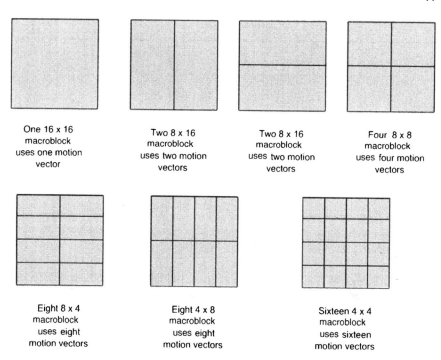

FIGURE 3.2   Block sizes supported by H.264/AVC

Macroblock Partitioning   The basic unit of the encoding or decoding process is the macroblock. Macroblocks are combined to form slices, which in turn form frames. Motion compensation used in previous versions of MPEG were limited to using 16 × 16 pixel arrays whereas the H.264/AVC standard provides an option to divide the macroblock into partitions. The purpose of this exercise is to provide enhanced support for motion-compensated prediction. Details of the seven different block sizes are illustrated in Fig. 3.2.

As shown, it is possible to send individual motion vectors for macroblocks as small as 4 × 4 pixels in size. This improves prediction and allows encoders to handle video motion at a very detailed level.

Multiple Frame Types   In addition to supporting I, B, and P frames, H.264/AVC has also added support for two additional frame types, namely, Switching I (SI) and Switching P (SP). These frames allow IPTVCD decoders to switch to a live stream from another stream at a particular location without using reference pictures.

Sophisticated Frame Referencing System   As discussed in our section on MPEG-2, the creation of a B frame is based on referencing other frames. In MPEG-2 these reference frames had to be located either before or after the specific frame. This limitation has been overcome in H.264/AVC and B-frames can access any frame in any part of the video timeline for motion-compensated predictions.

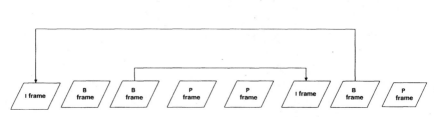

FIGURE 3.3   H.264/AVC frame referencing system

This feature helps to improve the efficiency of the compression process while being economical with bandwidth usage. A highly simplified graphical view of the frame referencing system is provided in Fig. 3.3:

Advanced Transform and Quantization Process   This module compresses the bulk of the video data. H.264/AVC makes use of an enhanced version of the DCT transform algorithm used by MPEG-2.

Deblocking Loop Filter   H.264 specifies the use of a filter to prevent the appearance of artifacts on the TV screen.

Specific Profiles   The initial H.264/AVC standard released in 2003 describes three "entertainment centric" profiles or feature sets:

(1) *Baseline*—Designed to support the robust transfer of content over a broad range of networking environments. Applications that use this profile include video conferencing and mobile TV. Specific technical characteristics associated with this profile include the ability to duplicate frame slices when network conditions are poor.

(2) *Main*—This is designed specifically for broadcast TV applications such as IPTV. Specific technical characteristics associated with this profile include the efficiency of its coding process and the ability of B frames to reference any frame in the entire sequence of frames.

(3) *Extended*—This profile combines the best characteristics from the two previous profiles into one. Not only can the devices that support this profile include the capability of operating in difficult networking conditions but they are also able to generate a high level of coding efficiencies.

Note that since 2003 a suite of four more profiles have been released that are designed for use within professional video environments.

All of the techniques outlined above in conjunction with many others, help to ensure that H.264/AVC performs at a very high level when used to compress video for delivery across an IP broadband network.

The practical use of H.264/AVC can be illustrated by the following example. Let us consider a sight we see relatively frequently in the Irish countryside, a farmer herding cattle across a field. From a technical perspective we can decompose the scene into the following objects:

(1) The field and hills in the background
(2) The sky
(3) The farmer and cattle running across the field
(4) The farmer's voice
(5) Noises emanating from the cows

H.264/AVC treats each one of these objects separately. Compression is then applied to each software object. The first two objects representing the field, hills, and sky are static. Thus, the motion prediction applied on these objects is straightforward. A more sophisticated motion prediction technique is however applied to the object representing the farmer and cattle in motion.

With regard to the audio objects, H.264/AVC encodes the farmer's voice with a high quality compression format, which helps to improve clarity. The noise from the cows is not overly important to viewers so a less precise compression mechanism is used on the fifth and final object. Therefore, the big advantage of separately compressing objects lies in the fact that some of the objects in the image do not have to be compressed as heavily as other objects within the same image, thus saving on bandwidth. This overall approach makes H.264/AVC a very suitable technology for use across IPTV networks.

3.2.2   VC-1

VC-1 stands for Video Codec 1 and is a next generation compression technology that was standardized by the Society of Motion Picture and Television Engineers (SMPTE). The specification for VC-1 was published in 2006 and can be found in SMPTE 421 M. In addition to this specification the SMPTE also published two more companion documents that detail the transportation of VC-1 content and conformance guidelines—SMPTE RP227 and SMPTE RP228. One of its most high profile implementations has been its adoption by Microsoft's Windows Media Video (WMV) 9 multimedia coding platform. In addition to Microsoft's support a number of other international standards including the high definition DVD formats HD-DVD and Blu-ray have also adopted VC-1. Some of the characteristics of the VC-1 specification include the following.

Deployable Across a Range of Platforms   VC-1 is supported by a wide range of IPTVCDs ranging from next generation DVD players and set-top boxes to portable media devices, and mobile phones.

Support for Three Separate Profiles   When integrated into an IPTV consumer device, VC-1 supports three profiles: simple, main, and advanced. Each of these profiles is suited to different types of applications.

The simple profile for instance is particularly suited to low bit rate Internet streaming type applications whereas the advanced profile is designed to accommodate the compression of HDTV content. Of the three profiles the main profile is the most suitable for use in an IPTV environment; particularly the delivery of video over DSL based networks. The main profile is further subdivided into levels. The purpose of applying levels to profiles is to identify some constraints that are applicable to the various parameters. For instance, the buffer size of the IPTVCD is one example of a level variable. Table 3.3 lists the levels associated with the main VC-1 profile.

ASF Support   The advanced systems format (ASF) is used to structure the IPTV stream.

Support for a Range of Block Sizes   In addition to supporting the coding of 8 × 8 pixel blocks, VC-1 has also added support for the coding of 4 × 4 pixel blocks and a number of other variations.

Additional Frame Type   In addition to supporting the coding of video into I, P, and B frames VC-1 introduces a new frame-type called BI, which is a variant of the I-frame but is not dependent on other frames. It is important to note that this frame type may not be used for predicting other frames.

Operates Across a Range of Network Transport Technologies   VC-1 is agnostic to underlying network transport protocols and is not tied to any particular mechanism of transporting the content across an IPTV network.

Although the market is dominated by MPEG-4 and VC-1 there are a couple of other video compression technologies who are vying for a piece of the IPTV encoding market. In China for instance, the country has developed a standard called audio video standard (AVS). The AVS compression mechanism is a China independent standard. The formal name is The Standards of People's Republic of China GB/T 20090.2 — 2006, Information Technology, Advanced Coding of Audio and Video, Part 2: Video. The efficiency levels achieved by this standard are

TABLE 3.3 List of Main Profile Levels

Level	Maximum Bits Rate (Mbps)	Resolutions Supported
Low	2	320 × 240 operating at a frequency of 24 Hertz. Typically used by computer displays.
Medium	10	Supports two resolutions: 720 × 480 at 30 Hertz and 720 × 576 at 25 Hertz. This level is well suited to the delivery of SD IPTV.
High	20	This level is capable of processing high resolution video content such as frames that contain 1920 × 1080 pixel densities.

quite similar to the performance levels achieved by AVC/H.264. The standard itself extends beyond compression and covers other areas such as digital copyright and content management. Ongoing development of the technology is the responsibility of the Audio Video Coding Standard Workgroup of China. Further information on this group is available at www.avs.org.cn/en/.

Note that while the standardization of video content is quite mature at the time of publication, a number of other proprietary and company specific compression technologies have been developed over the past number of years.

3.3 PACKETIZING AND ENCAPSULATING VIDEO CONTENT

The packetizing of video content involves inserting and organizing video data into individual packets. The term encapsulation is used to describe the process of formatting video content into datagrams. There are a couple of different approaches to encapsulating video content, namely, MPEG over IP and VC-1 over IP. Before discussing both these approaches, it is first helpful to understand a networking framework called the IPTV communications model (IPTVCM).

3.3.1 Overview of the IPTV Communications Model (IPTVCM)

The IPTV communications model is a networking framework composed of seven (and one optional) conceptual layers that are stacked on top of each other (see Fig. 3.4).

Video data is passed down the model from one layer to the next on the sending device, until it is transmitted over the broadband network by the physical layer protocols. The data arrives at the bottom layer of the IPTVCM resident on the destination device and travels up its IPTVCM.

Thus, if an encoder has video content to transfer to an IPTV consumer device, it must pass this material through the IPTVCM layered structure on both devices. Each layer in the communications model is generally self-contained and has specific responsibilities. Once its responsibilities have been met, the video data is passed to the next layer of the IPTVCM. Each layer adds or encapsulates some additional control information to video packets during processing. This control information includes specific instructions and is typically formatted as either headers or trailers. At the remote end, the data is passed up the communications model to the receiving application. Under this approach a logical communications connection between corresponding or peer layers is established. The seven and in some implementations eight IPTVCM layers can be broadly classified into two categories: upper and lower layers. The upper layers are concerned with dealing with specific IPTV applications and file formats, whereas the lower levels of the model deals primarily with the actual transportation of content.

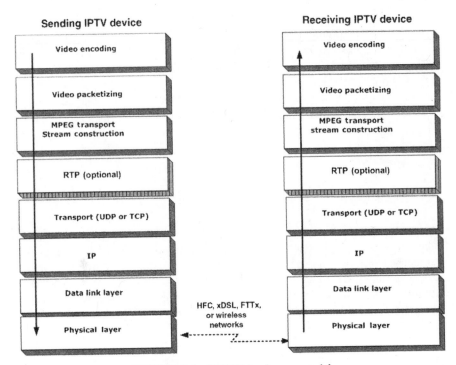

FIGURE 3.4   IPTVCM reference model

3.3.2 IPTVCM and MPEG Content Transportation

Figure 3.5 provides a more detailed insight into what happens to video content when compressed using the MPEG scheme in terms of encapsulation as it flows down through the IPTVCM protocol stack.

The following sections explain in detail the various encapsulations depicted in Fig. 3.5.

3.3.2.1 Video Encoding Layer   The communication process starts at the encoding layer where an uncompressed analog or digital signal is compressed and an MPEG elementary stream is outputted from the encoder. An elementary stream defines a continuous real-time digital signal. There are different types of elementary streams. For example, audio encoded using MPEG is called an "audio elementary stream." An elementary stream is effectively the raw output from the encoder. The stream is typically organized into video frames at this layer of the IPTVCM model. The types of information included in an elementary stream can include

- Frame type and rate
- Positioning of data blocks on screen
- Aspect ratio

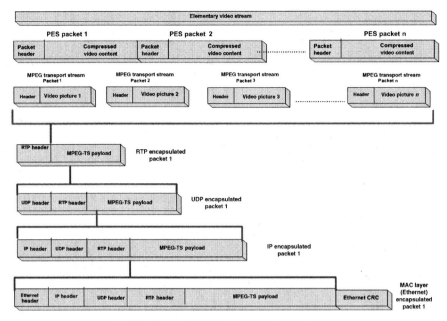

FIGURE 3.5   IPTVCM encapsulation layers

Elementary streams form the basis for the creation of MPEG streams. It is important to note that this layer is effectively divided into two sublayers by the H.264/AVC specification: Video Coding Layer (VCL) and Network Abstraction Layer (NAL). As discussed previously the VCL sublayer takes care of compressing the video content. The output from this layer is a series of picture slices. The bit stream at the NAL layer is organized into a number of discrete packets called NAL units. The format of an NAL unit is described in Fig. 3.6.

The diagram in Figure 3.6 depicts an NAL unit carrying a payload of video content. The term VCL-NAL is used to define this unit. It is also possible to include other types of payload in this field such as control information. These units are categorized as non-VCL units. NAL units are typically combined in a sequence to form an access unit. Note that the NAL used by H.264/AVC provides support for both IP and non-IP based networking infrastructures.

3.3.2.2 Video Packetizing Layer   In order for the audio, data, and video elementary streams to be transmitted over the digital network, each elementary stream is converted into an interleaved stream of time stamped Packetized Elementary Stream (PES) packets. A PES stream contains only one type of data from one source. A PES packet may be a fixed (or variable) sized block, with up to 65536 bytes per packet. This includes an allocation of approximately 6 bytes for the header with the remainder of the packet used to carry content. The elements of a PES header are illustrated in Fig. 3.7 and explained in Table 3.4.

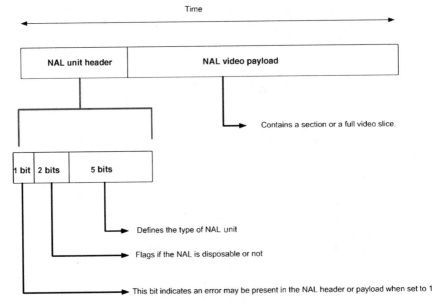

FIGURE 3.6   Structure of an NAL unit

Owing to the nature of networking the order or sequence of video frames outputted from the IPTV data center can be different to the order that they are received by the IPTVCD. Thus, to help synchronization, MPEG based systems often time stamp the various PES packets that are part of a particular video stream.

There are two types of time stamps that can be applied to each PES packet—Presentation Time Stamps (PTS) and Decode Time Stamps (DTS).

- *PTS*—A PTS is a 33-bit time value, which is set in the PES header field. The purpose of applying a PTS to each packet is to define when and in what order the video should be presented to the viewer.
- *DTS*—The purpose of applying a DTS to each packet is to instruct the IPTVCD decoder when to process the packets.

The concept of applying different types of timestamps to PES packets in an MPEG encoded stream is illustrated in Fig. 3.8.

As shown the order at which the packets are streamed over the network is different to the order, which is received at the input of the IPTVCD. As a result, the IPTVCD uses the PTS and DTS stamps to reconstruct the original video content. Besides carrying MPEG-2 compressed content a PES is also capable of transporting H.264/AVC units across an IPTV network. Details of the mapping involved with this process are illustrated in Fig. 3.9.

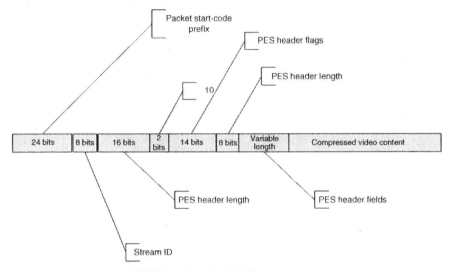

FIGURE 3.7 MPEG PES packet format

3.3.2.3 Transport Stream Construction Layer   The next layer of the IPTVCM communication model deals with building a transport stream, which consists of a continuous stream of packets. These packets, commonly called TS packets, are formed by breaking up the PES packets into fixed-sized TS packets of 188 bytes that are referenced to independent time bases. Using independent time bases versus identical time bases helps to reduce the potential of packet loss or noise corruption. Each TS packet contains one of the three media formats, video, audio, or data. Therefore, transport packets do not support a mix of media. Each TS packet comprises 184 bytes of payload and a 4 byte header. The elements of an MPEG TS header are illustrated in Fig. 3.10 and explained in Table 3.5.

The following data types may be encapsulated directly in the payload field of an MPEG transport stream packet.

- MPEG-1 Video (ISO/IEC 11172-2)
- MPEG-2 Video (ISO/IEC 13818-2)
- MPEG-1 Audio (ISO/IEC 11172-3)
- MPEG-2 Audio (ISO/IEC 13818-3)

Note, this layer also provides functionality that allows the creation of program streams. A program stream is a PES packet multiplex that carries several elementary streams that were encoded using the same master clock or system time clock. These types of streams are developed for error free applications such as storage of video content on optical and hard disks. Therefore, program streams are rarely if ever used in an end-to-end IPTV deployment.

TABLE 3.4   Structure of an MPEG PES Packet

Field Name	Description of Functionality
Packet start-code prefix	The PES packet starts with a prefix of 0×000001.
Stream identifier (1 byte)	This field identifies the type of payload carried in the packet. A bit pattern of 111x xxxx indicates an audio packet whereas a bit pattern of 1110 xxxx informs video processing equipment that the packet contains video content. The "X" values are used to represent the numbers of the MPEG streams.
PES packet length	This two byte field identifies the length of the packet.
Sync code	This field is used to synchronize audio and video content.
PES header flags	This 14 bit field contains various *PES Indicators* or flags that provide the hardware or software decoder resident in the IP set-top box with additional information. The types of flags included in this field follow:
	PES_Scrambling_Control—This informs the decoder whether the packet has been secured through the scrambling process.
	PES_Priority—This provides the decoder with information on the priority levels of the PES packet.
	data_alignment_indicator—This indicator determines if the PES payload commences with an audio or video bit.
	copyright information—When this bit is set the video content is protected by copyright.
	original_or_copy—This flag indicates whether or not this is the original content.
PES header data length	As the name suggests this field identifies the total number of bytes occupied by the various header fields.
PES header fields	This field contains a number of optional bits.
PES payload	The PES payload consists of the elementary stream audio and video data.

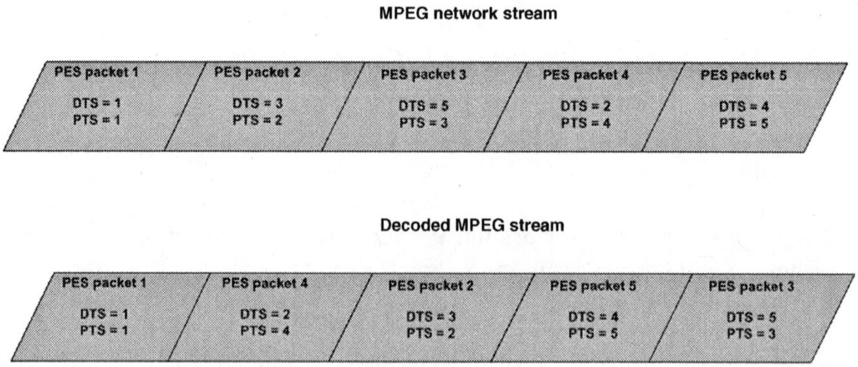

MPEG network stream

Decoded MPEG stream

FIGURE 3.8   Applying time stamps to MPEG PES packets

In addition to the compressed audio and video content, a transport stream includes a great deal of Program Specific Information (PSI) or metadata describing the bit stream. This information is carried in four PSI tables.

(1) *Program association table (PAT)*—The transmission of the PAT table is mandatory and is the entry point for the PSI tables. The PAT always has a Program Identifier (PID) of 0. This table gives the link between the program number and the PID of the transport packet carrying the program map table (PMT).

(2) *Program map table (PMT)*—The PMT is also mandatory and contains information regarding one particular program. The PMT lists all the PIDs for packets containing elements of a particular program (audio, video, data, and PCR information). Figure 3.11 graphically illustrates an example of the relationship between a PAT and a PMT.

So once an IPTVCD requests a particular program the PAT is consulted, which then interrogates the PMT to identify the PIDs for the audio, video, and data packets associated with that program. In this example the TV subscriber selects program 1 and the IPTVCD locates all transport stream packets with a PID of 36 for the video portion of the program, and packets with a PID of 3 for the audio portion of the program. If data is broadcasted with the program, then the PMT will also include details on where to locate the data transport stream packets.

(3) *Conditional access table (CAT)*—The CAT is an optional PSI table, which contains PIDs for EMMs (Entitlement Management Messages). The EMMs

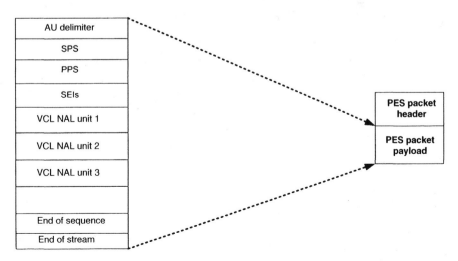

FIGURE 3.9 Mapping AVC access units into MPEG-2 PES packets

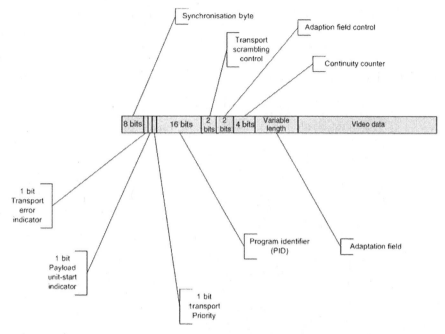

FIGURE 3.10   MPEG TS packet format

contain authorization level information for conditional access systems. The CAT is always carried in packets with PID of 1.

(4) *Network information table (NIT)*—The NIT is also an optional table that stores information such as channel frequencies and transport stream numbers. The set-top box uses this information to tune to specific programmes. The NIT table is rarely used for IPTV deployments.

Once the TS has been structured and formatted it is passed down the IPTVCM to either the transport layer directly or a layer that uses the Real-Time Transport Protocol (RTP).

3.3.2.4   RTP Layer (Optional)   This optional layer is used by a wide variety of IPTV applications. It acts as an intermediary between the H.264/AVC, MPEG-2, or VC-1 encoded content in the higher layers and the lower sections of the IPTVCM. The RTP protocol represents the core of this layer and is often the foundation block that supports the real-time streaming of media content across an IP network.

RTP delivers end-to-end streams of audio and video by encapsulating the content into a particular format called a packet. Each packet consists of a header and the payload IPTV data. To improve bandwidth efficiency, the payload typically includes more than one MPEG-TS packet.

TABLE 3.5   Structure of an MPEG TS Packet (188 Bytes) + 4

Field Name	Description of Functionality
Synchronization byte	The header starts with a well-known Synchronization Byte (8 bits). This has the bit pattern 0×47 (0100 0111) and is used to detect the start of the IPTV packet.
Transport error indicator	This single bit flag indicates an error in the associated transport stream.
Payload unit-start indicator	This flag indicates the start of the video payload.
Transport Priority	When set this flag identifies priority level of the video payload.
Program identifier (PID)	The most important field of the header is the 13 bits that define the program identifier. This uniquely identifies the stream that the packet belongs to. All packets belonging to this stream will have the same PID value. This information is used by the demultiplexer in the IPTVCD to distinguish between different packet types. Note that null packets are always assigned a PID value of 8191. Packets that have no PID values are typically discarded by the receiving IPTVCD.
Transport scrambling control	This two-bit field indicates the encryption status of the transport stream packet payload.
Adaption field control	This two-bit field indicates whether the associated transport stream packet header includes an adaptation field and payload.
Continuity counter	The continuity counter increments by one each time a transport stream packet with the same PID value is passed through the MPEG system. This helps to identify lost or duplicate packets, which could affect the quality of the video been viewed by the IPTV subscriber.
Adaptation field	This field may or may not be present in the transport header. The adaptation field contains a variety of data used for timing and control including the Program Clock Reference (PCR). The PCR is used to synchronize the IPTVCD clock with the source encoder clock. PCR values are 42-bits in length and increment according to a standard clock rate of 27 MHz. Once synchronization has taken place the decoding of the IPTV MPEG-2 stream can occur.

The header contains those functions that are essential for the successful transmission of real-time data across the network. An RTP header is identifiable with a value of 5004 in a User Datagram Protocol (UDP) header and contains quite a large number of fields. Details of these various fields are illustrated in Fig. 3.12 and described in Table 3.6.

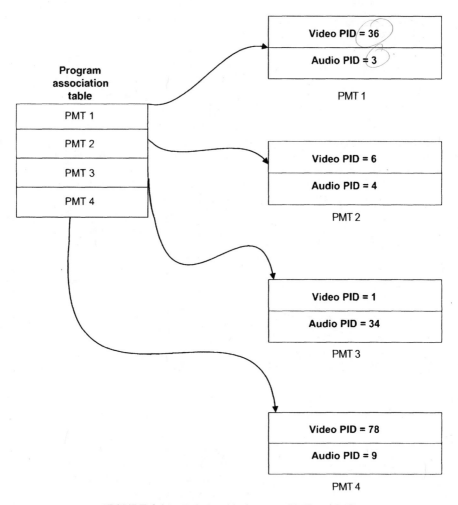

FIGURE 3.11   Relationship between PMT and PAT

It is also worth highlighting that RTP does not have a length field in its header because it relies on the underlying transport protocols to provide this type of information. As described in "Table 3.6", the two main benefits of inserting compressed video content into RTP packets are

(1) It adds a sequence number to the packet to help both the server and the IPTVCD to detect lost packets. Additionally, this number may also be used by the IPTVCD decoder to reorder packets that arrive from the IP network in the wrong sequence.

(2) The timestamp field helps to tackle issues such as jitter and incorrect clock synchronization between source and destination.

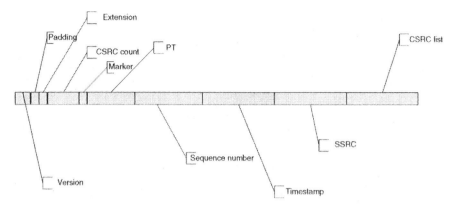

FIGURE 3.12 Typical format of an RTP header

Once the RTP header is added to the video payload, the RTP packet is sent to the TCP or UDP protocol for further processing.

In some instances it is possible to transport MPEG-2, H.264/AVC, or VC-1 compressed video, audio, and data content across an IP network by avoiding some of the upper IPTVCM layers and instead mapping the content directly inside an RTP packet. The following sections give a brief overview of using the RTP layer to carry these three different video compression formats.

RTP Payload Format for Encapsulating MPEG-2 Compressed Bit Streams   Rather than using UDP to directly carry MPEG-2 TS packets some IPTV systems use the RTP layer in addition to the UDP layer to transport the packets. The mapping of MPEG-TS packets into RTP packets is quite straightforward. In Fig. 3.13, the structure used in the transport of MPEG-2 based DVB content over IP is illustrated.

As shown the structure accommodates a header and a payload of MPEG2 TS packets. Each packet is 188 bytes in length and the standard practice is to use seven MPEG2 TS packets in each ETSI TS 102 034 compliant packet.

RTP Payload Format for Encapsulating H.264/AVC Compressed Bit Streams   The RFC 3984 provides specific recommendations on this method of transport for H.264/AVC content and defines three mechanisms for inserting NAL units into the RTP payload:

(1) *Single NAL unit packet:* This mechanism defines the mapping of a single NAL unit into a single RTP payload. The structure of the single NAL unit packet is graphically depicted in Fig. 3.14.

(2) *Aggregation NAL unit packet:* This mechanism defines the mapping of multiple NAL units into a single RTP payload. The structure of an aggregation unit packet is graphically depicted in Fig. 3.15.

TABLE 3.6 Structure of a RTP Based IPTV Packet

Field Name	Description of Functionality
Version (V)	This field identifies the version of RTP used in the IPTV packet.
Padding (P)	This field defines whether or not padding octets are present in the RTP packet.
Extension (X)	If this bit is set to 1 then the fixed header is followed immediately by an extension.
Contributing source (CSRC) count	This field contains information on the number of CSRC identifiers that are included in the packet.
Marker	Another one bit field whose functionality is defined by the RTP profile. Often used to define frame boundaries.
Payload type (PT)	As the name implies this field contains information about the format of the IPTV payload. For instance a value of 34 identifies a payload of video content that is encoded using H.263.
Sequence number	This has the same functionality as the sequence number field contained in a TCP packet. In other words it helps to detect lost packets on the network if they were to occur in a live IPTV environment. Additionally it helps the IPTVCD to reorder packets that arrive out of sequence, locate incorrect packet sizes, and identify duplicate packets. The value in this field is incremented by one every time an RTP packet is sent over the broadband network. Once the IPTV stream starts a random value is assigned to this field to reduce the risks of attacks from hackers.
Timestamp	This field holds the timestamp for the packet, which is derived from a reliable clock source. It is used to insert the audio and video packets within the correct timing order in the IPTV stream.
SSRC (synchronization source)	The purpose of this field is to identify the synchronization source on the IPTV network. This field is often used in conjunction with the sequence number field to rectify problems that may arise in the IPTV stream from time to time.
CSRC list	The purpose of this 32 bit field is to identify the various audio and video sources making contributions to the IPTV payload.

Aggregation NAL units were defined to accommodate maximum transfer units (MTU) of different types of distribution networking technologies. For instance, the MTU of Ethernet is 1500 bytes whereas ATM systems use an MTU of 54 bytes. The ability to use aggregation as a mechanism to map multiple NAL units into an RTP payload eliminates the need for transcoding and excessive packetization overhead when deploying IPTV across multiple platforms.

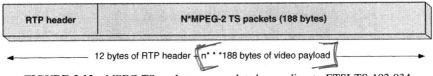

12 bytes of RTP header ⟮ n* * *188 bytes of video payload ⟯

FIGURE 3.13   MPEG-TS packets encapsulated according to ETSI TS 102 034

(3) *Fragmentation NAL unit packet:* As the name suggests this mechanism defines the mapping of a single NAL unit across multiple RTP payloads. The structure of a fragmentation unit packet is graphically depicted in Fig. 3.16.

It is important to note that NAL unit fragments need to be sent across the network in a sequential and consecutive order. This is enabled through the use of ascending numbers contained inside the RTP header. This mechanism has a couple of advantages for IPTV service providers. First and foremost it helps to deliver larger quantities of high definition IP based content and secondly it helps to improve the effectiveness of error correction techniques such as Forward Error Correction (FEC). The FEC topic will be discussed in greater detail later in the book.

RTP Payload Format for Encapsulating VC-1 Compressed Bit Streams   The delivery of VC-1 over IP is also facilitated by the IPTVCM. There are some

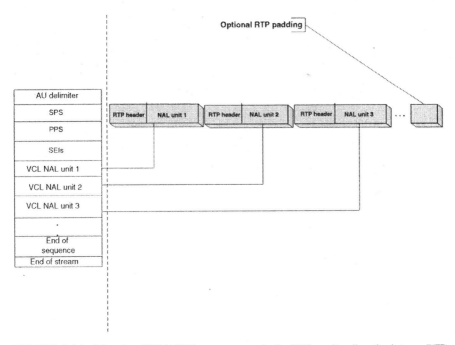

FIGURE 3.14   Mapping H264/AVC content as single NAL units directly into a RTP payload

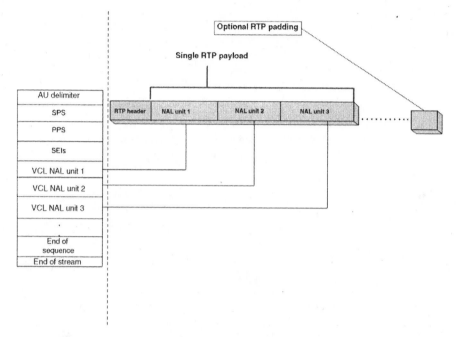

FIGURE 3.15 Mapping H264/AVC content as multiple NAL units directly into a single RTP payload

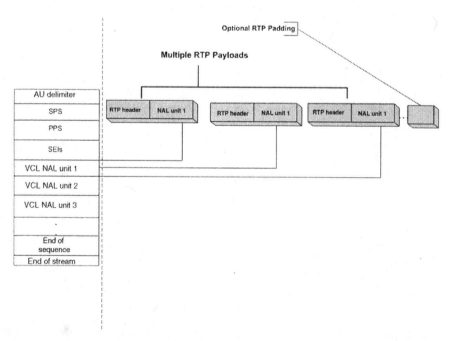

FIGURE 3.16 Mapping a single H264/AVC NAL unit directly into multiple RTP payloads

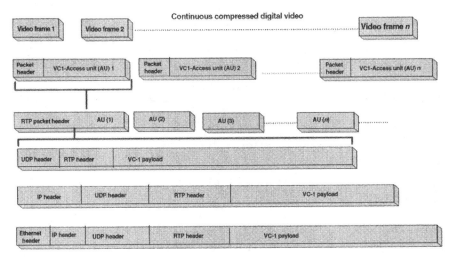

FIGURE 3.17 VC-1 encapsulation layers

variations in how the model is applied to transporting the VC-1 content. One of the main differences is that the video data is typically inserted directly into the RTP packets to avoid the use of MPEG encapsulation. Figure 3.17 provides a more detailed insight into what happens to VC-1 compressed content in terms of encapsulation as it flows down through the IPTVCM protocol stack.

As depicted, the encapsulation mechanisms are quite similar to the MPEG process. In fact, the lower transportation layers extending from RTP down to the physical layer are the exact same. There is however some slight differences in the upper layers of the IPTVCM that relate to data structures and the terminology used. For a start the transportation of VC-1 compressed bit streams over RTP is defined in RFC 4425. As depicted, the RFC transport mechanism involves encapsulation of VC-1 access units (AUs) inside a series of RTP packets. Each AU contains a header and variable length video payload.The structure of an AU packet header is illustrated in Fig. 3.18 and explained in Tables 3.7 and 3.8.

Note that RTP encapsulation mechanisms are generally deployed across networks that cannot guarantee sufficient levels of QoS for delivering IPTV services. Although RTP helps to improve the likelihood of streams arriving at their destination points in good order, it has not been designed to guarantee QoS levels. Thus, it is the responsibility of the service provider to ensure video traffic is treated as a priority as it passes across the networking infrastructure.

3.3.2.5 *Transport Layer* In general RTP packets form the input to the transport layer of the IPTVCM. It is important to note that it is also possible to map MPEG-TS packets directly into the transport layer protocol payload. This effectively avoids the RTP layer completely.

The IPTV transport layer has been designed to hide the intricacies of the IP network structure from the upper-layer processes. Standards at this layer provide for the reliability and integrity of the end-to-end communication link. If video data

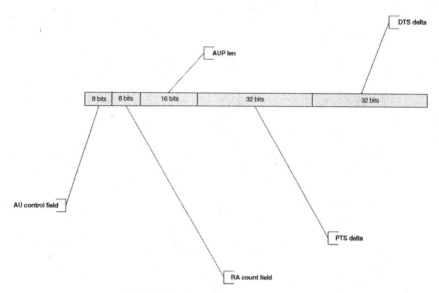

FIGURE 3.18   VC-1 AU packet structure

is not delivered to the IPTVCD correctly, the transport layer can initiate retransmission. Alternatively, it can inform the upper layers that can then take the necessary corrective action.

TCP and UDP are the two most important protocols employed at this layer of the IPTV communications stack. The following sections explain the functionalities of these transport layer protocols when used in an IPTV infrastructure.

TABLE 3.7   VC-1 AU Packet Header Field Descriptions

Field name	Description of functionality
AU Control (Mandatory)	This field consists of a number of flags that may be set to enable various functions. For instance the length present flag bit is set to one to inform the decoder that the access unit payload length field is included in the header. Further details of this important field are available in table 3.8.
Random access (RA) count (Mandatory)	This is the random access pointer counter field. This field gets incremented by a value of one each time the RA bit in the AU control field is set.
AUP len (optional)	AUP len is an acronym for access unit payload length. Thus, the field specifies the size of the video content payload contained in the AU.
PTS delta (optional)	The presentation time delta, abbreviated as PTS specifies the time as an offset (delta or difference) from the timestamp field included in the header of the RTP packet.
DTS delta (optional)	This field specifies the delta time between the presentation and the decode time values.

TABLE 3.8 AU Control Field Flag Descriptions

Flag Name	Size (Bit)	Description of Functionality
Frag	2	This flag indicates to the IPTVCD decoder where a complete video frame or a fragment is contained in the payload section of the packet. RFC 4425 recommends that an AU payload should contain a complete VC-1 frame. The RFC however acknowledges that there exists the possibility of RTP packet sizes exceeding the MTU of the network. Thus the RFC includes support for the following Fragmentation values: The payload is a complete frame. Contains a fragment in between the start and finish fragments. The first fragment of the video frame. The last fragment of the video frame.
RA	1	This indicates whether the AU is a random access point (RAP) or not. The RAP helps the IPTVCD decoder to restart processing in the event of lost packets across the network. Since the main profile is normally used for IPTV deployments an I-Frame is the RAP.
SL	1	This flag bit identifies changes in the sequence layer headers between consecutive AUs.
LP	1	A set LP or Length Present flag of 1 indicates to the decoder the presence of an AUP Len field in the AU header.
PT	1	A set PT or PTS Delta Present flag of 1 indicates to the decoder the presence of a PTS Delta field in the AU header.
DT	1	A set DT or DTS Delta Present flag of 1 indicates to the decoder the presence of a DTS Delta field in the AU header.
R	1	RFC 1445 recommends that this bit is set to 0 and ignored by IPTVCD decoders.

Using TCP to Route IPTV Packets In 1968, the U.S. Department of Defense (DoD) began an investigative research on network technology under the DoD advanced Research Agency (ARPA) that led to the development of TCP. The research team was working on network protocols and the technology that is now known as packet switching, when they came upon the concept of the TCP. Since its development back in the 1960s TCP has become synonymous with the transfer of data over the Internet and has been documented as the RFC 793 standard. TCP is a core protocol belonging to the Internet Protocol suite and is classified as being connection orientated. This basically means that a connection needs to be established between the IPTV headend servers and the IPTVCDs prior to transmitting the content over the network.

TCP complements IP's deficiencies and shortcomings by its ability to handle and deal with errors that occur during the transfer of content across the network. Lost, out of order, and even duplicate packets are the three main types of errors that are encountered in an IPTV environment. To cope with these situations, TCP uses a sequential numbering system to allow the sending device to retransmit video data that has been lost or corrupted. The sequential numbering system is implemented in the packet structure through the use of two 32 bit fields. The first field contains the starting sequence number of the data in that packet and the second field contains the value of the next sequence number the video server is expecting to receive back from the IPTVCD.

In addition to smoothing over the errors that can occur during the transmission of video data over an IP broadband network, TCP can also control the flow of data traversing the network. This is achieved through the use of the window size field in conjunction with an algorithm called the sliding window. As shown in Fig. 3.19, the value in this field determines the number of bytes that can be transferred over the network before an acknowledgement is required from the receiver.

In an IPTV environment the window size field value equates to the storage space of the buffers in an IPTVCD minus the amount of video content that is resident in

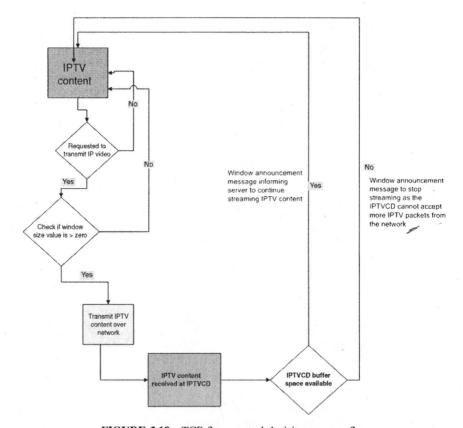

FIGURE 3.19   TCP flow control decision process flow

the buffers at a particular instance in time. The data in this field is kept current through the receipt of messages coming back from the IPTVCD called window announcements.

If the value of this field is zero, then the IPTVCD on the receiving end of the connection is unable to process the IPTV data at a fast enough rate. When this occurs TCP instructs the video server to stop or slow down the rate for sending data packets to the IPTVCD. This ensures that the IPTVCD does not become overwhelmed with incoming packets of data, which could result in the device crashing and causing serious disruption to the IPTV subscriber. Once the IPTVCD has processed some of the video data in the buffers, the IPTV server is informed of this fact, the value in the window size field increases and the video server starts transmitting content again. In an ideal IPTV environment the number of window announcements coming back from the IPTVCD telling the server that buffer space has been freed up will equate to the rate at which the IP video content is outputted from the video servers at the headend.

TCP Ports and Sockets   Each endpoint of an IPTV connection has an IP address and an associate port. So a typical connection will have four different identifiers:

(1) IP address of the video server
(2) Port number of the video server
(3) IP address of the IPTVCD
(4) Port number of the IPTVCD

The combination of an address with a port number allows a process on an IPTVCD to communicate directly with a process running on one of the servers located at the IPTV data center. A port is defined as a 16 bit number that identifies a direction to pass messages between network layers. Ports fall into two broad categories: well known and ephemeral.

- *Well known ports* have a value of between 1 and 1023. This category of ports are generally used by servers and managed by an organization called the Internet Assigned Number Authority (IANA).
- *Ephemeral ports* are established by IPTVCDs on a temporary basis when communicating back to the IPTV server. The ports are assigned by the embedded IP software stack. These values are normally greater than 1024 and less than 65535. IANA do not control or manage ephemeral ports.

Sockets are another key element of the IP communications model. A socket is essentially an application programming interface (API) and is used to facilitate communication between processes running on an IP device. A socket is formed by combining the IP address with the port number.

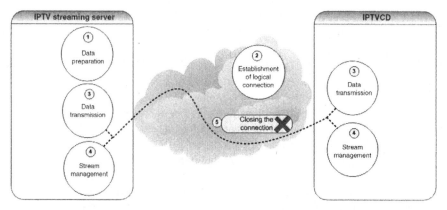

FIGURE 3.20   Inter-process communication over an IPTV network

Let us have a look at the relationship between IP addresses and sockets by taking a closer look at the sequence of steps that are involved with establishing a communications channel between a process running on an IPTVCD and a process running at the IPTV data center server. These steps are illustrated in Fig. 3.20 and described below.

(1) *Preparation of data:* The sending process running on the IPTV streaming server system prepares the content and calls on the TCP/IP communications module to transmit data to a process running on a remote client IPTVCD. The communication process commences and header information is added to the content whilst traversing down through the IPTVCM.

(2) *Establishment of a logical TCP Connection:* Both ends of the connection are identified by an IP address and a port number. This combination of an IP address and a port number is the socket. The addressing scheme for this communication link consists of the following variables:

- The protocol
- The IPTV server IP address
- The ID of the process running on the IPTV server
- The IPTVCD IP address
- The ID of the process running on the IPTVCD

(3) *Data conversation commences:* Communication starts via the sockets between both processes and the data is passed from the sending IPTV server to the receiving IPTVCD.

(4) *Managing the stream of IPTV content:* The TCP protocol takes care of managing the IPTV stream, while the connection is established.

(5) *Closing the TCP connection:* Once the transfer of IPTV content is complete the IPTVCD or the content server closes the socket and terminates the network connection.

TCP Segments TCP breaks a contiguous IPTV stream into smaller parts called "segments" and sends them separately across the broadband network. Once all these segments arrive at the destination IPTVCD, TCP puts them back together in its original sequence. Each segment has two components to it: a header and video data.

Header: This is the information attached to the segment that enables it to follow the route from the source to its final destination. The primary information given in the header is the port numbers of the source and the destination, the sequence number of the segment and a checksum. Port numbers ensure that the data reach to and come back from the right processes running on each IP device. The sequence number helps TCP to understand how to put back the data in the form it was in before being fragmented into segments.

Video data: In the context of an IPTV environment this field holds the actual video content.

A technical description of the TCP segment format is described in Table 3.9.

TCP maps the segment to the IP protocol after inserting all the necessary information in its header. As described above TCP provides a whole range of useful facilities for transporting data over an IP network. Let us now look at another popular protocol, which is also being used extensively by the IPTV industry to transport data and particularly video across an IP network—UDP.

3.3.2.5.1 Using UDP to Route IPTV Packets UDP is a protocol, belonging to the Internet's suite of protocols that enable streaming servers connected to a broadband network to send consumer acceptable broadcast quality television service to IPTVCDs. It is similar to the TCP but a more stripped down version, offering the minimum of transport services. UDP is connectionless protocol, meaning that a connection between the video server and the IPTVCD does not need to be established before video transfer across the network takes place. The video server simply adds the destination IP address and port number to the datagram and passes it onwards to the networking infrastructure for delivery to the destination IP address. Once on the network UDP uses a best effort approach to get the data to its destination point. Note that UDP uses data units called "datagrams" to transport content over the network. The next section reviews the basic structure of these data units.

About UDP Datagrams In the context of an IPTV networking environment, a UDP datagram consists of a header, which is 8 bytes long and associated video data. The key elements of a UDP based IPTV datagram are illustrated in Fig. 3.21 and described in Table 3.10.

Like anything, UDP has a number of advantages and drawbacks.

UDP Advantages Summing up the advantages of UDP in streaming of media, we can say, with UDP:

TABLE 3.9   Structure of a TCP Based IP Video Segment

Field Name	Description of Functionality
Source port	As the name suggests this field identifies the source port number of the sending IPTV application.
Destination port	Another 16-bit field used to identify the destination port number of the process or application running on the destination IPTVCD.
Sequence number	This field identifies the sequence of the TCP segment. This helps TCP to keep track of each IPTV packet that traverses the network.
Acknowledgment number	This field contains the number of the next sequence number that the sending device is expecting to receive. This will only contain this value if the ACK control bit is set.
Data offset	This short field indicates the location of where in the segment the video data commences.
Reserved	This field is typically reserved for future use and has a value of zero
Control bits	This field includes six control bits: URG: Urgent Pointer field significant ACK: Acknowledgment field significant PSH: Push Function RST: Reset the connection SYN: Synchronize sequence numbers FIN: No more data from sender
Window	This field identifies size of the window for delivering data between acknowledgements.
Checksum	The Checksum field is a simple error detecting measure used to protect data that is sent on the network. It ensures that the segment sent is the one that arrives at the destination. It consists of a number that is calculated by adding the octets of a TCP segment. This number is then added to the header so that the receiving system can do the same calculation and tally it with the number given in the header. A match ensures the integrity of the data, whereas a discrepancy would indicate an error happening somewhere along the network path. A flawed segment is sent back for replacement.
Urgent pointer	This field is only interpreted when the urgent pointer field is enabled.
Options and padding	Options included in this field are available in 8 bit multiples and can contain different types of variables. The options chosen will determine the length of the TCP segment. The padding is at the very end of the header and is composed of zeros.

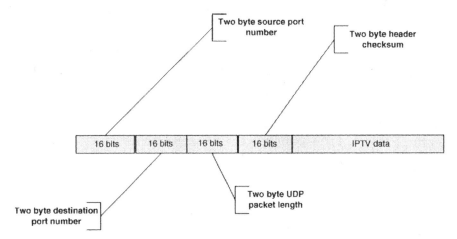

FIGURE 3.21 Typical format of a UDP based IPTV datagram

- *No pause in the delivery of IPTV content:* The delivery of IPTV content is not delayed even if there is delayed or damaged packets contained in the network traffic, whereas in using TCP, TV viewers are faced with a pause as they wait for a delayed packet or picture frame to arrive or wait for the damaged packet to be replaced.
- *Low overhead:* The size of the header is only 8 bytes when compared to the TCP header, which occupies 20 bytes of data.
- *Speedy connection setup:* The establishment and teardown of connections between IPTVCDs and IPTV data center networking components takes place in a very short time period. Therefore, the delivery of video packets using the UDP protocol is generally quicker compared to using the TCP protocol.
- *Supports one-way transmission:* UDP does not require a return path thus allowing companies like satellite operators to multicast IPTV content to their subscribers.
- *Easy implementation:* From a technical perspective UDP is pretty easy to implement because it is not required to keep track of video packets once they are sent onto the IP network.

Drawbacks of UDP Even though UDP is fast and efficient for those applications where time is of essence, it fails in certain areas.

- *Integrity of data:* For a start data integrity is not guaranteed in UDP since the only services it provides are check summing and multiplexing by port numbers. Any communication problem that may arise at either of the

TABLE 3.10   Structure of a UDP Based IPTV Datagram

Field Name	Description of Functionality
Source port	This field identifies the port number of the process that's sending the datagram. It is an optional port and if not used is typically filled with a zero.
Destination port	This field identifies the port number of the destination process running on the IPTVCD.
Length	This field helps the IPTVCD to identify the length and size of the incoming UDP datagram. The length field includes a value in octets, which includes both the header and the actual video data.
Checksum	This 2 byte field contains a predefined number that allows an IPTVCD to verify the integrity of the incoming UDP based IPTV datagram. If the incoming packet does not meet the integrity test and turns out to be corrupt or damaged the packet is discarded. Once discarded the IPTVCD makes no further attempt to request a new copy of the discarded IPTV datagram from the IPTV data center servers. It is also worth pointing out that in contrast to a TCP segment the Checksum field in a UDP datagram is optional and does not have to be used.
Video data	This portion of the datagram contains the video data. In the case of an IPTV environment the data in this part of the UDP datagram is formatted by the audio and video streaming protocol used at the IPTV headend.

communication ends need to be handled independently by the IPTV applications. Problems like retransmission, packetizing and reassembly, retransmitting lost packets, congestion, and flow control are all beyond the scope of UDP's error handling capacities.

• *Difficulty in penetrating firewalls:* Another drawback of UDP is that it fails to function effectively in the presence of certain network firewalls. Many of these firewalls block UDP information creating problems. This is not a major issue for IPTV service providers, however, it does have an impact on companies that offer Internet TV services.

In the context of IPTV, UDP is useful when a data center needs to send IP video content to multiple IPTVCDs and is the most popular transport level protocol employed by IPTV service providers.

3.3.2.5.2 TCP versus UDP for IPTV   When service providers are delivering IPTV content to their subscribers it is critical that the video content arrives on time and in good shape; in other words, the video packets are not corrupted. Therefore, they need to make sure to deploy a protocol that supports this level of delivery capabilities across their networking infrastructure. Although TCP provides applications with an extensive range of networking features in comparison to UDP, it is not a popular choice of transport protocol for service providers who are involved in the delivery of IPTV content. This comes down to the fact that IPTV is a real-time application and does not tolerate delays. TCP can often introduce latency into the delivery of IP video content due to fact that the protocol employs flow control mechanisms. The features and limitations of TCP that interfere with the real-time delivery of IPTV content are

Trade off between sensitivity and delays—IPTV is less sensitive to loss or corrupted packets than it is to time delays. Retransmitting packets improves the reliability of a connection between a server and IPTV access device; however, such activity increases delays. It is these delays that make TCP less desirable for transporting IPTV content.

TCP is a connection orientated protocol—As explained earlier TCP requires the establishment of a logical connection between the server and an IPTVCD before any transmission of IPTV content can take place. This causes unacceptable delays for end users who are changing from one channel to another in a live broadcast IPTV environment.

Support for error correction—As previously outlined TCP provides considerably more facilities, notably error correction and flow control. Ironically, the correction of errors that occur on an IP video network can in some cases degrade the quality of service delivered to subscribers. To understand why this is the case, we need to take a brief look at the basic characteristics of video transmission. First and foremost, video consists of a continuous sequence of images. Any interruption to the rate at which these images are processed and displayed by the IPTVCD causes impairments to the video display and affects the viewing experience of the end user. The typical processing time for each image is only a fraction of a second. Now, if TCP was used in this type of environment to fix a corrupted IPTV packet, the following would occur:

(1) The IPTVCD would flag that it has received a packet of data that contains errors.
(2) A message is sent to the IPTV server informing the application that one of the packets received was corrupt.
(3) Under the TCP regime the server will need to find and retransmit the same IPTV packet.
(4) The IPTVCD receives the new packet in its buffer and displays the video content contained in the packet.

While TCP cycles through the above steps the IPTVCD has either to wait for the retransmitted packet and incur a gap in the video flow or to discard the retransmitted packet when it arrives, thus negating the need for the TCP error correction mechanism.

Lost IP packets—The process of dealing with lost IPTV packets is more or less the same as fixing corrupted IPTV packets. Packets need to be transmitted, which in turn causes latency in the network, which has a negative impact on the delivery of IPTV services.

Support for flow control—In addition to the issue of error correction and establishing a logical connection, TCP's support for controlling the bit rate flow of the sending server can also cause difficulties in the transmission and reception of IPTV content. The difficulty occurs when the buffers in the IPTVCD start to fill up with IPTV packets or the network becomes congested. When these situations arise the server receives an instruction to reduce the rate at which the server is sending packets on to the network. If the slow down instruction is executed by the IPTV server, then there is a possibility that the rate will drop to a speed that will make it impossible to display a video picture on the TV screen. If a stop message is issued, then the IPTV service will completely shut down.

Lack of support for multicast—TCP is unable to scale effectively in a multicast environment. Therefore, UDP is a better choice over TCP when it comes to communications services like multicast IPTV applications and broadcast delivery since these do not exist on TCP.

Header Size—The TCP header is larger than a UDP header. The length of a TCP header is 20 bytes compared to 8 bytes used by a UDP header. Although a difference of 12 bytes appears in consequential within the context of transporting video data across high capacity ADSL and FTTH links it does add up in a network that may be delivering tens of millions of packets every day of the week.

These are the main reasons why TCP is rarely used in live IPTV environments. TCP is however extensively used by other applications such as e-mail and downloading Internet TV content for playback at a future date.

Although it is inferior to TCP in terms of reliability and error handling, UDP is used by most if not all IPTV implementations around the globe. Although UDP is the transport protocol of choice for the delivery of IPTV services it has couple of drawbacks, namely, its inability to deal with detecting and fixing error messages. This issue is overcome by building the error recovery functionality into the IPTV applications running across the network or inside the video stream itself.

3.3.2.9 IP Layer   The layer below the transport layer is called the IP layer (also known as the internetworking layer). Its primary objective is to move data to specific network locations over multiple interconnected independent networks called internetworks. This layer is used to send data over specific routes to its destination. IP is the best-known protocol located at the internetwork layer, which provides the basic packet delivery service for all IPTV services. The types of services vary from unicast systems where packets are sent from source to a single destination IPTVCD to multicast systems that send packets from a single encoder or streaming server to multiple IPTVCDs.

IP version 4, also frequently abbreviated IPv4 is the most commonly used protocol in IPTV networks today.The specifications for version 4 of the IP standard can be found in RFCs 791, 950, 919, and 922, with updates in RFC 2474. Note the most recent version of the protocol, IPv6 is published through RFC 2460. The main function of IP is to deliver bits of data segmented into packets from a source device to a destination device. IP uses a best effort mechanism for delivering data. In other words, it has no in-built processes for guaranteeing the delivery of information across a network. The foundation blocks of the IP protocol are segmentation of data bits into packets and addressing. The following sections give a brief overview of the IP video packets and associated addressing schemes.

IP Video Packets   An IP video packet is a unit that contains the actual video data and the details of getting the video from the IPTV data center servers to the destination IPTVCD. The key elements of an IPv4 video packet are illustrated in Fig. 3.22 and described in Table 3.11.

Understanding IP Addressing   In an IPTV environment, IPTVCDs and data center servers primarily use IPv4 addresses to identify each other. An IPv4 address is a series of four numbers separated by dots that identifies the exact physical location of a device such as an IP set-top box on the network. An IPv4 address is a 32-bit binary number. This binary number is divided into four groups of 8 bits ("octets"), each of which is represented by a decimal number in the range 0–255. The octets are separated by decimal points. An example of an IP address is 190.100.5.54. In binary notation this would be 10111110.01100100.00000101.00110110. The IP address is hosted in two sections:

(1) the network section identifies the broadband network that the IPTVCD is connected to,

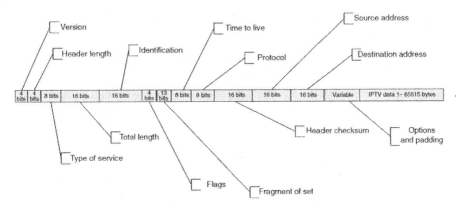

FIGURE 3.22   Typical format of an IPv4 video packet

TABLE 3.11 Structure of an IPv4 Video Packet

Field Name	Description of Functionality
Version	This identifies the version of IP used on the network—IPv4 or IPv6.
Header length	This field is used to describe the size of the header. This helps the IPTVCD to identify where the video data payload begins.
Type of service	This field also known as differentiated services code point (DSCP) is quite important when delivering IPTV content because it allows the service providers to set the type of content carried in the packet. This information is then processed by the IP routers located on the network. This allows routers to apply and adapt quality of service levels for different traffic types.
Total length	This field informs the IPTVCD of the entire length of the IPTV packet. This value is 16 bits in length, which means that a packet can have a maximum size of 65,535 bytes long.
Identification	This field is commonly used by routers to break a large packet into smaller fragments. Once fragmented the router uses this field to identify the various fragments of the original packet.
Flags	These fields identify different types of fragmentation, namely whether the packet is a fragment, whether it is permitted to be fragmented, and whether the packet is the last fragment, or there are more fragments.
Fragment offset	Once an IPTV packet is fragmented and delivered across a broadband network it is the function of the IPTVCD to reassemble the fragments in the correct order. This field numbers each of the fragments, which allows the IPTVCD to reassemble correctly.
Time-to-live	As an IP packet traverses a broadband network, the time-to-live field gets examined by each router along the way and the value inside this field gets decremented. This process continues until the value of the time-to-live field reaches zero. Once this happened the packet has expired and is discarded. The main function of this field is to eliminate IPTV packets on the network that are unable to reach their destination. This in turn reduces network congestion. A value of between 30 and 32 is typically used to transport IPTV content over a broadband network.
Protocol	This 8 bit field indicates the type of protocol encapsulated within the IP datagram. If the value is one then Internet Control Message Protocol (ICMP) is used, two represents Internet Group Management Protocol (IGMP), six is used for TCP traffic, and seventeen is used for UDP applications.
Header checksum	This field allows IPTVCDs to detect datagrams with corrupted headers. Corrupt packets are typically discarded or destroyed by the IPTVCD.
Source address	As the name suggests this is the IP address of the machine that sent the IPTV packet. In an IPTV environment this is typically a VoD server or an encoder.
Destination address	This is the IP address of the intended receiver(s) of the packet. In an IPTV environment this will typically be the address of the IPTVCD.
Options and padding	This field is used by IPTV vendors to provide additional features.
Data	This is the video content

(2) the host section identifies the actual IPTVCD.

It is also worth noting at this stage that the first few bits of the address identify how the remaining section of the address is separated between the host and network IDs. These bits are sometimes called highest order bits. For ease of use and administration, the IP addresses are broken up into the different classes. Further details of IPv4 address classes are available in Table 3.12.

The distribution of network and host address fields is shown in Fig. 3.23.

In addition to the above classifications, a number of IP addresses have been allocated and reserved for use by private networks. The IP address ranges reserved for private networks are

10.0.0.0 to 10.255.255.255

TABLE 3.12   IPv4 Address Classes

IP Address Class	Description
A	This describes a network whose first number in the IP address has a value of between 0 and 128. The remaining three numbers are used to identify an IPTVCD, server, or network device within the network. Thus, a class A address has a 7-bit network number and a 24-bit host address. The highest-order bit is set to 0. There are 126 class A network addresses in the world and each one of these networks has enough IP addresses to support more than 16 million unique network devices. All of these class A IP addresses have been licensed from InterNIC many years ago. InterNIC is a cooperative activity between the U.S. Government and a company called Network Solutions Inc.
B	A class B network has an address whose first number has a value of between 128 and 191. This equates to a 14-bit network number and a 16-bit local address. Values of 1 and 0 are assigned to the two highest order bits. There are approximately 16,000 class B networks on the Internet, each having the capability to support 64,000 unique network devices. Large organizations and Internet Service Providers have licensed most or nearly all of these addresses. For instance, in the class B address, 132.6.2.24, 132.6 identifies the network, and 2.24, identifies the host.
C	A class C network has an address whose first number has a value of between 192 and 223. This equates to a 21-bit network number and a 8-bit local address. Values of 1, 1, and 0 are assigned to the three highest-order bits. There are approximately two million class C addresses, each capable of supporting 254 addressable network devices.
D	The first portion of the address containing a value of between 224 and 239 is known as a class D address. These IP addresses are used for multicasting purposes.
E	Class E addresses range from 240 to 247 and are reserved for future use.

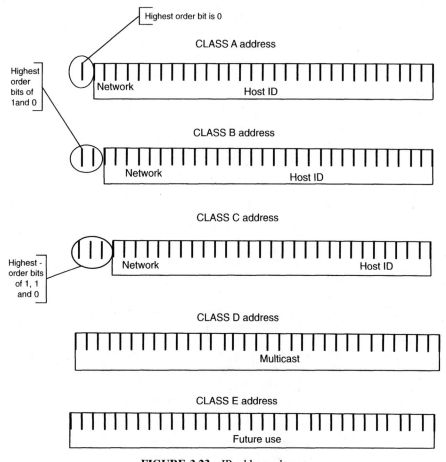

FIGURE 3.23   IP address classes

172.16.0.0 to 172.31.255.255
192.168.0.0 to 192.168.255.255

3.3.2.6.1 Subnetting an IPTV Network   In large IPTV networks with thousands of IPTVCDs spread across a large geographical area, IP based networks need to be subdivided into logical units called subnets. Subnetting an IPTV network allows service providers to identify and monitor individual sections of the organization's network without having to obtain new IPv4 addresses. Network operators also use subnet addressing to hide their internal private network structure against attacks from the public Internet. IPTV network administrators use special numbers called subnet masks to create subnets within an IPTV environment. Subnet masks are also a 32-bit IP address. The default numbers for Class A, B, and C addresses are as follows:

Class A—255.0.0.0
Class B—255.255.0.0
Class C—255.255.255.0

The subnetting process is implemented using hardware devices called routers. A router connects to more than one network and makes a decision on where to send information on a network. A router takes the assigned subnet mask and applies the number against the IP number of the IPTVCD. Applying in this context basically means that the router moves the demarcation line between *netid* and the *hostid* sections of the IP address. The result of this operation is a new number, which helps the router send an IP video stream to the correct IPTVCD.

3.3.2.6.2 Future of IP Addressing   When the Internet's IP address structure was originally developed in the early 1980s it was intended to meet the needs of current and future users. The 32-bit addresses used by the current version of IP (IPv4) can enumerate over 4 billion hosts on a possible 16.7 million different networks. The recent phenomenal growth of the Internet is rapidly exhausting the existing pool of IP addresses provided for by IPv4. The scale and speed of growth experienced by the Internet in recent years could not have been foreseen by the original developers of the IP protocol and the number of networks connecting to the Internet grows on a monthly basis. The global race to start deploying networks that support the triple play bundle of services is starting to put pressure on the IPv4 addressing scheme.

To find a solution to the limitations of the existing IP address scheme, the Internet Engineering Task Force launched IPv6 (Internet Protocol Version 6). IPv6 defines a 128-bit IP address that is compatible with current implementations of IP. This makes the transition to IPv6 an evolutionary rather than a revolutionary step. In addition to backward compatibility, IPv6 greatly increases the available address space. In fact the number of addresses possible under IPv6 will be four billion times four billion, which is a phenomenal amount of IP addresses. IPv6 also includes capabilities that will provide support for authentication, QoS integrity, encryption and confidentiality. The deployment of IPv6 across IPTV networks is particularly suited because of its in-built QoS mechanism and its ability to support a virtually unlimited number of IPTVCDs. Over the next five to ten years IPv6 is expected to gradually replace IPv4, with the two coexisting for a number of years during a transition period.

Why Use IPv6 When Deploying IPTV?   IPv6 scores over IPv4 in a number of major areas. The main reasons why an IPTV service provider should consider using IPv6 are as follows:

IMPROVED SCALABILITY   IPv4 has a 32-bit address while IPv6 has 128-bit address bit space making it four times more advantageous. This allows IPTV service providers

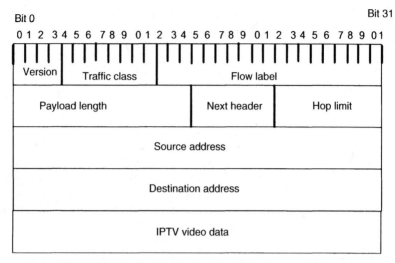

FIGURE 3.24   IPv6 header structure as defined in RFC 2460

to expand their reach into the management of numerous devices connected to in-home networks.

SIMPLIFIED HEADER STRUCTURE IPv6 reduces the size of the header down to fixed length of 40 bytes and greatly simplifies the structuring of its fields. The elements of an IPv6 header are illustrated in Fig. 3.24 and explained in Table 3.13.

IMPROVED SECURITY LEVELS IPv6 has two additional features in the basic specifications that ensure better security:

(1) Inclusion of an authentication header—which authenticates messages and verifies the packet sender.
(2) Encapsulated security payload—This feature ensures the integrity of the IPTV data and enforces confidentiality between the IPTV data centre servers and the various IPTVCDs.

BETTER FLOW OF REAL TIME TRAFFIC The flow labeling capabilities of IPv6 allows IPTV service providers to mark particular packets, which belong to specific services. So in a triple-play environment, routers are able to treat IP packets labeled with a video identifier "differently" to IP packets, which are labeled as carrying Web content.

AUTOMATIC CONFIGURATION The "plug and play" capability of IPv6 reduces the complexities of installing an IPTV service at a subscriber's home.

Because of the various benefits described above, IPv6 is starting to gain acceptance in the IPTV industry as the long-term solution for supporting the

TABLE 3.13   IPv6 Header Field Descriptions

Header Field Name	Size (Bits)	Description of Functionality
Version	4	Identifies version number.
Traffic class	8	Defines priority level of IPv6 packet.
Flow label	20	Defines specific treatment at routers.
Payload Length	16	Defines the length of the IPv6 TV packet.
Next Header	8	Specifies the presence or not of an extension header or header for another protocol, UDP for example.
Hop Limit	8	This replaces the TTL field in the IPv4 header and is modified by routers when traversing the IPTV network.
Source address	128	Identifies source address.
Destination address	128	Identifies destination address.

large-scale deployment of digital devices that are able to consume and render various types of IP based applications.

To conclude this section it needs to be noted that the main drawback of using the IP protocol is that it offers no certainty when it comes to delivering packets to their right destination, nor does it promise the correct arrival time. Even the order in which the packets arrive at the IPTVCD is not defined. Therefore, the IP layer works in conjunction with the transport layer protocols to ensure that packets arrive at the IPTVCD in an orderly and timely manner. IP can also introduce delays into the delivery of video content. As a result, most service providers deploy a quality of service (QoS) mechanism to minimize delays caused in this IPTVCM layer.

3.3.2.7   Data Link Layer   The link layer takes the raw data from the IP layer and formats it into packets that are suitable for delivery across the physical network. Note the data link layer varies between network protocols. The Ethernet technology is one of the more popular mechanisms used by IPTV systems. The types of functions carried out by the data link layer for Ethernet based networks include:

Encapsulation—This layer adds a header to the IPTV packets. Ethernet headers are the most common type of encapsulation that occurs at the data link layer of the IPTVCM. The elements of an Ethernet header are explained in Table 3.14.

Addressing—The data link layer deals with the physical addresses of IPTV consumer networking and server devices. The addressing schemes vary between network topologies. A media access control (MAC) addressing scheme for instance is used by Ethernet. MAC addressing allows each device connected to an IPTV network to have a unique identifier. The lengths of a MAC address is 48 bits and is often represented by 12 hexadecimal digit. Of the 12 hexadecimal digits the first six hexadecimal values identify the manufacturer of the IPTV device and the remaining digits identify the actual network interface.

TABLE 3.14 Structure of an Ethernet Header

Field Name	Size (Bits)	Description of Functionality
Ethernet destination address	48	This field identifies the address of the destination interface.
Ethernet source address	48	This field identifies the address of the source interface.
Type code	16	This field identifies the protocol used in the formation of the packet. For instance the "TCP/IP" type packets contain a hex value of "0×80 0×00".

Error checking—The error checking function is spread across several IPTVCM layers, including the data link layer. Corrupted packets are the most common type of error that occurs during the transmission of video content across an IP based network. An error correction scheme called cyclic redundancy check (CRC) is typically used in IPTV deployments to detect and discard these corrupted video packets. Under the CRC error checking mechanism, the transmitting IPTV device performs a mathematical computation on the packet and stores the resulting value in the packet. The same calculation is performed again by the receiving IPTV device when the packet is received. If the values are the same, then the packet is processed as normal. If however the values are different, then the packet contains errors and is discarded. A new video packet is subsequently built and retransmitted by the transmitting device. Informing the upper layers of the IPTVCM protocol stack when an error occurs is the main contribution made by the data link layer to the overall error checking mechanism used by end to end IPTV systems.

Flow control—Managing flow control is primarily dealt with at the transport layer. In the context of an IPTV networking environment the use of flow control ensures that the streaming server does not overwhelm a receiving IPTV consumer device with content. The data link layer executes any required flow control requirements in conjunction with the transport layer.

In addition to addressing, error checking and flow control this layer also takes care of encapsulating IP layer packets into a format that is suitable for transmission across the physical network. In most cases, packets are formatted into Ethernet frames. Table 3.15 provides further detail on the composition of a typical Ethernet frame that consists of a number of MPEG-2 TS packets. As a result of adding up the values listed in Table 3.15, the total number of bytes stored inside this Ethernet frame is 1370 bytes. In this example the size of the frame consisting of seven TS packets fits comfortably within the maximum limit size of 1518 bytes set out in the Ethernet standard.

3.3.2.8 Physical Layer The physical layer relates to topics that coordinate the rules for transmitting digital bits over the network. It is concerned with getting data

TABLE 3.15   Composition of a Typical Ethernet Frame used to Carry MPEG-2 Video

Description	Size
Seven MPEG transport stream packets	Each packet has a size of 188 bytes (184 bytes of video content plus 4 bytes used for header information). These seven MPEG-TS packets occupy 1316 bytes (10,528 bits) of the frame.
RTP header	This header occupies 12 bytes of the Ethernet frame.
UDP header	This header occupies 8 bytes of the Ethernet frame.
IP header	This header occupies 20 bytes of the Ethernet frame.
Ethernet header	This header occupies 14 bytes of the Ethernet frame.

across a specific type of physical network (such as DOCSIS, xDSL, and wireless). It defines physical network structures (topologies), mechanical, and electrical specifications for using the transmission medium.

Once the IPTV bit stream is transmitted over the network, the packets are passed back up the receiving device's IPTVCM and the encapsulation process is reversed. The data link layer for instance inspects the packet, removes from it the Ethernet header and the CRC fields. It next examines the type code field of the Ethernet header and determines that the packet needs to be processed by the IP protocol. Thus, the packets are passed upward to the network layer. The network layer subsequently inspects, removes the IP header and passes upward to the transport layer. This method of removing and stripping away headers at the various layers is called decapsulation. The other layers continue this process until the packets reach the top of the IPTVCM and the raw video gets displayed on the viewers TV screen.

TABLE 3.16   Limited Feature Comparison Between Different IPTV Compression Technologies

Technical Characteristic	MPEG-2	H.264/AVC	VC-1
Motion compensation block sizes	16×16 16×8	16×16 $16 \times 8, 8 \times 16$ 8×8 $8 \times 4, 4 \times 8$ 4×4	16×16 $16 \times 8, 8 \times 16$ 8×8
P frame referencing functionality	Single reference frame	Capable of referencing multiple frames	Single referencing of frames
B frame referencing functionality	Reference frames on either side	Capable of referencing multiple frames	Reference frames on either side

TABLE 3.17 Summary of IPTVCM Layers

Layer Number	Name	Overview
1	Physical	This layer defines the attributes of the network media responsible for carrying bits of IPTV data.
2	Data link	This layer handles the mechanisms used to access the network media. Error correction, synchronization and flow control are other functions that are provided by the data link layer.
3	IP	As the name suggests this layer deals with routing IP packets over the network. Mechanisms such as addressing and congestion control are typically used by this IPTVCM layer.
4	Transport	The primary function of this layer is to ensure that IPTV packets arrive at their destination point. TCP and UDP operate at this layer.
		Due to its low overhead UDP is a better choice over TCP when it comes to communications services like multicasts and broadcast delivery. UDP is the dominant transport layer protocol used by IPTV networking infrastructures.
5	RTP layer (optional)	Although UDP is the preferred transport layer protocol for delivering IPTV content across a broadband network it is not a reliable protocol and does not support error correction or dealing with packets that arrive at the IPTVCD out of order. As a result a number of service providers use RTP to address the deficiencies that are an inherent part of the UDP protocol. Note in addition to MPEG-2 it is also possible to use this layer to carry H.264/AVC NAL units and VC-1 packets as part of the RTP protocol payload.
6	Transport stream construction	This layer packetizes audio and video bit-streams. The resulting packets are typically 188 bytes in length.
7	Video packetizing layer	This layer creates a stream of time-stamped PES packets.
8	Video encoding layer	Audio and video elementary streams form the basis for this IPTVCM layer. The format of the streams employed at this layer depends on the compression algorithm used by the encoder.

SUMMARY

Video compression is a process used by IPTV service providers to reduce the amount of data contained in a video file down to a manageable size that may be transported across an IP broadband network. Different compression technologies are suited for different purposes. MPEG-2 for instance is a widely used video compression technology by digital television applications. MPEG-4 is the successor to MPEG-2 and includes a number of advanced features ranging from improved compression rates to support for interactive multimedia applications.

VC-1 is an encoding technology that has been standardized by an international organization called the SMPTE. It is published as SMPTE 421 M and is accompanied by two sister publications, which define the transport mechanisms and conformance guidelines for the technology. VC-1 is designed to compress video at a variety of different bit rates including those that are typically used by IPTV applications. Table 3.16 provides a limited feature comparison between the three major IPTV compression systems.

IPTVCM is a networking framework that is composed of seven and in some implementations eight layers, which each specify protocols and network functions that are used for communications between IPTV devices. The upper layers of the model implement various services such as encoding and packetizing of video content. The lower levels of the model are responsible for transportation-orientated functions such as routing, addressing, flow control, and physical delivery. Table 3.17 summarizes the IPTVCM reference model layers.

4

BROADCASTING LINEAR PROGRAMMING OVER IPTV

In addition to providing VoD applications, IPTV service providers are also starting to offer traditional broadcast linear TV programming to their subscribers. For VoD, a unicast communication system is used to establish individual links or sessions between the various IPTVCDs and the IPTV data center servers. This method of communications is not effective for delivering traditional broadcast style channels because the duplication of point-to-point sessions would overwhelm the network. Therefore, a technique called IP multicast is widely used to provide a single broadcast TV channel to multiple clients simultaneously. This chapter commences with a description of the video components used by end-to-end IPTV multicast systems. Coverage is also provided on a protocol called the Internet Group Membership Protocol (IGMP), which is used by multicast networks to select and control the delivery of TV channels to IPTV end users. There are currently three versions of the protocol with each new revision improving on the previous release. The chapter moves on to describe the various routing protocols used by multicast systems and examines the technologies used to deploy multicasting across IPv6 based networks. The primary purpose of multicast technologies is to ensure that end users are able to instantaneously and reliably switch channels during a TV viewing session. In fact, a number of telecommunication companies are marketing the fast channel-changing feature as a key benefit of IPTV over traditional pay TV services. The final section of this chapter examines some of the techniques that help to improve the channel-changing rate for multicast IPTV channels.

Next Generation IPTV Services and Technologies, By Gerard O'Driscoll
Copyright © 2008 John Wiley & Sons, Inc.

4.1   UNDERLYING VIDEO COMPONENTS OF AN END-TO-END IPTV SYSTEM

There are two main mechanisms for delivering IPTV content: multicasting of linear TV content and video on demand. Figure 4.1 shows the various hardware and software components as well as associated interfaces required to implement a complete end-to-end solution for IP based multicast TV and video on demand. The role of each component is briefly discussed in the following sections.

4.1.1   Integrated Receiver Devices (IRDs)

These digital devices are used to receive video assets from a number of different types of networks ranging from satellite links to dedicated video circuits and microwave links.

4.1.2   Real-Time Encoders

Central to an IP video-broadcasting network is the compression system whose job is to deliver high quality video and audio to end users using a small amount of network bandwidth. The main goal of the IPTV compression system is to minimize the storage capacity of information while maintaining the quality of the video

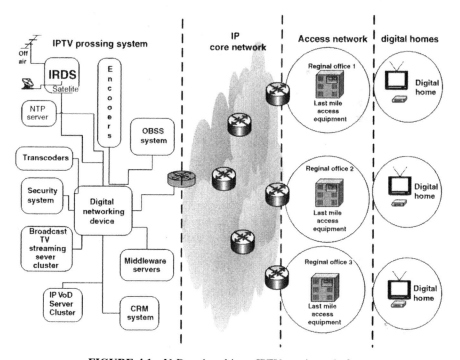

FIGURE 4.1   VoD and multicast IPTV services platform

and audio streams. This is particularly useful for service providers who want to "squeeze" as many IP digital channels into a broadband stream as possible. The compression system is made up of a number of real-time encoders, which are used to digitize and compress a range of audio, video, and data channels. Uncompressed analog or digital video content is received as input to the encoding device and compressed video content inside encapsulated video packets are outputted. Digital encoders allow IPTV operators to broadcast several high quality video programs over the same bandwidth that was formerly used to broadcast just one analog video program. Once the signal is encoded, a compressed video stream is transmitted to the IPTV distribution router. Note that most modern encoders feature an Ethernet interface for outputting the compressed content.

4.1.3 Broadcast TV Streaming Server(s)

These servers are typically configured into clusters for business continuity purposes and are responsible for streaming live IPTV content using selected protocols to end users.

4.1.4 An IP Transcoding System

The requirement of a transcoding subsystem at the IPTV data center is dependent on two factors:

(1) The format of the source video content
(2) The compression standard deployed on the IP distribution network

The addition of a transcoding system requires specialized hardware to fulfill this functionality. These devices are used to convert streams of MPEG-2, uncompressed Serial Digital Interface (SDI), or even analog material into another compression format such as H.264/AVC or VC-1. Transcoding does not involve reconstructing video signals, however, a change in either the bit rate or the resolution of the incoming video signal does occur. Other functions of transcoders include format conversion and adjustment of frame rates. Some of the more advanced transcoders also support encoding and decoding functionality. There are a variety of different methods of transcoding MPEG-2 content. For example, one method fully decodes the incoming video stream, processes it when the stream is uncompressed, and then proceeds to encode into the new compression standard at a lower bit rate.

4.1.5 An Operational and Business Support System (OBSS)

The OBSS, also known as a subscriber management system (SMS), is used in conjunction with other IPTV network elements to activate, fulfill, and provide IPTV services in real time to meet customer requirements. The types of information processed by an OBSS during the provisioning of a new service can include

- Subscribers name and address
- Billing and payment details
- IPTV multicast programs required
- IP-VoD assets required
- Network bandwidth usage required to provide a new service
- IP address allocation for a new service
- Subscribers desired time for installation and provisioning of new services

In addition to activating new services and connectivity options, the OBSS may include some if not all the following functionalities:

- Generating provisioning requests to the IPTV engineering and installation groups.
- Gathering usage information associated with particular types of IPTV services.
- Monitoring the status of provisioning requests and update accordingly.
- Storing a database of hardware and software resources owned or leased by the service provider.
- Monitoring and managing residential gateways connected over the broadband network to the IPTV data center.
- Managing and supporting content providers.
- Managing customer accounts, profiles, and invoicing.
- Providing a self-service system, which allow subscribers to order new products via a Web site.
- Providing an Internet based Web portal that allows IPTV end users to access information that will help them to resolve common technical problems.

A standard interface is used by the OBSS to communicate with external sub systems at the data center, including the customer relationship management (CRM) system.

4.1.6 IPTV CRM System

The use of a CRM system provides telecom operators with visibility into the sales of particular packages of services. A typical CRM may include a number of features and may be classified into three different modules. Table 4.1 explains these various CRM modules within the context of an IPTV networking environment.

4.1.7 An IPTV Security System

The output from the encoding system is fed into a security system for content protection. The purpose of the IPTV security system is to restrict access to subscribers and protect against the theft of IPTV content. The security system

TABLE 4.1   Functional Categories Used by IPTV CRM Systems

CRM Module	Functional Description
Interfacing with subscribers	This module ensures that detailed information both current and historical is available to help in improving the efficiency of interacting with subscribers. Contact management and customer care are two of the key features of this module.
Marketing IPTV products and services	This module manages various marketing and promotional campaigns. Typical functions of this module include organizing the distribution of marketing collateral and managing any resulting leads.
Selling IPTV products and services	This function is used as an aid to help the IPTV service provider generate sales. The CRM sales module interoperates closely with the marketing module in order to identify suitable sales channels and monitor the results from various marketing campaigns. Sales prospects and leads are also processed by this module. Up-selling is another integral part of the sales modules, which allows service providers to sell additional IP services to IPTV subscribers.

consists of two parts: Conditional Access (CA) and Digital Rights Management (DRM).

4.1.8   IP Video on Demand (VoD) Application Servers

As the name suggests the video servers store and cache the VoD files. Video servers are generally interconnected into a cluster, which provides for redundancy in the event of a server failure. VoD servers typically run a software application, which is required to support the management of VoD and the different types of multimedia data.

4.1.9   IPTV Headend Middleware and Application Servers

IPTV Middleware falls into two defined categories: client and server software. The middleware server software is implemented on a series of application servers that are running at the IPTV data center. IPTV Headend Middleware and Application Servers perform the following functions:

- Interacting with the backend OBSS and CA systems.
- Helping to manage the provisioning of new subscribers, billing, and overall management of video assets.

- Hosting software applications, which interface with the middleware clients embedded in IP access devices.
- Supporting the user interface for both IPTV multicast and on-demand services.

The client middleware resides at the IPTVCD and is used to isolate IPTV application programs from the details of the underlying broadband network. Chapter 6 provides a more detailed insight into IPTV client middleware software.

4.1.10   A Network Time Server

Many networking devices and servers use their own internal clocking systems to track time. Owing to the variations between different clock types, time drifts generally occur between the various devices over a period of time. As a result, IPTV data centers typically use a network time server to allow synchronization between network components. Interconnectivity with this server is facilitated through the Network Time Protocol (NTP)—an international standard published in RFC 1305.

4.1.11   IP Switching Infrastructure

An IPTV data center will typically include equipment for switching video and audio signals. Traditionally, large baseband switches were used to handle the routing of video signals between various types of content source devices. Drawbacks of baseband switches included their physical size and complexity. With the move to digital a new breed of switches are being installed in IPTV data centers. The size and complexity of these digital switches depends on how the video and audio signals are transported between the various IPTV data center components. With the move towards converting digital video signals into IP packets, a number of service providers have started to use standard IP networking equipment such as Ethernet routers and switches to perform the routing of signals. The use of standard networking equipment allows IPTV providers to consolidate various types of video, audio, and data signals onto one single network. This in turn reduces maintenance costs, simplifies network management, and increases the flexibility of the switching infrastructure.

4.1.12   A Distribution Router

The IPTV system architecture also includes a high speed distribution router, which typically resides at the service provider's headend and is responsible for delivering the interactive IPTV content to the distribution network. The router is directly connected to the IPTV core network. Note that an ATM switch may sometimes be used instead of an IP router to connect with SONET based networks.

4.1.13   IP Distribution Network

The IP distribution network consists of two parts, namely, the core and the access section.

(1) The *IP core* is responsible for aggregating all the IP video content in addition to other types of traffic in a triple-play environment.

(2) The *access or last mile* part of the system uses a mix of technologies such as DSL and WiMAX to bring requested services to the IPTV end user.

4.1.14   IPTVCDs

An abbreviation for IPTV consumer device, an IPTVCD is defined as a hardware device that typically terminates an IPTV connection. Examples of IPTVCDs include media center PCs, game consoles, residential gateways, and IP digital set-top boxes.

Table 4.2 summarizes the functions of each component used to process VoD and broadcast based IPTV services.

In a typical live IPTV network all of these components are tightly integrated in order to deliver a number of VoD titles and a large number of standard definition and high definition channels.

4.2   DIFFERENT APPROACHES TO STREAMING IPTV CONTENT

The network traffic patterns of real-time IP based TV differs to patterns generated by other IP services such as VoIP and high speed Internet access. For instance, video traffic sustains high throughput levels, whereas Internet traffic fluctuates between high and low levels of activity. Three different techniques are used to accommodate the unique patterns associated with video traffic.

4.2.1   Unicast

In unicasting, every IPTV video stream is sent to a single IPTVCD. Therefore, if more than one IPTV end user desires to receive the same video channel, each IPTVCD will need a separate unicast stream. Each of these streams will flow to the destinations across the high speed IP network. The principle of implementing unicast over an IP network is based around delivering a dedicated stream of content to each end user. From a technical perspective this configuration is quite easy to implement; however, it does not make effective use of the bandwidth on the network. The diagram presented in Fig. 4.2 shows an example of how IP connections are established when five IPTV subscribers access a specific IPTV broadcast channel over a two-way high speed network.

As shown, when multiple IPTV end users decide to access the same IPTV channel at the same time a number of dedicated IP connections are established

TABLE 4.2   IPTV and VoD Headend Components

Component Name	Description	Involved in the Processing of Which Category of IPTV Service
IRDs	IRDs interface with the source content provider to receive video content from various sources.	Both VoD and broadcast centric IPTV services.
Encoders	This component receives analog or digital video content, compress the content, and output a compressed digital video stream.	Broadcast centric IPTV services.
An IP transcoding system	This component may be used to change the compression format of video streams in real time.	Broadcast centric IPTV services.
Broadcast TV streaming server	These servers interface with the encoders and stream TV programming to a number of different types of IPTVCDs.	Broadcast centric IPTV services.
IP VoD servers	These servers host and stream VoD files to IPTV viewers.	VoD centric IPTV services.
Headend middleware and application servers	This cluster of servers stores the backend middleware applications that are required to support front-end functionality of the middleware platform running on the various IPTVCDs. It integrates with most of the video components running in the data center.	Both VoD and broadcast centric IPTV services.
OBSS	The OBSS contains the various items of information required to administer and manage an IPTV business, including billing details and information on which services the subscriber is allowed access.	Both VoD and broadcast centric IPTV services.
CRM	The CRM system is used to efficiently manage interactions between the service provider and subscribers.	Both VoD and broadcast centric IPTV services.
IP switching infrastructure	Standard networking switches are typically used to route IP video data between the various components located at the IPTV data centre.	Both VoD and broadcast centric IPTV services.

TABLE 4.2 *(continued)*

Component Name	Description	Involved in the Processing of Which Category of IPTV Service
CA system	The CA system is responsible for encrypting and managing access to both IPTV broadcasting and on-demand content. It verifies whether or not a subscriber is authorized to view a particular channel or VoD title and grants or refuses access based on the rights of the subscriber.	Both VoD and broadcast centric IPTV services.
IP distribution router	This component takes the IP stream from the VoD and broadcast streaming servers and forwards onwards to the IP distribution network.	Both VoD and broadcast centric IPTV services.
IP distribution network	The IP network is in some cases subdivided into two sections, namely, the core and access networks.	Both VoD and broadcast centric IPTV services.
IPTVCDs	There are a number of different devices that are capable of accessing IPTV content. IP set-top boxes are however by far the most popular option used by service providers to deploy IPTV services. The purpose of an IP set-top box is to take the incoming IP stream and convert it to a format that is suitable for viewing on a standard television.	Both VoD and broadcast centric IPTV services.

across the network. In this example the server needs to provide an active IP connection to every IPTV subscriber that requires access to Channel 10, with a total of five separate streams originating from the content server and connecting to the destination router. These five connections subsequently get routed to their destination points. The connections are spread across two regional offices, with three dedicated IP connections to regional office 1 and two dedicated IP connections to regional office 2. More connections are subsequently set up between the routers at the regional offices and the residential gateways (RGs) installed at the five digital homes. In this unicast environment the need for multiple IP connections leads to a requirement for very high capacity network links. This method of transporting IP

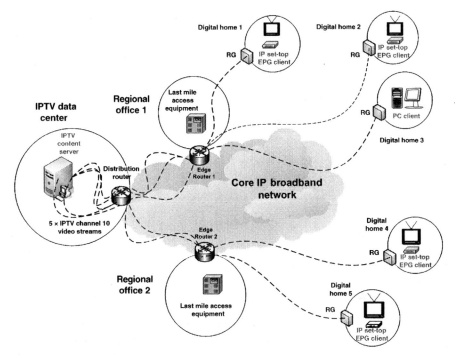

FIGURE 4.2   Multiuser unicast IP connections for single IPTV broadcast channel

video is well suited to on-demand applications such as network based digital video recording (NDVR) and VoD, where each subscriber receives a unique stream.

4.2.2   Broadcast

IP networks also support broadcasting functionality, wherein the same IPTV channel is streamed to every IPTV access device connected to the broadband network. When a server is configured to broadcast, an IPTV channel is sent to all IPTVCDs connected to the network regardless of whether they requested the video stream or not. This is a major issue because the resources of IPTVCDs are tied up in processing unwanted video packets. Also the other issue that makes broadcasting totally unsuitable for IPTV applications is the fact that this communication technique does not support routing. Since most IPTV networks make extensive use of routers, the use of broadcasting is prohibitive. This is because the network and the various IPTVCDs become overwhelmed when all channels are sent to all recipients.

4.2.3   Multicast

Groups and membership form the basis of how multicasting operates. In the context of an IPTV deployment, each multicast group is a broadcast TV channel and its members equate to the various IPTVCDs that are "tuned" into and viewing that channel. Thus, each IPTV channel is only streamed to the IP set-top boxes that

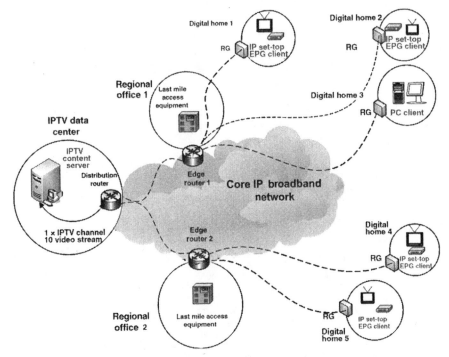

FIGURE 4.3   IP Connections used by the multicasting technique

want to view the channel. This keeps bandwidth consumption relatively low and reduces the processing burden on the server to a small fraction of that found under the unicast one-to-one communication system. Figure 4.3 illustrates the affect of using multicast technology on the example of five IPTV subscribers simultaneously accessing IPTV Channel 10.

As shown, only a single copy is sent from the content server to the distribution router. This router makes two copies of the stream and sends them to the routers at the regional offices via dedicated IP connections. Each one of the routers will subsequently make copies of the streams for digital homes connected to their interface ports that want to view the stream. Note that this approach significantly cuts down the number of IP connections and video streams traversing the network. This approach is typically used by service providers to broadcast live IPTV programming and is an efficient technique for utilizing an existing IP infrastructure. Note that multicast does not utilize an upstream path for communication between IPTVCDs and the broadcast server. The multicasting of IPTV content is considered to be more sophisticated when compared to using the unicast or broadcast communication models. The following sections provide a more detailed insight into how multicasting is applied to the transmission of IPTV content.

4.3   MULTICASTING ACROSS AN IPTV NETWORK

Multicasting refers to the technique of transmitting a single video signal simultaneously to multiple end users. So just like in traditional broadcasting, all viewers receive the same signal at the same time but there are no separate streams for each recipient. It provides an efficient way to support high bandwidth, one-to-many applications on a network. As shown in Table 4.3 a number of applications take advantage of IP multicasting.

In addition to the above applications, IP multicast is also widely used in the delivery of broadcast TV services over IP networks. There are a number of reasons for this. First and foremost, it significantly reduces the amount of bandwidth required to transmit high quality IPTV content across a network. This is because only a single copy of every video stream needs to be sent to a router, which in turn makes a copy of that stream for the requesting devices. Not only does multicast reduce the bandwidth requirements of the network but the processing power of the content server can also be kept relatively low because it only transmits one copy of an IPTV stream at a time. In contrast, a unicasting based networking environment is required to simultaneously support the transmission of multiple video streams to multiple end users and only high performing servers can perform this task. Multicasting does however have some drawbacks including

VCR type controls are unsupported—Multicast does not allow subscribers to rewind, pause, or fast-forward the video content as per their wishes.

Limited flexibility—Similar to current pay TV services, when IPTV subscribers turn on their TVs they can only join the viewing when the particular channel is already in progress.

Routers need to support multicast—Service providers who want to deliver live IP broadcast television over their networking infrastructure need to ensure that all routers between the data center and client IPTVCDs are multicast enabled. Routers that do not provide this level of functionality need to be replaced.

Increases the workload and processing requirements of routers—Routers play a prominent role in the transmission of IPTV content across a network. In addition to forwarding traffic to the correct output ports, routers also need to

TABLE 4.3   Categories of IP Multicasting Applications in use Today

Multicasting Application	Category
Video conferencing	Real time
e-Learning	Real time and non-real-time
Stock updates	Real time
Online weather reports	Real time
Database replication	Non-real-time
Automobile traffic updates	Real time

handle additional tasks such as replicating video streams and keeping track of multiple copies of video packets. Processing the various tasks associated with IP multicasting adds a significant burden to the workload of IP routers.

Support needs to be consistent between source and destination—All of the components between the IPTV content source and the IPTVCD are required to support IP multicasting technology. This is problematic for Internet TV providers who have limited or no control on how devices are configured on the public Internet.

Blocking of IP multicast traffic—Security devices such as firewalls are often configured by IT network managers to block IP Multicast applications. This is not a major issue for operators that own the networking infrastructures; however, it should be borne in mind when deploying IPTV services across the public Internet.

4.4   IPTV MULTICASTING NETWORKING ARCHITECTURE

The deployment of a multicast system is based on a distributed networking architecture. The logical and physical components needed to deliver IP multicast services can be categorized as follows:

- IGMP devices
- Multicasting groups and addressing
- IPTV multicasting protocols
- Multicast transport architecture technologies

These are described in greater detail in the following sections.

4.4.1   IGMP Devices

A multicast-enabled host is configured to send and receive (or only send) multicast data. There are two categories of devices used in an end-to-end IGMP communications transaction:

- An *IGMP* host is any client or server connected to an IPTV network. A set-top box, a mobile phone, or a standard PC are all examples of IGMP hosts.
- *Multicast routers* also called IGMP routers are a key component of an IPTV networking infrastructure. Routers on an IPTV network fall into two broad categories, distribution and aggregation routers. The distribution router(s) normally reside at the IPTV data center or headend and directly interface with the source content servers. All of the IPTV channels are available at the distribution router. The aggregation router is positioned downstream in the network and closer to the end user. Only the channels that are viewed by IPTVCDs connected to the aggregation router are available at this point on the network. Multicast IP routers are an integral and essential component

used in the delivery of linear IPTV broadcast channels. They are connected to the IPTV transmission network and support the following functions:

Receiving multicasted IPTV content—On an IPTV network many of the router interfaces operate in multicast mode. This means that the router analyses incoming packets to see if they require further processing or not. When a multicast router receives packets that have a specific bit in the packet header set to one it sends the packet through the IP communication protocol for further processing. Multicast routers also use complex algorithms to handle forwarding of IPTV multicast traffic.

Managing and processing IGMP messages—Multicast routers also need to be capable of receiving, processing, and managing various types of IGMP messages.

Keeping the routing tables current—Multicast routers use the same tables that are used by any underlying unicasting protocols that are configured on the router. These tables are kept current by the routers operating system.

Replication of IPTV streams—Replication is defined as the ability of a router to take an incoming IPTV video stream and copy this stream via a single or multiple router ports to individual IPTV subscribers.

All of the tasks described above add to the processing requirements of the router. Thus, the deployment of multicasting technologies has a serious impact on the performance of the routers and often requires an upgrade to support the effective delivery of IPTV content.

4.4.2 Multicasting Groups and Addressing

Multicasting over an IPTV network works by sending video packets to a group of IPTVCDs that have expressed an interest in receiving a particular IPTV channel. The group is identified by a unique class D IP address. As shown in Fig. 4.4, a class D IP multicast address begins with 1110 as the first four bits and the address range extends from 224.0.0.0 to 239.255.255.255. This equates to a total address space of 268,435,456 multicast groups.

The multicast address range is subdivided further to support different types of applications (see Table 4.4).

All of the layer three IP addresses shown in Table 4.4 need to be converted to a hardware MAC address when used to deploy IPTV multicast services.

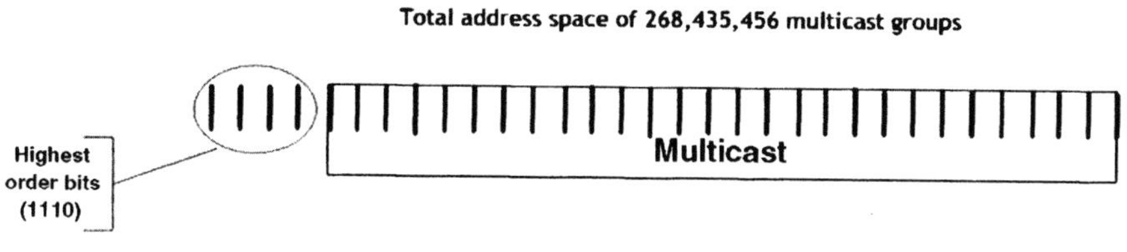

FIGURE 4.4   Structure of a class D multicast address

TABLE 4.4 IP Multicast Address Ranges and Associated Applications

Address Block Range	Address Type	Description
224.0.0.0 through to 224.0.0.255	Permanent	These addresses have been reserved by IANA for use by single hop multicast applications. The IP address 224.0.0.2 is an example of a reserved address that identifies all routers connected to the local network. 224.0.0.22 is another example, which is used by IPTV enterprise applications. These group addresses will always exist even if the group membership has dropped to zero.
224.0.1.0 through to 238.255.255.255	Globally scoped (Internet wide)	Addresses used in this range can be used to multicast IP video content across the public Internet. These groups will only exist as long as there are available members.
239.0.0.0 through to 239.255.255.255	Administratively scoped	This multicast range is used for IP multicasts, which are limited to a local group or organization.

This process can be slightly problematic due to the fact that only 23 of the original 32 bits available in the IP address are used to form the multicast MAC address.

4.4.3 IPTV Multicasting Protocols

IPTV multicasting employs specialized protocols for distributing and replicating IPTV content.

4.4.3.1 IGMP Versions 1, 2, and 3   Internet Group Management Protocol (IGMP) is an integral part of the IP communication model, which is used by IPTVCDs to join or leave a particular multicast group address. There are three different main flavors of IGMP in use today, namely, IGMP version 1 (IGMP v1), IGMP version 2 (IGMP v2), and IGMP version 3 (IGMP v3). The following sections give an overview of the three IGMP versions when used across an IPTV infrastructure.

IGMPv1   The specification for IGMPv1 was published in 1989 and can be found in RFC 1112. It is the original version of the protocol and was implemented on a variety of UNIX and Microsoft based operating systems. Some of the characteristics of the IGMPv1 specification include

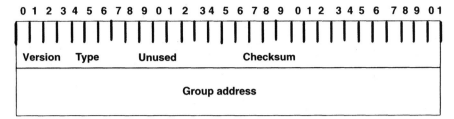

FIGURE 4.5   IGMPv1 message format

Message types—It supports two message types, namely, *host membership query* messages and *host membership report* messages. The query messages allow multicast routers to probe group membership details. The membership report messages allow local IP set-top boxes or PCs to inform their local router of the desire to join a particular IGMP group.

Sending multicast traffic—The specification extends the IP protocol by including a provision to send multicast IP datagrams.

Receiving multicast traffic—The specification extends the IP protocol by including a provision to join IP multicast groups.

Ethernet encapsulation—Support is provided for transmitting multicast traffic across an Ethernet network by mapping multicast addresses directly into the destination field of Ethernet packets.

Specific message formatting—The key elements of a version 1.0 IGMP message are illustrated in Fig. 4.5 and explained in Table 4.5.

Version one is rarely used nowadays. In fact, most if not all IPTV deployments use either version 2 or the latest version 3 of the IGMP protocol.

TABLE 4.5   Structure of an IGMPv1 Message

Field Name	Description of Functionality
Version	This field identifies the version of IGMP.
Type	This field identifies the type of message used. For version one this will be one of two options: a query or a host membership report.
Unused	Generally populated with zeros.
Checksum	Used to protect the integrity of the data.
Group address	The value of this field will vary between the message types. For instance in the case of a query message this field contains zeros when being sent whilst the report message holds the IP group address.

IGMPv2   Owing to the lack of support for leaving a particular hosting group in version one combined with a number of other technical factors meant that the Internet Engineering Task Force went back to the drawing board and produced a second version of the protocol called IGMPv2 in 1997. Version 2 of IGMP improves on the original version and is described in RFC 2236. It is the default version of the protocol at present for selecting to receive a particular IPTV broadcast channel. The key characteristics of IGMPv2 include

✓ *Improved leave latency*—IGMPv2 reduces the time taken for a multicast router to learn that there are no longer IPTV subscribers accessing or viewing a particular broadcast channel. This feature eliminates any unwanted streams, which in turn reduces the possibility of congestion on the network and the delivery of poor quality video to end users.

Widely supported—It is supported by all modern versions of Microsoft operating systems. Modern releases of Linux and Unix also provide strong support for IGMPv2.

Backward compatible with IGMPv1—In addition to providing additional functionality, IGMPv2 is also backward compatible with version one of the protocol.

Message types—IGMPv2 adds new message types that were unavailable in IGMPv1. Table 4.6 provides further information on the various types of messages used by IGMPv2.

Support for IPTV—Unlike its predecessor, IGMPv2 supports the multicasting of IPTV traffic. This is due to the fact that IGMPv2 can explicitly make an instruction that it wants to leave a particular channel.

Integral part of IPv6—IGMPv2 multicasting functionality has been tightly integrated with Internet Protocol Version 6.

Message formatting—All IGMPv2 messages are encapsulated inside IP datagrams. An identifier of 2 is used in one of the IP header fields to indicate that the datagram contains an IGMPv2 message. The key elements of an IGMPv2 message packet are illustrated in Fig. 4.6 and explained in Table 4.7.

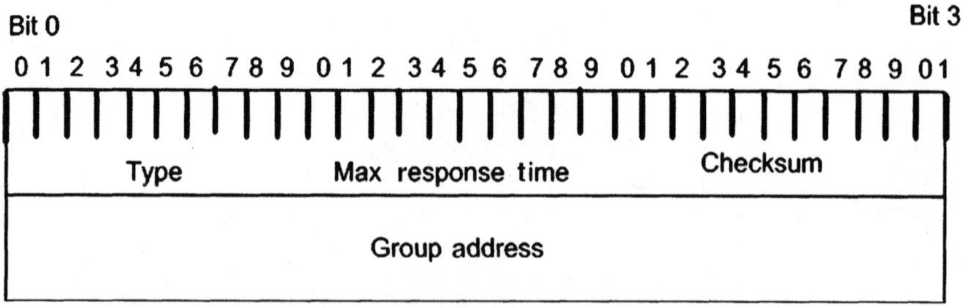

FIGURE 4.6   IGMPv2 message format

TABLE 4.6 IGMPv2 Message Types

Message Type	Message Description	Hex Value in Type Field
Membership query	These messages are periodically sent onto the network by the multicast router. Only one router is allowed under the rules of the protocol to send query messages on a network segment. There are two types of query messages: general and group specific. General queries are used to determine which IPTVCDs are viewing which multicast streams. The group specific query messages are used to find out if any subscribers are watching a specific multicast IPTV stream.	0×11
Version 1 membership report	This message type is used by IGMPv1 clients to join or connect to a particular IPTV channel. These records are also used to respond to queries and are often sent unsolicited when a request is received to change IPTV channel.	0×12
Version 2 membership report	This message type has the same function as IGMPv1 membership reports except that they are intended for IGMPv2 supported routers.	0×16
Leave group	This message type is not available in IGMPv1. The purpose of this message is to reduce the latency associated with stopping a multicast router from forwarding IPTV streams when there are no longer members in the group. In other words, when all subscribers on a particular network segment have stopped watching a particular IPTV channel. When this message is received the router will automatically stop transmitting copies of the broadcast packets. This will ensure that the network bandwidth becomes available to other uses.	0×17

It is worth noting that it is possible to create IGMPv2 messages that are greater than the 8 octets described in Table 4.7.

Querier election process—Under the IGMP protocol only one router can send query messages onto a network segment. Version 2 of the IGMP protocol

TABLE 4.7   Structure of an IGMPv2 Message

Field Name	Description of Functionality
Type	Identifies the type of message used. For version two this will be one of four options, namely membership query, v2 membership report, leave group, and v1 membership report. The leave message is new in IGMPv2 and is sent by the IPTVCD when leaving a particular group or broadcast TV stream. This message helps to reduce the latency associated with stopping the unnecessary transmission of video streams when compared to version one of the standard.
Max response time	This field is only relevant to membership query messages. It specifies a maximum time period for the delivery of a responding report. The value of this field is measured in increments of a tenth of a second.
Checksum	Used to protect the integrity of the data and check for any errors.
Group address	The value of this field will vary between message types. For instance, in the case of a query message this field contains zeros when being sent whilst the report or leave group message holds the IP group address inside this field.

supports an election type process for identifying the router responsible for sending out query messages. This is based on the principle that routers with lower IP addresses are elected to the position of "querier."

Whenever an end user changes a channel, the IPTVCD responds by sending two commands to the central equipment:

(1) To leave the existing video multicast.
(2) To join the desired new video multicast.

We explore these commands in the following subsections.

IGMPv2 LEAVE PROCESS   Owing to the limited bandwidth capabilities of many DSL based access networks many IPTVCDs can only receive one channel at any particular instance in time. Therefore, it is critical that an efficient mechanism is available to IPTV clients that allow them to communicate their intentions to leave an IP multicast group (in other words switch channels). This mechanism is supported by IGMPv2 and is the main difference between versions 1 and 2 of the protocol. This is facilitated through the leave group message. Figure 4.7 shows an example of how the IGMPv2 leave process is executed in an IPTV networking environment.

The steps involved include

Terminate access to an existing IPTV stream—The IP set-top box in the digital home receives an instruction from the viewer to change viewing from

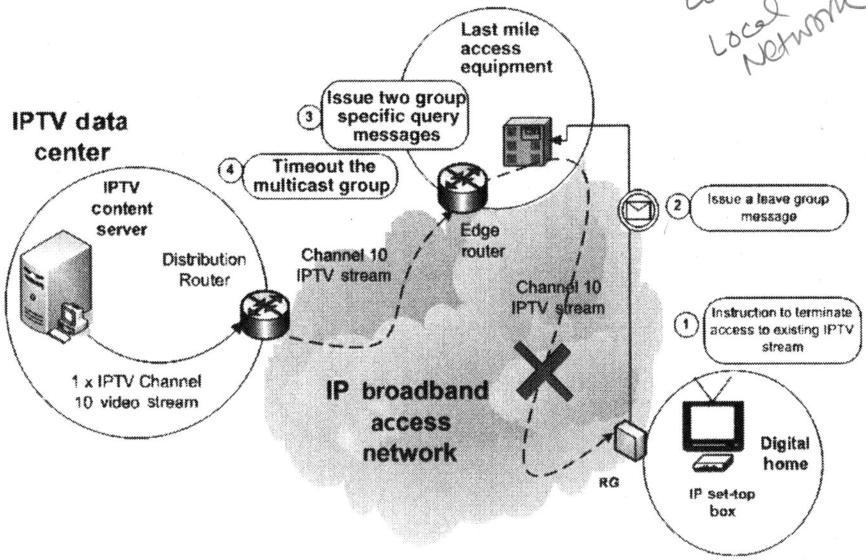

224-0.0.2 → all routers connected to the Local Network.

FIGURE 4.7   Protocol flow—IGMPv2 leave process

Channel 10 to another. In technical terms this equates to terminating access to the Channel 10 IP multicast stream and joining another stream.

Issue a leave group message—The leave group message, which contains the IP address of the TV channel to leave, is sent to all multicast routers on the multicast address 224.0.0.2. In this case the message is received by the edge router at the regional office.

Issue group specific query messages—The edge router typically sends out two group specific query messages to determine if there are any more IPTVCDs interested in receiving this specific stream or broadcast TV channel.

Further router processing—If there are no replies to the group specific query message, the router proceeds to time out the multicast group, removes the entry from the IGMP interface group table and stops sending traffic out to the appropriate interface.

This speedy and effective approach to instructing a multicast router to stop streaming a specific TV channel ensures that network bandwidth is freed for other IPTV applications.

IGMPV2 JOIN PROCESS   For an IP set-top box to start receiving multicast packets from an IPTV content server it must cycle through the IGMPv2 Join Process. The term "Joining" is used to describe the fact that a particular IGMP client has indicated that it wants to receive a specific IPTV broadcast channel. Figure 4.8 shows how an IGMPv2 join process is executed when a request has been issued by an electronic

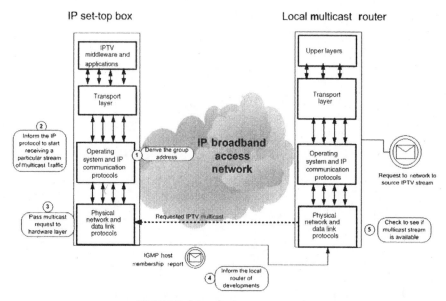

FIGURE 4.8   IGMPv2 join process

programme Guide (EPG) application running on an IP set-top box to change to a particular broadcast channel.

The steps involved include

Deriving the group address—The OS or middleware software running on the IPTV set-top box derives the group address of the requested broadcast channel from a hard-coded multicast address or an address derived from a universal resource locator (URL) string.

Informing the IP protocol to start receiving a particular stream of multicast traffic—Once the group address is determined, the IP communication protocol is informed to start receiving IPTV multicast traffic on a particular IPv4 or IPv6 address.

Passing multicast request to hardware layer—The network adaptor, which in most cases is Ethernet is instructed to listen and respond to multicast medium access layer (MAC) addresses that correspond to the multicast address of the requested broadcast channel.

Informing the local router of developments—The IP set-top box sends an unsolicited IGMP Host Membership Report or "Join" message to the local router. This message tells the local router, typically located in the regional office of the IPTV service provider that it is listening for multicast traffic for a specific IP address, which translates to a particular broadcast TV channel. It is also worth noting that the IP set-top box in the example can also gain access to an IPTV channel by responding to query messages originating from the multicast router.

Checking to see if multicast stream is available—The local router needs to determine if it is already receiving the requested IPTV multistream. If so, it simply replicates the stream and sends the stream back to the IP set-top box via the appropriate interface. If the stream is not available, the router sends out a request to the network to source the stream. The network responds with the stream, which is in turn is copied by the local router and sent onward to the IP set-top box.

As this example illustrates the process of joining an IP multicast is relatively complex and can be time consuming, particularly in situations where the requested stream is not available at the local router.

Having established a use for query messages within the joining and leaving process, the following section covers this message type in greater detail.

IGMPv2 QUERY PROCESS The multicast router is responsible for periodically broadcasting IGMP query messages onto an IPTV network segment. The purpose of these messages is to identify what IPTVCDs belong to what multicast groups. In the context of an IPTV environment, this equates to the IP video channels that subscribers are watching. Additionally, the query process is used to determine if an error has occurred during a "joining" or a "leaving" process. The unplugging of an IP set-top box during the execution of one of these processes is an example of a typical error that occurs an IPTV networks. As discussed previously, under version 2 only one multicast router is allowed to send out query messages. An election mechanism is used by IGMPv2 to assign this functionality to a particular router. An example of how to identify a router to issue IGMP query messages (the querier) is depicted in Fig. 4.9.

As shown the process consists of a small number of steps:

(1) When an IP Multicast router is originally connected to a broadband network, it assumes by default that it is responsible for managing and broadcasting IGMP query messages.

(2) It starts to send out IGMP query messages. The interval between sending out messages has a default value of 125 s on most routers and the messages are sent out to the 224.0.0.1 reserved multicast address.

(3) While connected to the network, it may receive IGMP messages from another router.

(4) It examines this message and if the IP address contained in the message is numerically a lower value when compared to its own address, it stops sending out query messages. If no query message is received within a particular time period (approximately 255 s), then the router becomes the "querier" and continues to manage and process IGMP query messages.

Reducing the number of querier multicast routers on an IPTV network helps to reduce the number of queries and reports traversing the network, which in turn reduces bandwidth consumption.

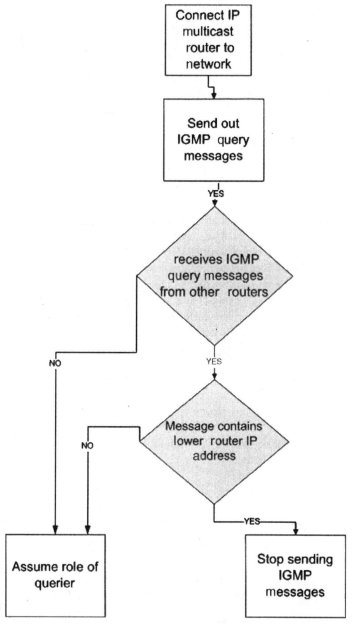

FIGURE 4.9   Election mechanism for identifying a router to manage IGMP query messages

IGMPv3   The IGMPv3 protocol is specified in RFC 3376. It was published in 2002 and considered to be a major revision to the protocol at the time of its publication. It builds upon previous versions of the standard and includes the following key characteristics.

SUPPORT FOR SSM   Under the previous versions of IGMP, an IPTVCD sends out a message indicating that it wants to join a particular group (i.e., start viewing a particular TV broadcast channel). The message contains the destination IP address of the group or requested broadcast channel, and the channel is streamed to the requestor. The IP address of the multicast group is typically obtained from the EPG. While receiving the broadcast channel, the IPTVCD is also configured to listen to all traffic in that multicast group. This approach often called the "Any Source Multicast" (ASM) has been refined in IGMPv3. IGMPv3 provides support for a new approach called source specific multicast (SSM). This additional IGMPv3 feature enables an IPTVCD to explicitly specify which broadcast channels it wants to receive and the IP addresses of the source of those IPTV channels. So instead of specifying the group IP address only, IGMPv3 messages contains the multicast group IP address and the unicast IP address of the content source. This combo identifier is represented as (S, G) where S is the unicast IP address of the video content server and G is the group address of the broadcast TV channel. IPTVCDs get access to a channel by subscribing to an (S, G) multicast channel. Note that the address range 232.0.0.0 through 232.255.255.255 has been reserved by IANA for the purpose of deploying SSM applications and protocols.

The source IP address is a useful feature in an IPTV networking environment because it allows IPTV access devices to instruct network components that they only want to receive packets from a particular source and not any other source. Additionally, IGMPv3 also allows IPTV access devices to specify that there are certain source addresses that they will not accept content from. Both of these functions are implemented in the protocol through the use of two filter modes, namely, include and exclude:

(1) *Include mode*—When an IPTVCD interface is configured with a filter mode of include, the device will only receive broadcast streams from specified content sources.

(2) *Exclude mode*—When an IPTVCD interface is configured as exclude mode, the network filters out particular streams and does not pass onto the access device.

This new SSM model of addressing IP multicast groups is quite a useful feature for those involved in the deployment of IPTV broadcast services. With the traditional ASM model, all IPTVCDs joined to an IPTV video stream could send content to all channel/group members. The main drawbacks of this approach are increased security risks and network traffic loads. Alternatively, the SSM approach supports a number of different viewers accessing the same stream but only allows one source content provider. The deployment of SSM does however have one main drawback that is its ability to interoperate with legacy IP applications that do not support subscriptions to (S, G) channels.

IMPROVED FORMATTING OF THE MEMBERSHIP REPORT MESSAGE IGMPv3 has similar features to version 2; however, the message types are slightly different. The biggest difference between both versions is the introduction of a new version 3 membership report message. This message type is primarily used to "Join" and "Leave" IP multicast broadcast TV groups. They are also sent in response to queries and provide information such as their multicast reception states. The key elements of a IGMPv3 membership report packet are illustrated in Fig. 4.10. As illustrated the packets contain one or more group records. This is the case when an IPTVCD sends out an unsolicited report to change channel. In that case the message would contain two records—one providing details on the "leave" and the second record providing details on the requested "join." Details of the formatting of these records are available in Table 4.8.

The query messages are similar to IGMPv2. The only enhancement in this version is the creation of a new type of message called the "Group-and-Source-Specific Query." As per previous versions, IGMPv3 messages are also encapsulated inside IPv4 datagrams.

THE USE OF LEAVE GROUP MESSAGES HAS BEEN DISCONTINUED IGMPv3 does not use leave group messages, as this functionality is provided through the source address filtering system.

INTEROPERABLE WITH PREVIOUS VERSIONS OF THE STANDARD IGMPv3 is designed to be interoperable with IGMPv1 and IGMPv2.

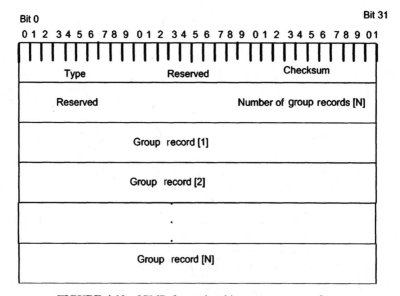

FIGURE 4.10 IGMPv3 membership report message format

TABLE 4.8   Internal Format of a Group Record

Field Name	Description of Functionality
Record type	There are three different record types used by IGMPv3: Current-State Record (CSR) — This provides a response to a particular query and provides information to the IP Multicast router on the reception status state of a particular IPTV access device. The filter mode of operation is also included in the response. Filter-Mode-Change Record (FMCR) — This is a record sent out when the status of the interface filter mode changes. This will have two states; either CHANGE_TO_INCLUDE_MODE or CHANGE_TO_EXCLUDE_MODE. Source-List-Change Record (SLCR) — This record is triggered when the source list is changed. The type of information in this record can include a list of additional IPTV channels that the access device wishes to hear or a list of video streams that the device no longer wishes to process.
Aux data length	As the name suggests this field contains details on the amount of auxiliary data available in the record. Auxiliary data is rarely used and this field is often set to zero.
Number of sources (N)	A figure for the number of IPTV source content addresses present in the record are listed here.
Multicast address	This is the multicast address of the specified broadcast TV channel.
Source address [N]	This is a series of unicast addresses that identify the locations of the various IPTV broadcast channels. Note that N is the value in the record's number of sources (N) field.
Auxiliary data	If available this field will store any additional information that is relevant to the group record. RFC 3376 does not define data for this field and therefore is often ignored in a live IPTV implementation.

IMPROVED BANDWIDTH UTILIZATION AND SECURITY   The provision in IGMPv3 to specify an IP address of a video content server is an important enhancement because it prevents IPTVCDs from receiving traffic from other devices connected to the same IGMP multicast group. This is a desirable feature because it prevents content originating from unwanted sources, which diminishes the amount of bandwidth available on the network. The lowering of bandwidth availability to IPTVCDs can sharply degrade the reception quality of the IPTV channel being broadcasted by the group. SSM also helps to prevent against other risks including denial of service attacks. This is important because under previous versions of IGMP the IPTV access devices subscribed to a particular channel were exposed to attacks from both inside and possibly even outside the core IPTV distribution network.

SUPPORT FOR NEW ROUTING PROTOCOLS New multicasting protocols have been adopted by IGMPv3 to support the deployments of source specific IP multicasting technologies. These protocols will be discussed later on in the chapter.

ADOPTED BY THE CABLE INDUSTRY CableLabs has included support for IGMPv3 in the latest version of its DOCSIS 3.0 communication standard.

IGMPv3 JOIN PROCESS As shown in an example illustrated in Fig. 4.11, there are differences between how IGMPv3 joins an IP multicast when compared to previous versions.

For a start the request to change a TV channel can be executed with a single request. Previous versions required the IPTVCD to issue two separate commands, namely, "Join" and "Leave". The ability to half the number of requests required to change an IPTV channel speeds up the rate at which subscribers can zap through their channel lineups. Additionally, all of the IGMP messages are sent in an out of band (OOB) control channel identified by the group IP address 224.0.0.22. This contrasts with previous versions of the protocol where the packets were addressed to each IPTVCD who was a member of a particular broadcast TV group.

As discussed previously IGMPv3 also enables IPTVCDs to send additional information to a multicast router, including the specification of the content source(s) combined with the group IP address.

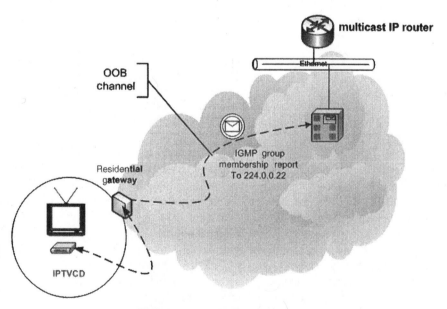

FIGURE 4.11 IGMPv3 join process

IGMPv3 QUERY PROCESS The increased sophistication of the IGMPv3 protocol has also increased the complexity of the querying function. For instance, the querier has to keep track of much more activity when compared to previous versions including status details on each multicast group or IPTV channel and processing the various record types that are an integral part of IGMPv3.

One other difference with previous versions is the fact that only one host per multicast group was required to respond to a query. The objective of this method of responding to queries was to minimize traffic, which worked quite well. However, it had one main drawback and this was the fact that the querier or multicast router received no information on the IPTVCD that responded to its queries. Version 3 addresses this issue by ensuring that all devices provide detailed information. An example of how a query is processed by IGMPv3 devices is presented in Fig. 4.12.

As shown the membership report is sent by the multicast router to the all-systems multicast address —224.0.0.1. This address is also used by IGMPv2 compliant multicast routers. Each of the three client IP set-top boxes subsequently respond to the query by sending IGMP membership reports to an IP address destination of 224.0.0.22. The multicast router listens on this address specifically for IGMPv3 response messages.

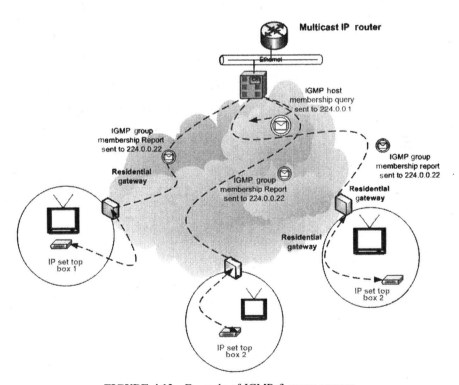

FIGURE 4.12 Example of IGMPv3 query process

IGMPv3 LEAVE PROCESS Keeping the channel changing time to a minimum is a
critical part of any IPTV deployment. In IGMPv2 a channel change required the
execution of four separate steps before the channel change could occur:

(1) The IP set-top box sends a leave group message.
(2) The multicast router at the regional office or central IPTV data center
    responds with a query message.
(3) The IP set-top box responds to the query with a membership report.
(4) The IP set-top box interprets the membership report and reconfigures its
    interface to comply with the instruction contained in the report.

In IGMPv3, the number of steps required for the same action has been reduced
to two, namely, the issuing of a membership report that contains a change of state
record and corresponding actions undertaken by the multicast router. Assuming that
there are no issues, such as network congestion, this adjustment to the IGMP
protocol typically improves the rate at which IPTV subscribers change channels.
The other main enhancement to the leave process is the fact that an IGMPv3 client
is able to specify that it wants to stop receiving content from a specific IPTV server
within a particular group. In previous versions a leave group message is sent to the
router to instruct the router to stop sending the video stream. In IGMPv3 however
the IPTVCD sends a membership report addressed to 224.0.0.22.

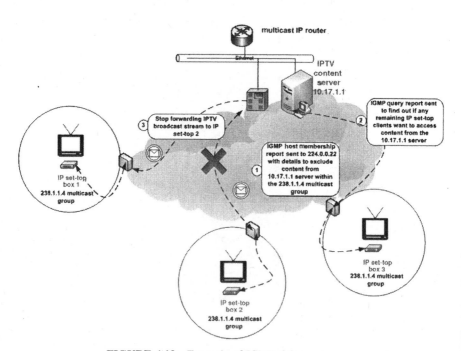

FIGURE 4.13   Example of IGMPv3 leave process

As illustrated in the example shown in Fig. 4.13, the membership report message originating from IP set-top box 2 contains a state change record that instructs the router to exclude or stop receiving content from the 10.17.1.1 server connected to the 238.1.1.4 multicast group. Once the message is received by the multicast router, a query message is immediately forwarded to the group requesting details on whether any of the other IP set-top boxes in the group want to view content from the 10.17.1.1 server. If no response is received within a particular time period, then the router will stop forwarding the IPTV broadcast stream.

Although IGMPv3 has a huge amount of benefits, there are a couple of drawbacks that we should mention before concluding our coverage of this topic. For a start, the devices that support IGMPv3 need to be able to examine group records that are part of the membership report messages. This often involves a mechanism called deep packet inspection, which can require additional hardware resources to meet this requirement. Extra hardware resources such as processing speed or additional memory are expensive. So due to hardware limitations on some networks, it might not be possible to run IGMPv3 when deploying IPTV services. Another factor to bear in mind when using IGMPv3 is the fact that the EPG needs to also be configured to take advantage of the improved features of the protocol.

4.4.4   Multicasting Transport Architecture

The delivery of video over an IP networking infrastructure generally uses a number of advanced routing protocols and technologies. This section presents details on some of the routing technologies that are used by IPTV networks:

- Multicast distribution trees
- Multicast distribution protocols
- Multicast forwarding techniques

4.4.4.1 Multicast Distribution Trees   A multicast distribution tree is used to efficiently deliver linear TV services from the streaming server at the IPTV data center or one of the regional offices to the various IPTVCDs in the field.

An IP multicast router uses information obtained from the IGMP protocol and other sources to create lists of nodes that map out the route or path that packets of IPTV content need to travel in order to reach their destinations, which are typically IPTVCDs. These node lists or route maps are called multicast distribution trees. The two basic types of multicast distribution trees are source trees and shared trees.

Source Trees   A source tree is a relatively straightforward implementation. It is based on the principle of identifying the shortest path through the network from source to the destination IPTVCD. Owing to the fact that source trees identify the shortest path, they are also referred to as *shortest path trees* (SPTs). A new source tree is generally configured when new source servers are added to the IPTV network. Figure 4.14 shows a simplified example of a source based IPTV multicast distribution tree.

In the example shown, the source distribution tree is made up of an IPTV server, R1 that serves as the root and routers 2, 3, 4, and 5. Any of the routers used to pass a video stream to an IP set-top box client, are part of the tree and routers that are not delivering streams are not included in the tree. In addition to identifying routing paths, these trees are also used by multicast routers to efficiently manage the process of replicating IPTV streams in different locations within the network.

In this example R2 replicates the incoming IPTV stream because two of its interfaces have active viewers—the paths or "branches" that connect to routers 3 and 5. The branch connected to router 4 has no IPTV viewers of Channel 10, therefore it is not part of the multicast distribution tree. Once a subscriber connects to R4 and requests to see the Channel 10 stream then this branch gets active, the tree is dynamically modified and extended by the router, and the requested traffic is forwarded down the tree to the IP set-top box. So the structure of the tree is continuously changing as the IPTVCDs join and leave the network at various intervals throughout a normal day. When all IPTVCDs leave a multicast group, the routing branch leading down to that subnet of IPTVCDs is "pruned." In other words, the local multicast router stops forwarding multicast traffic associated with a specific group down that particular branch or subnet. The routers on the IPTV network are responsible for keeping these trees constantly updated. Note also that

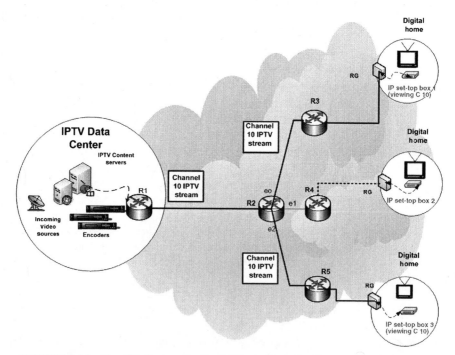

FIGURE 4.14   Simplified example of a IPTV multicast distribution tree (type source)

the extending and pruning of multicast trees also occurs when new access networks join or leave the core networking infrastructure.

Shared Trees   The configuration of a shared tree is different to a source tree in the sense that the shared multicast distribution tree locates the root at a chosen point on the network called a rendezvous point (RP). The RP acts as an intermediate device between IPTV sources and IPTVCDs. This contrasts with source trees that locate their routes at the source of the IPTV content. Multiple RPs may be required to support IPTV implementations that broadcast hundreds and even thousands of channels across a network. Figure 4.15 shows a simplified example of a shared multicast tree for Channel 10 (Group IP address of 224.5.5.5).

In the example shown above, the RP is located at router 5 and this contains the details of both sources. The video streams from the source servers in the regional headend and the central IPTV data center travel to the root of the tree, which is router 5 in this example. From router 5 the content continues to travel down through the shared tree and ending their journeys at the two IPTVCDs (IP set-top boxes one and two). The identification of the shared tree is derived from the (S, G) notation, which were discussed earlier in this chapter. In a shared tree environment the wildcard "*" is assigned to the "S" variable and is used to represent all source servers. Therefore, the shared tree illustrated in Fig. 4.15 has an identifier of (*, 224.5.5.5). The use of

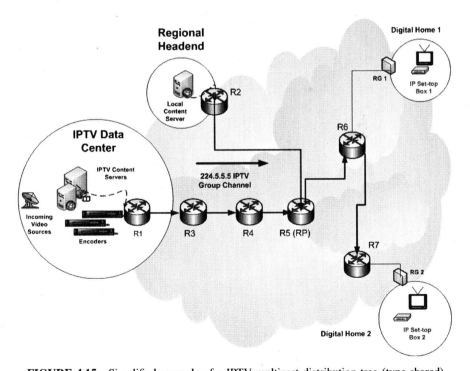

FIGURE 4.15   Simplified example of a IPTV multicast distribution tree (type shared)

the wildcard negates the need for the IP set-top boxes to know the IP address of the IPTV content source server because these details are available on the RP.

Both types of multicasts have their advantages and disadvantages when used to support IPTV services. Source trees for instance create paths through the network that keep latency to an absolute minimum, which is ideal for delivering IPTV services. However, high processing and memory resources are required to meet the demands of creating and operating source tree multicast structures. Shared trees, on the contrary, consume less hardware resources when compared to source trees; however, they can on occasions introduce latency on some of the routing paths because all traffic is forwarded to the RP. The acceptability of this latency delay will vary from network to network.

4.4.4.2 Multicast Distribution Protocols   In some of the large IPTV deployments, multiple multicast routers are required to deliver multiple TV broadcast channels to a spatially diverse and large number of IPTVCDs. Protocols such as Protocol Independent Multicast (PIM) are used to build multicast distribution trees that route IPTV video content through a high speed broadband network. PIM defines a small collection of multicast routing protocols that are optimized for delivering different types of services. There are four PIM variants:

PIM dense mode (PIM-DM)   operates on the principle of flooding multicast packets to all routers on the network. Routers that do not have multicast IP group members connected to their interfaces send a leave or prune message to the source of the packets. Once the source receives the prune/leave message, the transmission of multicast packets to that part of the network ceases. This flooding of unwanted multicast traffic consumes valuable bandwidth. Therefore, this PIM variant is rarely deployed across an IPTV network and is typically used to support IP applications that operate across enterprise and local area networks (LAN).

PIM sparse mode (PIM-SM)   The most recent version of the PIM-SM specification was published in 1998 and can be found in RFC 2362. It is designed to route IP multicast traffic over routes that operate across wide area networks (WANs). In particular, it defines how routers interact with each other to build and maintain different types of multicast distribution trees. As the name suggests, PIM-SM is based on the assumption that viewers of a particular channel are sparsely distributed throughout an IPTV network. The delivery of TV channels under the PIM-SM protocol is done so using the pull mode content delivery mechanism and only IPTVCDs that have explicitly requested to view a channel will be forwarded the associated IP video traffic. This is a useful technique for preserving bandwidth; however, the time taken to join a broadcast stream may cause a slight delay since IGMP instructions need to be issued to the nearest upstream multicast router that is processing the requested channel. PIM-SM scales quite well for large network infrastructures. When implemented across a live IPTV network, PIM-SM is enabled on the various router interfaces.

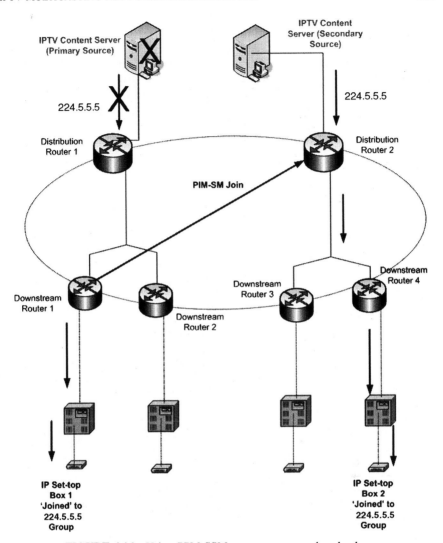

FIGURE 4.16   Using PPM-SSM to support network redundancy

This helps to form a dedicated video multicasting network infrastructure. When used with IGMP, PIM-SM can provide support for the detection and recovery of video infrastructure components such as encoders and streaming servers. Figure 4.16 illustrates how PIM-SM provides support in the event of a failure. In this example, both servers are sending to the same multicast address associated with two separate multicast trees. Thus, in the event that IPTV content server 1 fails, the PIM-SM routing protocols detect the failure and recalculate the routing paths. As a result, PIM-SM sends a new join request to the content server 2, and the stream gets rerouted to the IP set-top box that was affected by the server failure.

PIM source specific multicast (PIM-SSM) is a routing protocol that operates at layer 3 of the IPTVCM and has been derived from PIM-SM. As the name suggests it supports the deployment of SSM, a delivery model supported by IGMPv3, which allows IPTVCDs to explicitly specify the channels they want to receive. The operation of the protocol is relatively straightforward and uses the SSM addressing notation to support the execution of PIM Join and "Prune" commands.

Bidirectional PIM (BIDIR-PIM) is slightly different to PIM-SM. Lack of support for encapsulation and source trees are two of the main differentiators of this PIM based routing protocol. BIDIR-PIM is considered to be quite a useful protocol in cases where scalability is required. It does however suffer from instances of delays when deployed across large IP networks.

Of the four PIM variants PIM-SM is the most popular multicast distribution protocol used to support deployments of IPTV.

4.4.4.3 *Multicast forwarding techniques*

Within a multicast network, IP routers are used to forward video packets between networks. There are two approaches used by routers to forward IP packets:

Unicast—In unicast routing, the router examines the destination IP address of the video packets and looks up its routing table. The routing table provides information of various IP devices connected to its port in addition to information about remote networks. In an IPTV environment these tables are kept up to date through the use of various types of routing protocols. Once the routing table has been interrogated, the video packets are forwarded to the next router on the path toward the destination point. This process continues until the packets arrive at their destination point. So the methodology behind unicasting is based around the final destination of the video packet.

Multicast—In contrast to unicasting, the process of multicasting is more concerned with the origin or source of IPTV packets rather than the final destination point. When using multicast to send video content from an IPTV data center server to a remote IPTVCD, the video packets are first passed to the local distribution router. This is a similar approach to unicasting however the similarities stop at that point. This is because multicasting packets contain a group address, which will in most cases require transmission across multiple interfaces. As a result, multicast routers employ specialized distribution protocols such as Reverse Path Forwarding (RPF).

RPF Reverse Path Forwarding is an integral part of the PIM-SSM multicast routing protocol. The principle of RPF routing is that a multicast packet will only get forwarded onward if it is received on the interface that is closest to the IPTV source content server. In other words, it will drop IP packets that are not received by an upstream interface. An example of how RPF operates is shown in Fig. 4.17.

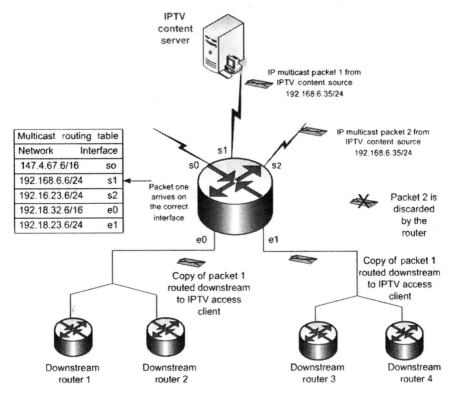

FIGURE 4.17 Example of RPF verification tests

As shown two multicast packets are received by the router. Multicast video packet 1 arrives on serial interface port S1. The routing table is looked up; the packet is copied and forwarded down through the appropriate interfaces or downstream paths. In this example, the appropriate interfaces are Ethernet interfaces e0 and e1. This first packet gets processed correctly because it arrived on the interface leading back to the IPTV source content server. The second multicast packet from source 192.168.6.35/24 is received on s2. The router references the routing table and finds out that it has arrived on the wrong interface and is subsequently discarded. This concept of only forwarding packets that have been received on the upstream interface eliminates the possibility of routing loops.

4.4.5 IGMP Snooping Functionality

Till this point in the chapter our focus has primarily been on the IGMP client device and the multicast router. During a typical IGMP communication session, the IGMP report messages often pass through other intermediate devices such as residential gateways, network switches, DOCSIS 3.0 CMTSs, or DSLAMs. These reports are normally monitored and analyzed by the intermediate devices in a process called IGMP snooping. Therefore, the intermediate devices are completely transparent

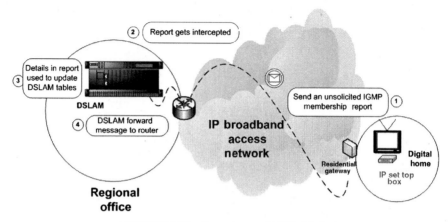

FIGURE 4.18 Snooping an IGMPv2 join

while IGMP messages are traversing the network. The information gathered from the various IGMP messages gives the intermediate device(s) a greater understanding of the nature of the IPTV multicast network traffic. The gathering of this information takes place when a particular interface is configured for IGMP snooping. Once enabled the snooping monitor examines the packets as they flow between IPTV access devices and multicast routers. The packets are unaltered and never modified during the examination process. Once examined the IGMP snooping agent records in its multicast tables the IPTV access devices that have joined and left the various multicast groups. Consider the simple example in Fig. 4.18.

In this case, the IP set-top box has sent a channel change request. This request takes the form of an unsolicited IGMP Membership report. It first passes through the residential gateway and across the network. Once the packet arrives at the DSLAM in the regional office, it gets intercepted. The DSLAM inspects the message and updates its interface tables with the new information contained in the packet. The IGMP message is then forwarded onward to the IPTV multicast router.

The main benefit of IGMP snooping is that it prevents the DSLAM from flooding multicast video packets to all ports and instead limits the streaming of packets to ports that have IPTV clients requesting a specific channel. It is also used as a basis to implement IGMP fast leaves. As suggested by the name, the fast-leaving process allows the DSLAM to remove an interface from the tables without sending out a group message. This speeds up the termination of channel streams and helps to improve bandwidth management. Although fast leaves speed up the termination of IPTV streams there is however the risk that other IPTVCDs connected to the same port may still be requesting to view the stream. The DSLAM or a DOCSIS 3.0 compatible CMTS, which is the case for HFC networks will be unaware of this fact because a group message was not been sent out. Thus, the stream gets terminated and the other devices that are viewing the stream will be left without a signal. Therefore, the use of fast leaves is generally limited to deployment in scenarios where there is only a single device making requests. So, for example, the introduction of the fast-leaving feature will work just fine for a scenario where a single IP set-top box is connected to a DSL connection. In this case the IGMP requests are only originating from one source, and the DSLAM or CMTS is guaranteed that no other devices will be affected by its actions. The use of fast leaving becomes more complicated when there are multiple devices connected to the last mile broadband connection. It is important to note that the latest version of DOCSIS has removed the requirement of IGMP snooping, which was part of previous DOCSIS specifications.

4.4.6   IGMP Proxy Functionality

IGMP proxy functionality allows devices such as DSLAMs to issue IGMP host messages on behalf of downstream IPTVCDs that they have retrieved from their IGMP enabled interfaces. So when the IGMP proxy function is enabled, it assumes the following responsibilities:

- When queried by the upstream routers it sends out group membership reports.
- It sends out an unsolicited group membership reports on behalf of IPTVCDs to join particular multicast groups.
- It replicates requested channel streams and forwards downstream to the requesting IPTVCDs.
- It sends out an unsolicited group membership report to 224.0.0.2 when the last of the IPTVCDs connected to its interfaces leaves a group.

Thus, by enabling the IGMP proxy feature on a DSLAM or on a CMTS in a HFC network, these devices act as an IGMP server to the downstream IPTVCDs and as a client to the upstream edge and distribution routers. The operation of IGMP proxy typically involves the establishment of two separate IGMP signaling paths: one path between the IPTVCD and the terminating device at the regional office (generally a DSLAM or a CMTS) and another path between the regional equipment and the upstream router. This concept is depicted in Fig. 4.19.

4.5   MULTICASTING IPTV CONTENT ACROSS IPV6 NETWORKS

With the exponential growth of next generation services, such as VoIP and IPTV, an increasing number of telecommunication companies are examining the possibility of deploying IPv6 across their broadband networks. Such a deployment requires service providers to use a multicasting signaling protocol called Multicast Listener Discovery (MLD) when delivering broadcast TV channels to IPTV end users. Note that in order for an IPTV networking environment to take full advantage of MLD, both routers and IPTVCDs must support the IPv6 protocol stack.

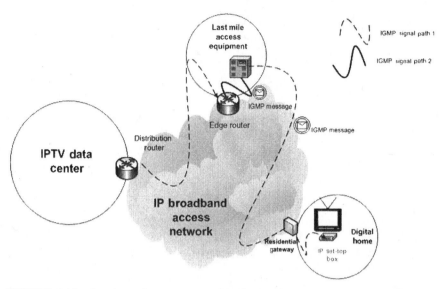

FIGURE 4.19 Creation of two separate IGMP signal paths to support IGMP proxy functionality

Two versions of MLD were developed at the time of writing, MLDv1 and MLDv2.

(1) The specification for *MLDv1* is derived from IGMPv2 and is found in RFC 2710.

(2) The specification for *MLDv2* is derived from IGMPv3 and is published in RFC 3810.

The following sections give a brief overview of versions 1 and 2 of the protocol when used in an IPTV infrastructure.

4.5.1 MLDv1

MLDv1 was published in 1999. The primary purpose of the specification is to provide IPv6 enabled routers with a protocol for discovering the presence of network devices that are interested in or actually receiving IPTV multicast traffic. This information ensures that the correct IPTV streams are delivered to their correct destination IPTVCDs.

As is the case with IGMP the core functionality of MLD is achieved through the use of messages. The key elements of an MLDv1 message are illustrated in Fig. 4.20 and explained in Table 4.9.

MLDv1 uses three different types of ICMPv6 (IP Protocol 58) messages when deployed in a networking environment—queries, reports, and done messages:

MLDv1 *Query* messages come in two variants: general and multicast-address-specific.

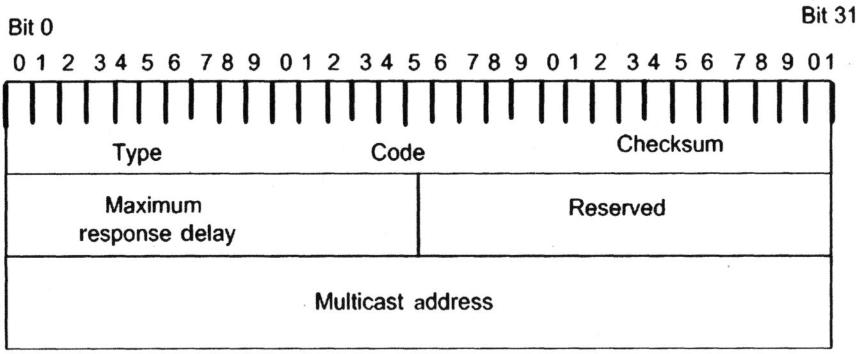

Bit 0   **Bit 31**

```
0 1 2 3 4 5 6 7 8 9 0 1 2 3 4 5 6 7 8 9 0 1 2 3 4 5 6 7 8 9 0 1
```

Type	Code	Checksum
Maximum response delay		Reserved
Multicast address		

FIGURE 4.20   Generic MLDv1 message format

TABLE 4.9   Structure of a Generic MLDv1 Message Format

Field Name	Description of Functionality
Type	As the name suggests the message type is identified by this field. A query message for instance has a value of 130.
Code	This field is normally set to zero, and is generally ignored by IPTVCDs in most cases.
Checksum	Similar to IGMP, this field is used for error correction across the entire MLD message.
Maximum response delay	This defines the maximum time allowed in milliseconds before a responding report message is sent. This is only relevant to MLD query messages. Note that in all other MLD message types, this field is set to zero and ignored by receiving IPTVCDs. Keeping this value as low as possible can improve the leave latency times for IPTVCDs.
Reserved	Typically set to zero and reserved for future uses.
Multicast address	Contains the group multicast address.

- The *general* query message type is used to identify which multicast addresses have IPTVCDs listening on a directly attached subnet. Note that the multicast address field is set to zero for general query message types. The information gathered by general query messages is used to update internal router tables.
- The *multicast-address-specific* query message type directs the query at a specified address to find out more about associated IPTVCDs listening on a directly attached subnet.

Note that both of the above message variants are identified by a decimal value of 130 contained in the 8-bit type field.

MLDv1 *Report* messages generally respond to query messages and are identified by a decimal value of 131 contained in the 8-bit type field.

MLDv1 *Done* messages are the equivalent of IGMP Leave messages and are identified by a decimal value of 132 contained in the 8-bit type field. Thus, when an

IPTVCD wants to stop receiving a particular streaming channel, it sends out a Done message to the router at the local regional office.

The operation of MLDv1 is relatively straightforward. At the IPv6 router a list is kept for each interface of which multicast addresses are used by IPTVCDs to access broadcast TV channels and related timing information. Similar to IGMP, MLDv1 makes extensive use of query, report, and done message types to provide join and leave functionality to IPv6 enabled IPTVCDs.

4.5.2   MLDv2

MLDv2 is based on the IGMPv3 protocol and is fully interoperable with MLDv1. One of the main enhancements of MLDv2 when compared to version one is its support for SSM. This allows an IPTVCD to explicitly specify which broadcast channels it wants to receive and the IP addresses of the source of those IPTV channels. Alternatively, it allows an IPTVCD to exclude receiving multicast content from particular source addresses. Similar to IGMPv3 the Include and Exclude filter modes are used to implement this type of functionality. MLDv2 adds two new message types that are unavailable in MLDv1: MLDv2 multicast address and source specific query messages and updated report messages.

4.5.2.1   Overview of MLDv2 Multicast Address and Source Specific Query Messages   The key elements of MLDv2 query message are illustrated in Fig. 4.21 and explained in Table 4.10. Note that the shaded areas in the table and the figure identify the fields that are common to both MLDv1 and MLDv2.

4.5.2.2   Overview of MLDv2 Report Message   MLDv2 report messages are sent by IPTVCDs to neighboring IPv6 enabled routers. The key elements of an MLDv2 report message are illustrated in Fig. 4.22 and explained in Table 4.11.

Note that each multicast record contains information about the IPTVCD sending the report message.

4.6   INTRODUCTION TO CHANNEL CHANGING

One of the important aspects of delivering broadcast live content over an IP network is the speed at which end users can change channels during a TV viewing experience. This entails leaving one TV channel and joining another channel immediately. On an IPTV network, the process of channel changing takes place on a server instead of the set-top box, which is the case for traditional broadcast TV networking environments. End users expect to be able to quickly change channels on their TV sets. Owing to the fact that the set-top box interacts with the network during a channel change, latencies in the viewing experience are a possibility. In addition to delays from equipment at the digital home, the network and the IPTV data center can introduce potential delays into the channel changing process.

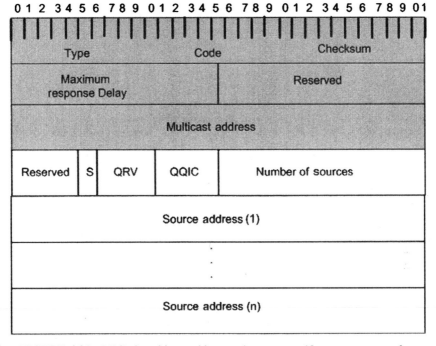

FIGURE 4.21 MLDv2 multicast address and source specific query message format

Delays introduced in these various subsystems are illustrated in Fig. 4.23 and explained in the following subsections.

4.6.1 At the IPTV Data Center

At the data center a number of systems can affect the times taken to change a broadcast TV channel:

Encoding system—The encoding of incoming and processing of source content increases latency times for newly requested broadcast TV channels.

CA and DRM systems—Functions like encrypting premium broadcast channels to protect copyright, generating keys, and sending encryption keys over the network can all add to delay the channel changing function. The use of DRM systems can further complicate the security function and subsequently introduce further delays.

The other systems that operate in the IPTV data center such as the middleware and application servers generally have no impact on time taken to execute a change channel request.

TABLE 4.10   Structure of an MLDv2 Query Message

Field Name	Description of Functionality
Type	As the name suggests the message type is identified by this field. A query message for instance has a value of 130.
Code	This field is normally set to zero, and is generally ignored by IPTVCDs in most cases.
Checksum	Similar to IGMP, this field is used for error correction across the entire MLD message.
Maximum response delay	This defines the maximum time allowed in milliseconds before a responding report message is sent. This is only relevant to MLD query messages. Note that in all other MLD message types, this field is set to zero and ignored by receiving IPTVCDs. Keeping this value as low as possible can improve the leave latency times for IPTVCDs.
Reserved	Typically set to zero and reserved for future uses.
Multicast address	Contains the group multicast address.
S	This flag determines whether multicasts routers that receive the message along the route should or should not suppress the normal timer updates that they execute.
QRV	Short for Querier's Robustness Variable, the QRV field is generated by the router acting as the querier on the IPTV network.
QQIC	Short for Querier's Query Interval Code, the QQIC value is expressed in seconds of the query period.
Number of sources	This field defines the number of source addresses present in the query message. This value is set at zero for general and address specific queries. However it includes a non-zero value for multicast address and source specific messages.
Source address (n)	This field contains the actual IP source addresses up to a value of n. The number of addresses supported in a message will depend on the maximum transfer unit of the physical network. For Ethernet networks for instance the value is 1500 bytes, which limits the number of source IP addresses contained in a message to approximately 89.

4.6.2   In the Network

Once the broadcast channel is prepared for transmission it gets forwarded to a distribution router that takes care of routing video traffic to the core of the IPTV distribution network. Once live on the distribution network the various routers, physical links, and regional office devices (DSLAMs for example) can all add time to a response to a channel change request. The speed at which the DSLAMs and the routers reply to IGMP Leave and Join messages is critical to ensuring that latency times are kept low. Although the Real Time Streaming Protocol (RTSP) protocol is generally used to set up a unicast connection, it can in some cases be used to access a multicast channel. The establishment of this connection needs to be negotiated via various request and response message round trips. This can also introduce delays

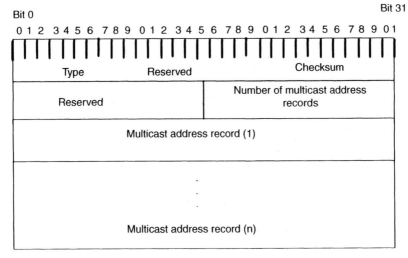

Bit 0   Bit 31

0 1 2 3 4 5 6 7 8 9 0 1 2 3 4 5 6 7 8 9 0 1 2 3 4 5 6 7 8 9 0 1

FIGURE 4.22   MLDv2 report message format

TABLE 4.11   Structure of an MLDv2 Report Message

Field Name	Description of Functionality
Type	As the name suggests the type of message is identified by this field. A report message for instance has a value of 143.
Reserved	This field is typically set at 0 when the message is transmitted over the IP broadband network.
Checksum	Used by IPv6 for error correction.
Number of multicast address records	Identifies the number of multicast address records present in the report message.
Multicast address record	This is the actual multicast address records.

into the process. All of the links used to interconnect the network devices need to be able to support bandwidth capacities of the various IPTV channels. Issues with any of the network components can result in the introduction of jitter, which can impact directly on channel zapping times. Other factors such as the multicast distribution protocols chosen and the version of IGMP supported by the network also come into play with regard to the channel changing response times.

4.6.3   At the Digital Home

The following lists five major elements that occur inside the digital home that can have an affect on the time it takes to change a channel:

Packet processing at the RG: The RG typically sits between the IPTVCD and the NGN distribution network. In a DSL environment, the RG includes a

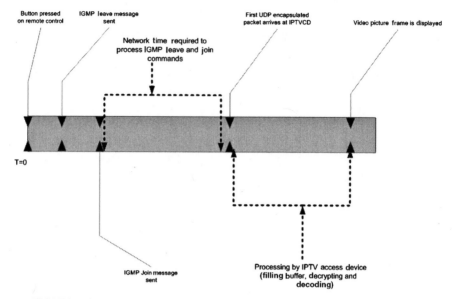

FIGURE 4.23   Actions that may have an affect on the channel changing process

modem and interfaces to the in-home network. The fact that a channel change gets processed by the RG means that an increase in the Join/Leave latency time is an obvious outcome.

Decoding: The decoding of the incoming compressed signal at the IPTVCD also plays a role in any delays that are experienced during the channel changing process. For instance, when using the MPEG compression algorithm, the IPTVCD needs to wait for an I-frame in the incoming stream before decoding can commence. The I-frame defines a group of pictures (GOP) and contains all the necessary information required to reconstruct the encoded picture. It has no dependencies on previous frames, which cut down on the affects of lost frames. The waiting period for the I-frame will depend on the number of I-frames broadcasted by the encoder per second. A high frequency will enable the IPTVCD to start decoding almost immediately. However, because I-frames contain the most amount of data of any frame type a large amount of network bandwidth is required to support a high density of I-Frames on a network. In some cases an I-Frame can take up to 50% of the available bandwidth with the remainder been used up by P and B frames, so they are sent as rarely as possible. In an IPTV environment I-frames are generally 15 frames apart, although this can vary significantly from network to network. The actual display panel used to present the IPTV channel has no impact on the time taken to process an IPTV channel.

Decrypting: If the multicast channel is encrypted, then the keys to decrypt the packets need to be acquired from the incoming stream. This security information is often carried inside a table within the IPTV transport stream. The frequency at which these tables are available within the stream will determine how soon an IPTVCD can start decrypting the incoming stream.

The keys in some of the advanced CA systems are periodically updated so the IPTVCD needs to ensure that the keys it retrieves from the stream are always up to date. The delay introduced by decryption can vary significantly between 0 for unencrypted channels and 1500 ms. On some networks it can even take longer than 1500 ms because the IPTVCD will have to wait for the latest set of decryption keys to become available in the stream.

Again, the service provider needs to make a trade off between the frequency of these tables and the bandwidth availability of the IPTV distribution network.

Buffering: Owing to the variable bit rate nature of video there is normally a variation between the time IPTV packets arrive relative to the just-in-time value. This time difference is called jitter and can originate on the server or may be introduced on the network. Therefore, a de-jitter buffer is typically utilized to ensure that a supply of data is always present for processing and display. In other words, there is no "under-run," which would result in the display of black screens. In addition to coping with jitter the buffers are also used to deal with retransmission of packets if necessary. So the use of buffering is critical to the smoothing of IPTV streams. However, these buffers need to fill prior to decoding commencing, which introduces delays. Additionally, the size of the memory buffer also has an impact on the channel changing time. For instance larger buffers, which are typically found in IP set-top boxes due to the eratic nature of IPTV streams compared to standard MPEG-2 traffic, can increase the time required to fill. Long buffer filling times can slow down the rate at which the channel change request is met.

Processing power and software architecture: The hardware architecture in terms of CPU, decoder, and memory performance are all factors that can affect the rate at which a channel change is achieved. This will generally be a trade off between performance levels and costs associated with purchasing the IPTV consumer devices. The structure of the software installed on the IPTV access device can also make either a positive or a negative impact on channel-change times.

4.7 FUNDAMENTALS OF CHANNEL CHANGING

There are two different types of interactions that subscribers have with IPTV multicast streams:

(1) Using the remote control to select a broadcast channel.

(2) Changing from one broadcast channel to another.

We explore these interactions in the following subsections.

4.7.1 Selecting a Broadcast Channel

On a traditional cable or satellite TV network, the selection of a channel is relatively straightforward because all the channels are available on the different frequencies and the set-top box tunes to the selected channel. Choosing an IPTV channel is more complicated because all the channels are not available simultaneously on the network. The steps involved with selecting a channel will depend on the networking infrastructure deployed. Selecting a broadcast channel takes a finite amount of time. The length of time required to complete this function will be impacted by the location of the requested video channel. For instance, if the channel is available at a nearby networking switch or router, then the delay will be negligible. However, if the stream needs to be accessed from a server at the central IPTV data center, then a longer period of time will be required to deliver the stream to the IPTVCD. Consider an example of a request that originates from an IP set-top box and the channel is available at the distribution router located at the IPTV data center. Figure 4.24 illustrates some of the possible steps that may occur across a DSL based IPTV network.

(1) Once a subscriber wishes to change channel they press a channel number on their remote control or select from an EPG application. These commands are received by the infrared receiver.

(2) The IP set-top box accepts the channel changing command and sends this instruction in the form of an IGMP join request to the RG. The RG sees the request and will either pass onward to the DSLAM or else examine the request to see if the requested channel already exists at one of its ports. If this is the case, then the RG simply copies the stream and sends to the requesting device. The only time that this would occur is when someone else inside the home is also watching the same channel. It is also worth noting that the RG will need to support IGMP snooping to facilitate this functionality.

(3) If the channel is not available at the RG or IGMP snooping is unsupported the request is forwarded to a DSLAM, in the case of DSL networks. It is worth

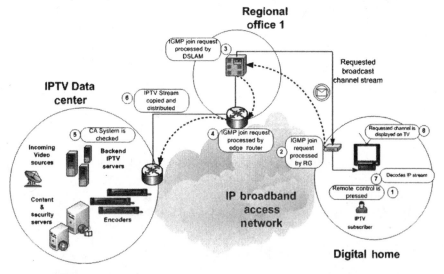

FIGURE 4.24   Possible steps involved in selecting an IPTV broadcast channel

noting that a CMTS is used for cable broadband networks. The next action will depend on the level of sophistication supported by the DSLAM or CMTS. Support of IGMP proxy functionality and snooping will allow the device to see if the channel already exists at one of its ports, if so the stream is copied and sent downstream to the IPTVCD. If however these advanced features are unsupported or the stream is unavailable, the request is sent to the upstream routers.

(4) When the router located at the regional office receives a request that has not been serviced by the various downstream network components, it also has two options, namely, to copy the stream to the correct interface or pass the request upstream to the distribution router in the event that the channel is unavailable at its downstream interface.

(5) The request for the channel finally ends up at the IPTV data center where all broadcast channels are available. It is important to note that the channel is generally identified by an IP address. This corresponds to a frequency and a channel on a traditional RF network. The CA system is checked to verify that the end user is authorized to view the particular channel.

(6) Once authorization is completed, the IP address of the subscriber's IP set-top box and associated details get added to the multicast list. The channel is then copied and sent onward to the IP set-top box.

(7) The set-top receives the new IP stream and additional security information such as CA tables and encryption keys. It then buffers in memory, and waits for an I-frame to arrive before decoding starts. The I-frame contains all of the necessary information required to reconstruct the original picture frame. Once the I-frame is received the IP set-top box, which can take between a half and two full seconds, can start the process of displaying the channel.

(8) The first picture frame of the new channel is displayed on the TV display.

Note that the ITU-T FG IPTV group is recommending that the time taken to acquire a broadcast channel should not exceed 2.5 s.

4.7.2   Changing from One Broadcast Channel to Another

As illustrated in Fig. 4.25, the process involved in changing channels in the middle of the TV viewing experience is quite similar to selecting a channel. There is however a couple of extra steps:

(1) When a viewer wants to change to another IPTV signal the remote control button is pressed and the instruction is accepted by the IP set-top box.

(2) The set-top box issues an IGMP Leave message to terminate the stream associated with the old channel.

(3) The termination of the stream can take place in one of three places, namely, the RG, the DSLAM, or CMTS located at the regional headend or the edge router at the regional office. This depends on what level of IGMP support is

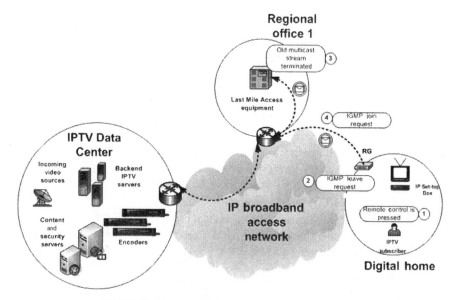

FIGURE 4.25   Changing IPTV broadcast channels

available on each of these devices. For instance, if the RG has an IGMP proxy function, then it can carry out the termination of the channel stream. This is assuming that no other IPTVCDs inside the digital home are watching the channel. If IGMP functionality is unsupported then the request is sent upstream to either the DSLAM or the edge router who takes care of terminating the multicast stream. In this case the DSLAM supports IGMP snooping and terminates the old channel stream. It is important that the process of terminating the old channel takes place at a very fast pace. Otherwise the requested stream may arrive at the interface of the IPTV set-top box before the previous stream has been terminated. This could double the bandwidth requirements, which may not be supported by the connection. This can result in packet loss, increased jitter, and ultimately a degradation in the quality of the picture displayed by the new incoming stream.

(4) A join message is then sent to start viewing the next channel. The steps described above with regard to selecting an IPTV broadcast channel are then executed.

4.8   TECHNIQUES FOR SPEEDING UP IPTV CHANNEL CHANGING TIMES

For IPTV to be a commercial success the time required to change channel on an IPTVCD should be at least as quick as channel switching on a cable TV network, namely, 500 ms. There are at least five approaches that may be used by service providers to improve the channel-changing time:

- Implementing IGMP proxy functionality inside the DSLAM.
- Increase the number of I-frames generated by the encoder at the IPTV data center.
- Increase the frequency of CA tables for encrypted channels.
- Reduce the size of the IP set-top box buffer.
- Predictive tuning and static joins.

The following sections give a brief overview of these approaches when used in an IPTV networking environment.

4.8.1 Implementing IGMP Proxy Functionality Inside the DSLAM

One technique used by some DSL service providers to speed up channel-changing times of their IPTV offerings is to configure the DSLAMs in the regional offices to act as IGMP proxy devices. An IGMP proxy is best defined as a networking device that issues IGMP host messages on behalf of IPTVCDs that are connected to its interfaces. Any device that acts as an IGMP proxy assumes the roles of both IGMP client and IGMP router. Implementing IGMP proxy functionality in the DSLAM has a couple of benefits.

(1) First and foremost, if a group of IPTV subscribers are watching a particular IPTV channel and a new subscriber wants to join the group and watch the same channel, the IGMP proxy device can immediately start forwarding the requested stream to the new subscriber. This is because the IGMP proxy negates the need to propagate the IGMP message report upstream to the source server. As a result, the new IPTV subscriber experiences a minimal delay in changing to the new channel.

(2) Proxy devices can also be used to keep an eye on which IPTVCDs are leaving the subnet, which the IGMP proxy resides on. In the event that all IPTVCDs have "tuned" away from a particular channel, then the proxy sends a leave message on behalf of all the subnet devices to the upstream multicast router to stop sending that particular channel.

If it is impractical or cost prohibitive to implement IGMP functionality in the DSLAM, then the next option is to configure the edge router to support IGMP Proxy Functionality . When this occurs the edge router needs to have enough processing power to swiftly and effectively process a high volume of IGMP leave and join requests, which are received from the various DSLAMs and CMTSs.

4.8.2 Increase the Number of I- Frames Generated by the Encoder at the IPTV Center

Because the IPTVCD cannot start decoding until it receives an I-frame, it is critical that IPTV service providers configure there encoders to insert I-frames into the stream as frequently as possible. The problem with this approach is that I-frames consume a relatively high amount of bandwidth and in some circumstances, particularly for MPEG-2 transport streams; consecutive I-frames are located within 10 frames of each other. This in turn increases channel-changing latency times. Thus, service providers need to balance the demands for lowering channel change times versus the economic demands of driving down bit rates and bandwidth requirements of delivering IPTV multicast channels to their subscribers.

4.8.3   Increase the Frequency of CA Tables for Encrypted Channels

As previously discussed, the IPTVCD needs to wait for encryption keys and CA tables to arrive before the decryption process can start. Increasing the frequency of this information within the stream can help to speed up channel changing times. This obviously increases the bandwidth requirements of the IPTV stream.

4.8.4   Reduce the Size of the IP Set-Top Box Buffer

As already discussed IPTVCDs that have large buffer sizes are more likely to output better quality video compared to devices with smaller buffers. This is particularly true for networks that suffer from jitter. The one main drawback of using large buffers is that it takes time to refill the buffer during the processing of a live multicast channel, which has an affect on the channel changing times. For networks that do not suffer from jitter then the use of smaller buffers is another technique used to speed up channel changing times. This reduces the time necessary for the buffer to fill.

4.8.5   Predictive Tuning and Static Joins

The use of predictive technology is a widely used feature in the mobile communication sector and may also be applied to an IPTV environment. The technology examines previous viewing habits of the IPTV end user and makes a prediction as to what are the most likely channel(s) that will be tuned to next. Once the prediction is made the network copies the stream and sends to the IPTVCD. This approach can also be used to make previous and next channels available at the IPTVCD. This is quite a good technique for speeding up channel changing; however, there may not be enough bandwidth coming from the regional office to the IPTVCD to support the delivery of multiple IPTV streams. Predictive tuning is a technique, which is particularly suited to NGN network technologies such as VDSL and FTTX. A static join is very similar to predictive tuning. Static joins are based around the same concept where a particular channel is always available at a particular point on the network, this could be the DSLAM or even at the RG in some cases. As with predictive tuning the main benefit is the fact that channel-changing latency is kept to an absolute minimum. The drawback is that the channel is always available and is using a section of the networks bandwidth even when nobody is watching the TV channel. So as with predictive tuning a trade off needs to be made between delay levels and bandwidth usage. Various statistical models are available to aid service providers in identifying the benefits or drawbacks associated with implementing a network design that supports either predictive tuning or static joins.

4.8.6 Implement a QoS Across the Network

In addition to locating the IGMP functionality as close to the IPTV end user as possible, increasing program information frequency levels within the stream, and reducing buffer sizes, it is also important to apply effective QoS mechanisms across the network. In addition to helping to increase the video signal quality delivered to the end user, a properly implemented QoS mechanism can also speed up channel changing times. Channel-changing latency is a biproduct of transporting video content over an IP network.

In addition to the backend techniques identified above, most providers display a dialogue box on the television screen to help in making the channel changing process as transparent as possible to IPTV end users. This dialogue box typically shows various types of information including the program name, brief description, and current time.

4.9 DISCOVERING CHANNEL PROGRAM INFORMATION

On a traditional RF network the digital set-top boxes generally read the frequencies associated with each channel from PMT and PAT tables on the network and tune to the appropriate channel when requested by the TV viewer. On an IPTV network the retrieval of channel and IP service details is different. For a start the IPTVCD needs to know the IP addresses of the various IPTV source servers and encoders connected to the network. The means of getting this information can vary between implementations. For instance, a network that is sending DVB content over IP will use a standard called service discovery and selection (SD&S), whereas the SDP protocol is the more likely choice for networks that use native RTP. We explore both of these approaches in the following sections.

4.9.1 SD&S as Defined in ETSI TS 102 034

A mechanism called SD&S is defined in section 5 of the ETSI TS 102 034 standard, which describes the discovery, selection, and delivery of DVB service information over an IP broadband network. The standard itself, also known as DVB over IP-based Networks (DVB-IPI), specifies a process for discovering IP services that are available on a DVB network.

SD&S basically involves the IPTVCD retrieving channel information from a particular multicast address and port number. The ETSI standard uses XML records and schemas to represent this information. The standard describes a number of records that may be used to describe a range of IPTV services including video on

TABLE 4.12   IPTV Service Provider XML Discovery Record Structure as Defined in ETSI TS 102 034

XML Attribute	Description	Comments
@DomainName	This field contains the registered domain name of the DVB-IPTV service provider.	Mandatory attribute
@Version	This field gives the version number. Any change in this value indicates to the IPTVCD that a modification has occurred in the details of the record.	Mandatory attribute
@LogoURI	This points the IPTVCD to the location of the service provider's logo.	Optional attribute
Name	The name of the DVB-IPTV provider is held here. The standard allows the display of this information in multiple languages.	Mandatory attribute
Description	Further descriptive details of the service provider may be inserted here.	Optional attribute
Offering list type	This section of the record includes a number of entries that define the IP addresses and port numbers of where to find further information about IPTV services offered by the network operator. The standard refers to these points as DVB-IP offering locations.	A mix of mandatory and optional attributes.
Payload list type	This final section indicates the type of service information available at the DVB-IP offering location. This value generally varies between multicast TV or on-demand.	Optional attributes

demand and multicast TV. Rather than covering all of these records, Tables 4.12 and 4.13 provides information on two XML records that are particularly relevant to this chapter.

It is important to note that the discovery record file structure can be enhanced to carry IPTV service information from multiple DVB-IPTV providers. To achieve this objective the full specification needs to be downloaded and reviewed. Accessing this record on a live IPTV networking environment may be done by an IPTVCD specifically retrieving the file or else the service provider multicasting the XML record onto the network.

The second XML file format that is particular to this chapter provides the IPTVCD with information required to select and find multicast channels on the network that have embedded SI information. This record is derived from another

TABLE 4.13   Multicast Discovery Record Structure as Defined in ETSI TS 102 034

XML Attribute	Description	Comments
textualidentifier@ DomainName	The domain name identifies the service provider.	Optional attribute
textualidentifier@ ServiceName	This identifies the name of the channel	Mandatory attribute
DVBtriplet@OrigNetID	This provides the NetID to the IPTVCD.	Mandatory attribute
DVBtriplet@TSid	This provides the MPEG transport ID to the IPTVCD.	Mandatory attribute
DVBtriplet@ServiceId	This provides the service ID or program ID to the IPTVCD.	Mandatory attribute
MaxBitRate	This provides the bit rate of the stream.	Optional attribute
Service location type	This section of the record provides information on channel location and includes details such as The IP multicast source address The IP multicast destination address The IP multicast Port Number RTSP address if unicast mode is to be used to access content. Details of audio and video compression algorithms used on the stream.	A mix of mandatory and optional attributes.

record called the DVB IP offering record and one of these XML records is created for each channel available on a DVB network. Further details are available in Table 4.13.

The standard also includes details on a record structure for networks that do not use DVB-SI tables to carry program information. Delivering the XML based channel information across a multicast network involves the use of a new protocol called DVB SD&S Transport Protocol. Full details of this protocol are available in section 5 of the ETSI standard. The standard also specifies the use of the HTTP protocol for transporting channel information over unicast connections.

4.9.2   SAP/SDP

When RTP is used to deliver IPTV services, the channel information is typically carried across the network by the session announcement protocol (SAP) —an international standard published in RFC 2974. The information transferred over the SAP transport protocol is formatted in compliance with the Session Description Protocol (SDP) format defined in RFC 2327.

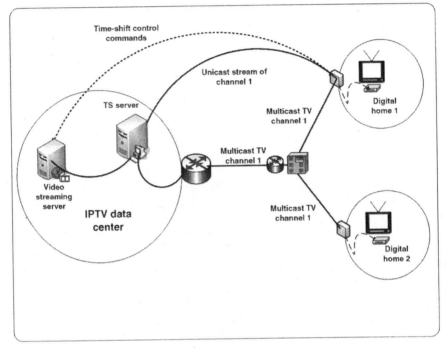

FIGURE 4.26   TSMIPTV services network architecture

4.10   TIME-SHIFTING MULTICAST IPTV

In addition to providing a linear multicasting service, it is also possible to implement a networking infrastructure that supports the delivery of time-shifting multicast IPTV (TSMIPTV) services. Figure 4.26 briefly illustrates the network components required to support the delivery of TSMIPTV services.

As illustrated the main feature of this architecture is the addition of a time-shifting (TS) server at the IPTV data center. The principle of operation is relatively straightforward from a technical perspective. In this example, digital home 1 has subscribed to a TSMIPTV and is therefore able to receive a unicast stream of channel 1 at a time, which is after the original multicast of the channel 1 stream. The delivery of the service involves the rerouting of the multicast stream through the TS server and onward to the IPTVCD. As the stream is passed through the server it gets copied to a storage device. This copy is subsequently available to other IPTVCDs that are permitted to access a time-shifted version of the particular channel. Storing sections of the stream onto a storage device allows endusers in digital home 1 to use trick mode functionalities such as pause, rewind and fast forward to multicast TV channels. It is important to note that the TS server may not be a separate piece of physical hardware and the time-shifting functionality in some configurations may be embedded directly into the IP streaming server at the IPTV data center. End users in the digital home 2 are only able to watch channel one in real-time and are unable to avail of the convenience provided by TSMIPTV services as they have not subscribed to the service.

4.11   CHANNEL-CHANGING INDUSTRY INITIATIVES

The telecommunication industry has placed a significant importance and focus on promoting rapid channel changing as one of the reasons why TV viewers should migrate away from their existing TV provider and move to providers that offer IPTV services. To further develop this IPTV feature an industry group called ISMA has developed a specification that addresses the technical issues surrounding the changing of TV channels over an IP network. The specification examines the factors that can contribute to a delay between the request for a particular IPTV stream and the subsequent rendering of that stream on the TV screen. It goes on to propose some techniques that may be used to minimize the perceived delay experience by a viewer when changing a channel or moving to viewing an on-demand video title. The document concludes with a set of recommendations. The full text of the specification is available at www.isma.tv.

SUMMARY

IP multicasting is a popular technique used to broadcast live TV channels. From a technical perspective it is a hybrid of unicast and broadcast technologies. With an IPTV multicast, only a single copy of each video channel is transported across the network irrespective of the number of people viewing the channel. This is an excellent mechanism for reducing the traffic requirements of the network and helps to conserve bandwidth. IGMP devices fall into two broad categories: client hosts and routers. Client IGMP devices are responsible for issuing messages to view and leave a particular broadcast channel, whereas routers fulfill the requests to view new IPTV channels. In addition to fulfilling requests from various types of IPTVCDs and forwarding network traffic, multicast routers have a number of other capabilities ranging from managing distribution trees that control the path that IP multicast traffic takes through to the network to replicating streams, processing multicast protocols, and managing IGMP messaging. The destination address of an IP multicast address is always in the range of 224.0.0.0 through to 239.255.255.255. A specific IP multicast address is also known as a group address. There are three versions of IGMP in use today: IGMP v1, IGMP v2, and IGMP v3. The main difference between IGMPv1 and IGMPv2 is the leave group feature and the reduction in time it takes for a multicast router to determine that there are no longer people watching a particular IPTV stream. Additionally, IGMPv3 adds support for a new multicasting paradigm called SSM. This new paradigm allows an IGMP client to explicitly define the source of the broadcasted content. IPTVCDs must support IGMPv2 because it is the earliest version of the protocol that supports multicasting.

TABLE 4.14   Feature Comparison Between Different Types of IGMP Systems

Parameters	IGMPv1	IGMPv2	IGMPv3
Number of message types supported	Two	Four	Two
Compatible with previous versions	Not applicable	Yes	Yes
SSM support	No	Yes	Yes
Security levels	Low	Fair to medium	High

Table 4.14 provides a feature comparison between the three versions of the IGMP protocol.

A multicast distribution tree is used to pass IPTV packets across a network. There are two types of distribution trees, namely, shared and source. PIM is a family of multicasting protocols that are used to distribute traffic throughout the core of a high speed IP core network. PIM-SSM is generally used to support IPTV applications and provides support for source and shared multicast trees.

In a multicast network, routers are responsible for replicating source content and forwarding it to multiple IPTVCDs. The purpose of IGMP snooping is to watch and monitor IGMP membership reports.

MLD is the IPv6 version of IGMP. Two versions exist for deployment across an IPTV networking infrastructure, namely MLDv1 and MLDv2. The main differentiator between both versions is MLDv2s ability to support filtering of source multicast addresses.

The main goal of the channel-changing process within the context of an IPTV environment is to ensure that the experience is at least on par or better than the current channel changing process on a standard RF based digital TV service. Delays between the request to access an IPTV stream and actual rendering of the stream on the screen is introduced by a number of factors such as packet processing at the RG, decoding, decrypting, and buffering. There exists a number of techniques and solutions that serve to minimize the affect of the delay factors associated with accessing an IPTV stream. Note that the DVB-IPI specification includes standardization guidelines on providing channel and IPTV service information across a DVB based IP network.

5

IPTV CONSUMER DEVICES (IPTVCDs)

IPTV enables the delivery of a wide variety of video content across a range of different consumer devices. Of the various IPTV consumer devices that are vying for a share of the emerging marketplace, the bulk of deployments today use residential gateways and IP set-top boxes. This chapter provides a thorough analysis of the hardware, software, and standards used by both of these types of IPTV consumer devices (IPTVCDs). Apart from the strong focus on IP set-top boxes and residential gateways, this chapter also provides a brief overview of other types of IPTVCDs such as game consoles and media servers.

5.1 ABOUT RESIDENTIAL GATEWAYS

An Residential Gateway (RG) sits between the IP broadband access network and the in-house network to allow multiple digital devices in a home to share a single IP connection. In addition to high speed Internet access applications, RGs are also capable of intelligently routing IPTV services to specific digital televisions, flat panel displays, and mobile devices. (see Fig. 5.1).

Given the fast-changing nature of the IPTV industry, it is helpful to examine the different generations of residential gateways that have been deployed over the past number of years.

Even though the term "residential gateway" is a relatively new one to most people, RG devices already exist in many households around the world. For example, most households already own or lease a couple of first generation

Next Generation IPTV Services and Technologies, By Gerard O'Driscoll
Copyright © 2008 John Wiley & Sons, Inc.

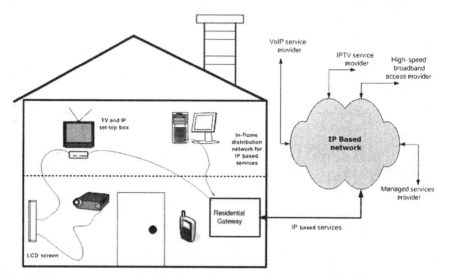

FIGURE 5.1   Using an RG to provide IP connectivity to a range of in-home digital devices

residential gateways—a set-top box for receiving television, a meter from the utilities company, and a modem that allows people to connect with the Internet. Second generation residential gateways include advanced features such as broadband connectivity, home networking interfaces, and VoIP capabilities.

Powerful third generation RGs are now commercially available that are capable of supporting multiple broadband and home-network interfacing technologies. The hardware architecture of these appliances is modular in design. This support for modularity means that RGs are starting to evolve into a type of application server that consumers are using to distribute broadband, IPTV, and VoIP services throughout their homes. From an IPTV service provider's perspective, RGs are an important strategic technology. They allow network operators to introduce in-home networking products that differentiate their services from their competitors and provide additional revenue opportunities.

An RG needs to have reliable hardware and software systems to accommodate the delivery of multiple services, be secure, and interoperable. An RG typically combines the functions of a modem, a router, and a hub to access IPTV services. There are different types of residential gateways available based on the gradients of functionality required. They can be categorized as

Simple—Simple or "dumb" RGs have limited routing, connectivity processing power and applications. In this category RGs support bridging at layer two of the IPTVCM. This is not a popular design because the network traffic is sent to all device interfaces.

Complete—Most RGs provide capability for advanced data, voice and video routing within the home. A complete RG contains a modem and networking software and does not depend on a PC to operate. Routing functionality is supported that provides Layer 3 connectivity between private IP addresses assigned to IPTVCDs connected to the home network and the broadband access network.

5.2 RG TECHNOLOGY ARCHITECTURE

Since broadband technology is relatively new and continuously evolving, the residential gateway is also evolving in its functionality. However, in its basic form, it contains and combines three distinct technological components:

(1) A digital modem
(2) Home networking chipsets
(3) Relevant software

Much of the configuration flexibility is brought about by a gateway's support for a range of different types of modems. These modems provide connectivity to the following types of broadband access networks:

(1) Direct broadcast satellite
(2) Wireless Wi-Fi and WiMax systems
(3) Two-way hybrid-fiber cable TV networks
(4) DSL, ADSL, and VDSL systems
(5) FTTH networks

Most RGs also include a console port that allows IPTV engineers to troubleshoot the RG if problems arise.

At the home networking side of the RG, a chipset is available that provides the interface to the particular technology running on the home network. The types of home technologies supported by RGs include

- HomePNA
- HomePlug
- MoCA
- GigE
- Wi-Fi

In addition to the various types of broadband and home networking chipsets, all gateways contain computing resources that support the software required to operate the device. The software running on the RG enables the smooth interoperation of digital consumer devices and IPTV services within the home. The software ensures that the complexities of the system elements are hidden from the IPTV subscriber.

The location of the RG differs between households. Some consumers connect their home networks to RGs that reside on the outside of their houses. These device types are weather proofed and housed in a durable casing. The main benefit of having the gateway located on the outside of the residence is the ease of access provided to the IPTV service provider for maintenance and upgrades. RGs that are designed for internal use allow consumers to interact with the gateway and are cheaper than external units to install. As with similar products, such as satellite TV receivers or cable modems, RGs can be installed either by the consumer or by a professional technician.

5.3 RG FUNCTIONALITY

The RG acts as a centralized point that provides a number of functions that are particularly relevant to delivering IPTV content to a digital home network.

5.3.1 Bridging and Routing Network Traffic

An RG is used to bridge network traffic between logical topologies known as virtual LANs (VLANs) and ATM based Private Virtual Circuits (PVCs) with in-home digital consumer devices connected to RG home networking ports. Note further details of VLANs and PVCs are available in Chapter 9. The bridging function is normally implemented at layer 2 of the IPTVCM. The types of encapsulations associated with bridging are listed and explained in Table 5.1.

In addition to bridging traffic, most if not all RGs used in IPTV deployments also support routing of packets into and out of the home network.

TABLE 5.1   RG Encapsulation Methods

Physical Media Type	Interface Category	Encapsulation Header Type
Copper pairs	Broadband link	Multiprotocol encapsulation over ATM adaptation layer 5
RG-59 and RG-6 coaxial cable	Broadband link	DOCSIS versions 1, 2, and 3
Wireless	Home networking link	UWB, IEEE 802.11a, b, g, and *n*.
Category 5e or 6 cable	Home networking link	Fast Ethernet and Gigabit Ethernet (GigE)
Powerline	Home networking link	HomePlug AV and power line communications (PLC)
Phoneline	Home networking link	HomePNA
RG-6 or CT-100 coaxial cable	Home networking link	MoCA, HomePNA, and IEEE 1394

5.3.2   Enforcing QoS Functionality

The DiffServ QoS architecture is often deployed on IP networks to support the effective delivery of telecom services including IPTV. Note that DiffServ is a mechanism used to improve the performance of time-sensitive applications such as IPTV. The RG can typically play a role in DiffServ by marking packets that originate in the in-home network and are destined for the IP broadband access network. From a technical perspective the marking of packets generally involves assigning a label called a differentiated services code point (DSCP) to each IPTV packet. This process helps to improve the end-users experience of the various IPTVCDs connected to the home network. The use of DiffServ in a home networking environment is not mandatory because it may cause issues that impact on performance. For example, DiffServ labels will increase the overall size of network packets, which can cause difficulties for some types of basic networking switches and hubs. As a result, a number of other approaches are used to maintain adequate QoS levels within a home network and seperate IPTV traffic from other IP services, including:

5.3.2.1   By Physical Port   Figure 5.2 illustrates a typical RG traffic separation model based on physical ports in operation.

As depicted the network traffic on the wide area network interface are typically separated into distinct 802.1Q VLANs or in some cases ATM based VCs. DSCP packet markings are typically used to segregate the IP traffic into separate VLANs. Within the digital home, each network device is generally part of a particular ecosystem or IP subnet. So PCs are associated with the high speed Internet access service, whereas the phones are used to access VoIP services, and any IP set-top boxes are used to view IPTV services. The method of separating IP services based on physical ports works well when there is no ambiguity between the functionality

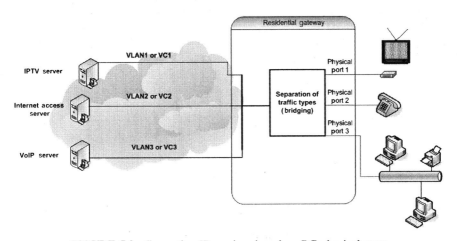

FIGURE 5.2   Separating IP services based on RG physical ports

provided by each device connected to the home network. However, newer digital consumer multifunctional devices have increased in popularity over the last few years, which can cause difficulties for RGs who are tasked with separating services for transport to the various VLANs.

5.3.2.2 By MAC Hardware Address   Another approach used by RGs to separate traffic types is by hardware address. Under this approach the RG maintains a table in which each entry includes the hardware address of the digital consumer device, the port that it is connected on, and the associated VLAN or ATM based VC.

5.3.2.3 By IP Address   It is also possible to use either the source or the destination IP address contained in the packet header to classify traffic traveling upstream across the RG and onward to the IP distribution network.

5.3.2.4 By Packet Length   This approach uses the packet length to determine actions in terms of queuing, scheduling, and packet dropping policy. For instance, the length of a VoIP packet is typically smaller than a packet used to carry Web content. Thus, the RG can add the smaller packets to a queue that avoids excessive jitter and latency.

5.3.2.5 By Protocol Type   Separation by transport protocol type is another configuration, which may be used by RGs. This is quite a general approach and classifies packets as either TCP or UDP traffic types. From this classification the RG can then implement appropriate treatment policies.

Once the traffic is classified using one of the mechanisms described above, the packets are assigned to an appropriate queue for delivery onward to its destination point.

5.3.3 IP Address Management

Most modern RGs will support a relatively sophisticated approach to managing multiple IP subnets operating across an in-home digital network. To accomplish the management of IP addresses in a triple-play environment, the components shown in Fig. 5.3 are generally supported.

The role of each component is discussed in the following sections.

DHCP Clients   A separate DHCP client is required for each service. Each of these clients communicates across a VLAN to a DHCP server in the headend. In this case, the IPTV DHCP client in the RG communicates over VLAN1 back to the server in the data center responsible for managing the IPTV address pool. The server uses the VLAN1 connection to assign a unique "external IP address" to the DHCP client resident in the RG. As depicted the DHCP clients for the Internet access and VoIP services are also assigned unique IP addresses. Note that point-to-point Protocol over Ethernet (PPPoE) clients may be used instead of a DHCP client for the high speed Internet service.

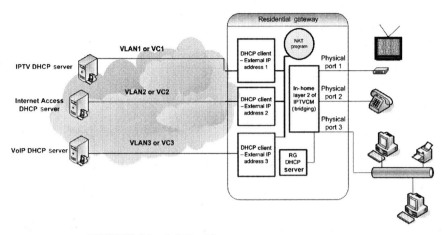

FIGURE 5.3 RG IP address management components

DHCP Server The purpose of the RG DHCP server is to allocate addresses to local in-home digital consumer devices. These addresses are used to create a single subnet that allows devices connected to the in-home network to communicate with each other.

Network Address Translation (NAT) Program The main function of the NAT program is to alter the IP address information contained in the IP header. Alterations include the replacement of the source address as the IPTV packets pass through the RG. The main benefits of incorporating NAT functionality into an RG include the following:

- It allows multiple IP devices inside a home network including an IP set-top box to access servers at the IPTV data center by using a small number of RG external IP addresses.
- It ensures that IP addresses of the various IPTVCDs connected to the home network are not made available to users on the public Internet. This is particularly appropriate when accessing Internet TV channels.

It is important to note that the use of NAT in an RG can cause some difficulty when used to deliver multicast TV services to a home network. To overcome these difficulties, some RGs use features such as IGMP snooping to avoid the execution of NAT on particular types of multicast traffic.

5.3.4 Facilitating Remote Administration

Most basic RGs include a software client that allow IPTV support engineers to remotely monitor, manage, and troubleshoot the device. In addition to remote support, all RGs include a local port that allows an end user to make configuration changes. The application used to make these changes is typically browser based.

5.3.5 Managing Congestion

As discussed previously, an RG is a single device that interfaces between two network types, namely, the IP access network and the internal home networking infrastructure. Both network types generally differ in terms of network media, protocols, and bit rate throughput. The bit rate throughput characteristic is particularly relevant to RGs because variations between both network types can lead to congestion on either the RG upstream or the downstream interfaces. Thus, congestion occurs when the RG is receiving traffic faster than it can be transmitted. The interfaces that are susceptible to congestion are shown in Fig. 5.4.

As shown network congestion can occur at three different points of the RG.

(1) Upstream intersection point between RG and the IP access network—As depicted the rates at which the home network operates are quicker than the bandwidth capacity of the access network. In this example, Gigabit Ethernet (GigE) and powerline based equipment with the potential to operate at several 100 Mbps has being installed in the home. However, the ADSL2 + broadband connection used to provide connectivity to the regional office can only operate at a maximum upstream data rate of 1 Mbps. Under these circumstances congestion will occur unless the RG is able to constrain the rate at which home networking devices send traffic upstream.

(2) Downstream intersection point between RG and the IP access network— The level of congestion at this point depends on the type of technology used by the access network. For instance, the potential for congestion is greater

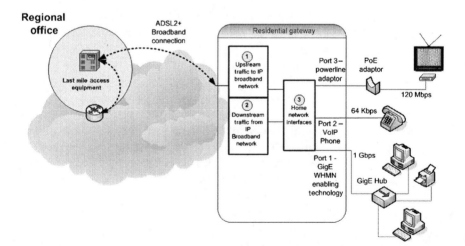

FIGURE 5.4 RG interfaces that can potentially suffer from congestion

for FTTH networks than it is for a network that uses ADSL2 technologies. In general, the congestion levels are normally low at this point because RGs are typically designed for a specific type of IP access network.

(3) Intercommunication between different home networking technology types— Rate mismatches can also occur between different home network enabling technologies. For example, an RG may be required to facilitate communication between a wireless 802.11b enabled notebook and an IP set-top box, which is connected to the RG using a HomePlug AV powerline bridging adaptor. As a result, the RG has to deal not only with the bit rate mismatch between both networking types but also with congestion that might occur due to IP access traffic and local home networking traffic competing for the same interfaces.

Once congestion occurs at any interface, a queuing mechanism is applied to alleviate the situation. Queuing involves storing the packets temporarily until bandwidth becomes available to start transmitting the queued IP packets. The types of queuing mechanisms supported by RGs are described in Table 5.2.

The emptying of the various queues above will depend on the availability of sufficient physical layer network bandwidth to ensure that traffic flows smooth.

5.3.6 Enforcing Various Security Mechanisms

An RG normally combines a number of components to create an integrated security system, including

A firewall—Most RGs include an integrated firewall. The purpose of the firewall is to monitor and inspect packet headers of all network traffic going upstream from the home network to the access network and vice versa. In addition to monitoring traffic, the firewall is also responsible for enforcing rules that minimize the risk of security breaches occurring.

Internet content filtering—Specific IP ports on the gateway can be set to block particular types of traffic emanating from the Internet. With this capability it is possible to block applications such as FTP, instant messaging, and e-mail traffic entering the home network. In addition to port blocking, some gateways support the definition of groups with specific access rights. For instance, an IPTV installation engineer may be requested to establish a group for the homeowner's children. Once the group is defined, lists of full or partial Web addresses that need to be blocked are then assigned to this group.

Encryption keys—In some security configurations public and private keys are stored in the RG for use in encrypting and decrypting messages. It is important to note that some RG security mechanisms also inspect the payload in addition to the headers, which can add latency to the delivery of IPTV streams.

TABLE 5.2   Types of Queuing Mechanisms

Type	Description
First-in-first out (FIFO) queuing	FIFO queuing is based on the principle that the first packet to arrive at the queue at one of the RG interfaces is the first packet to get transmitted across the network media. This is not used as a mechanism for processing time sensitive network traffic such as IPTV.
Priority queuing	As the name suggests this mechanism places IP packets into one of a multiple of queues. The number can vary between RGs. At a minimum most RGs will support three queues—namely, one for each of the triple play services. Under this mechanism each queue is assigned a particular level of priority. As a result, the packets in the higher priority queue(s) get processed, while the packets stored in the lower priority queues need to wait until the higher priority queue(s) are emptied. This obviously causes delay issues for the traffic traversing the RG via the lower priority queues. IPTV and VoIP are typically stored into the higher priority queue because of their sensitivity to delays and network jitter.
Round robin queuing	This type of queuing adds packets to all of the queues. Removal and subsequent processing by the RG is done on a round-robin fashion.
Weighted round robin Queuing (WRR)	WRR assigns a weight or a variable to each queue. The processing of the queue is based on round robin, however, the amount of data taken from the queue during each schedule is dependent on the weighting assigned to the queue.

5.4   RG INDUSTRY STANDARDS

Industry standards are crucial for enabling mainstream consumer adoption of residential gateway products. This has been proven for a number of different markets such as VCRs, cellular phones, set-top boxes, and DVD drives. The same model applies to the residential gateway market. Standards and interoperability conformance enable the following:

- Protection of home networking users from obsolescence.
- Improvement of consumer confidence in the purchase of residential gateways.
- Encouragement for vendors to build interoperable products—products that are compatible among different brands, as well as with other devices connected to a home network.
- Increase in the widespread availability of consumer equipment from a number of different vendors in retail stores and from hardware manufacturers.

Developing and adhering to standards is critical to the success of residential gateways. The current definition of a residential gateway has its beginnings in a white paper developed by the RG Group, a consortium of companies and research entities interested in the residential gateway concept. The RG Group determined that the residential gateway is "a single, intelligent, standardized, and flexible network interface unit that receives communication signals from various external networks and delivers the signals to specific consumer devices through in-home networks." Since the publication of the RG Group's white paper in 1995, a number of technologies are vying to become the defacto standard for residential gateways. A discussion of these technologies is provided in the following sections.

5.4.1 ISO/IEC 15045-1

The International Organization for Standardization and the International Electrotechnical Commission (ISO/IEC) are proposing a standard called the ISO/IEC 15045-1, which specifies the minimum functional requirements of an RG. The standard specifies what an RG should do in order to deliver services in a suitably safe, secure, and future proof. Some of the characteristics of the ISO/IEC 15045-1 specification include the following:

- It supports communications among devices within the premises, and among systems, service providers, operators, and users outside the premises.
- In addition to supporting the delivery of IPTV services it also allows telecom operators to provide their customers with a range of advanced IP based services such as home appliance control and preventive maintenance, remote metering, and energy management.
- In addition to supporting the IP protocol, 15045-1 compliant gateways can also support a number of other in-home networking protocols, including MoCA, HomePNA, HomePlugAV, IEEE 1394 over Coax, and Wi-Fi.
- When operating in an IP environment, security is a major concern for service providers and consumers alike. The 15045-1 standard specifies security measures to ensure the integrity of information that may pass through the residential gateway.

5.4.2 OSGi

The Open Services Gateway Initiative (OSGi), an international alliance of companies, was set up in 1999 and has developed technical specifications that allow residential gateways to support the provision of IPTV services into a subscriber's home. Some of the characteristics of the OSGi specification include the following:

- The software used in the specification uses industry standard languages such as Java, HTML, and XML.
- The major identified components of a complete end-to-end OSGi framework model as defined in the specification include

  Services gateway

  Service provider

  Service aggregator

  Gateway operator

  WAN

  Local devices and networks

- OSGi allows devices to be directly connected to an RG using hardware links such as parallel or serial connections or indirectly connected by means of a local transport network such as IEEE 1394, or wireless.

5.4.3 TR-069

Produced by the DSL Home-Technical working group the TR-069 specification defines a protocol, which is used to remotely manage a DSL enabled RG. In addition to the various technical features, TR-069 also includes support for business models that allow IPTV service providers to sell gateways through retail distributors. A typical DSL based networking architecture using TR-069 compliant equipment is presented in Fig. 5.5.

As shown the TR-069 client is an application running on the RG and it interacts using a suite of different protocols with a configuration server at the IPTV data center. A brief overview of the protocols used by TR-069 are described in Table 5.3.

Other protocols such as TCP and IP are also used by the TR-069 management protocol. A connection to an OBSS allows service providers to apply various management policies to the RG. Characteristics of the TR-069 specification include

- *Provisioning capabilities*—TR-069 allows the automatic provisioning of DSL based RGs once the device is connected to the broadband access network. DHCP in conjunction with the Domain Naming System (DNS) are generally used by the RG to identify the address of the configuration server.
- *Software updates*—IPTV providers can manage the download of software update image files to DSL enabled gateways. Advanced download features

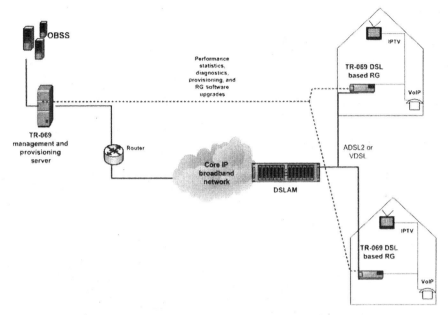

FIGURE 5.5 Controlling an RG via TR-069

such as digital signatures, version identification, and management notifications are also supported.

- *Performance monitoring*—The delivery of gateway status information and performance statistics back to a network management system is also facilitated by this protocol.
- *Diagnostics*—The ability to diagnose problems on a gateway is an essential part of managing an IPTV network. TR-069 allows IPTV providers to test and diagnose connectivity problems when they arise.

5.4.4 Home Gateway Initiative

Launched in December 2004 by a group of telecommunication companies, the Home Gateway Initiative (HGI) is an open forum whose primary objective is to define a common set of technical requirements to support the deployment of next generation home gateway applications. In 2006, the group published a set of requirements in a document titled the "Home Gateway Technical Requirements: Release 1." The HGI specification addresses a number of separate topics associated with the standardization of RGs including.

5.4.4.1 Network and Service Requirements In an effort to enforce the practicalities of RGs, the HGI defines business requirements associated with a typical deployment of an RG in a family household.

TABLE 5.3   TR-069 Protocol Descriptions

Protocol	Functionality
Sockets layer (SSL)	The purpose of the SSL cryptographic protocol is to combat snooping of confidential information transferred between the server and the TR-069 gateway.
Transport layer security (TLS)	TLS is a successor to SSL and is sometimes used instead of SSL by TR-069 equipment manufacturers.
Remote procedure call (RPC)	TR-069 enabled servers use the RPC protocol to read and write parameters to the RG. Additionally, it enables IPTV service providers to download software upgrades to RGs in the field.
HTTP (hypertext transfer protocol)	HTTP is the communications protocol used by TR-069 clients and servers. When a request is received, the TR-069 server opens a connection to the RG and sends the requested file. After servicing the request, the configuration server returns to its listening state, waiting for the next HTTP request.
Simple object access protocol (SOAP)	TR-069 combines SOAP with HTTP to exchange RPC messages across a network. CPE WAN Management Protocol (CWMP) is the term used by the DSL Forum to describe the combination of SOAP and HTTP protocols.

5.4.4.2 HGI Hardware Reference Architecture   The architecture as defined by the HGI consists of the following components:

- Home Network Connectivity Ports—IEEE 802.11b/g and Fast Ethernet were chosen as the home networking technologies for HGI compliant RGs. Note the next release of the specification due out in 2008 is likely to include support for the IEEE 802.11n WLAN standard.

- RG and home network designs need to align—The topology of the home network determines whether the design of an RG is centralized or distributed. In a centralized design, all of the traffic is processed by the RG because all the IPTVCDs and other device types are connected directly into the RG. In contrast, an RG built to support the distributed model requires fewer interfaces because most of the layer 2 networking is done by an external switch or hub.

5.4.4.3 Management Architecture The management architecture section of the HGI defines all the tasks that are required to manage an RG, including

- Managing home networking devices
- Managing RG software upgrades
- Managing RG security
- Managing RG QoS

Other areas such as performance monitoring and diagnostics are also supported by the HGI specification. Most of the HGI management tasks leverage much of the work completed by existing home technology standards bodies. For instance, the management architecture makes extensive use of the TR-069 standard. In addition to the core standard, a number of related standards are also used to support different levels of functionalities (see Table 5.4).

5.4.5 CableHome

CableLabs, a research and development U.S. based consortium of cable television system operators, has developed a specification for cable based RGs called CableHome. The first revision of the technology, known as CableHome 1.0, was issued by CableLabs in 2002 and has been through one revision over the past 5 years. The key historical milestones of how CableHome has evolved since the project was originally kicked off are set forth in Fig. 5.6.

Table 5.5 summarizes the features supported by the first two generations of CableHome technologies.

As shown in the table the second version of CableHome builds on the baseline features of its predecessor. It gives cable operators the ability to provide managed home networking services to their subscribers. In addition to improving home networking functionalities, CableHome 1.1 also puts a strong emphasis on security and QoS. CableHome consists of a number of specifications that allow cable

TABLE 5.4 DSL Forum Standards Used by the HGI Management Architecture

DSL forum standard	Description
TR-069	A remote management protocol.
TR-111 (Part 1)	This allows the remote home management server at the IPTV data center to identify an RG that is associated with a particular device connected to a home network.
TR-111 (Part 2)	This allows a remote home management server to establish a direct communications session with a home networking device that is connected to the access network through an RG.

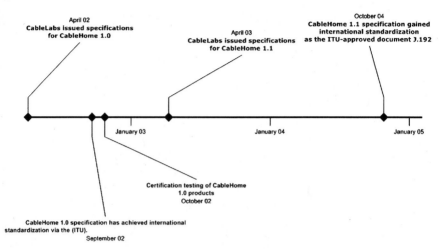

FIGURE 5.6   Evolution of the CableHome specification

TABLE 5.5   Key Features of CableHome Compliant RG Devices

	CableHome 1.0	CableHome 1.1
Automatic RG provisioning and authentication	✓	✓
Secure remote management and configuration	✓	✓
Support for local address translation	✓	✓
Support for connectivity testing	✓	✓
Support for firewall functionality	✓	✓
Local domain naming service	✓	✓
Firewall reporting capabilities		✓
Parental control		✓
Port forwarding and support for virtual private networks (VPNs)		✓
Messaging to network management consoles		✓
QoS functionality		✓
Discovery of devices connected to a home network		✓
Support for Wi-Fi networks		✓
IP statistical information		✓

operators to provide their subscribers with a rich suite of advanced home networking, telephony, and IPTV services. Table 5.6 lists some of the specifications developed as part of the CableHome project over the last couple of years.

The major components of a full CableHome deployment are shown in Fig. 5.7.

Managed IP backbone and servers—The headend servers store the IPTV content and the IP backbone is responsible for carrying the IP video content to the various regional sites.

Cable modem termination system (CMTS)—The CMTS takes the IPTV traffic coming in from the core backbone and routes onward to the HFC RF based network and vice versa.

CableHome compliant RG—The CableHome gateway interacts with IP based in-home devices.

Whole home media network—The WHMN component is used to deliver IPTV content to a variety of different consumer electronic devices almost in any part of the home.

From a technical perspective the CableHome specification addresses four separate areas of functionality:

5.4.5.1 Discovery and Provisioning CableHome enabled RGs are easy to install and configure. They use the DHCP networking infrastructure to acquire an IP address

TABLE 5.6 List of CableHome Specifications

Description	Specification
CableHome 1.1 specification	CH-SP-CH1.1-I11-060407
CableHome™ QOS MIB specification	CH-SP-MIB-QOS-I05-050408
CableHome 1.0 specification	CH-SP-CH1.0-I05-030801
CableHome CAP MIB specification	CH-SP-MIB-CAP-I08-050211
CableHome CDP MIB specification	CH-SP-MIB-CDP-I08-041216
CableHome CTP MIB specification	CH-SP-MIB-CTP-I06-040409
CableHome PSDEV MIB specification	CH-SP-MIB-PSDEV-I10-060407
CableHome Security MIB specification	CH-SP-MIB-SEC-I07-040806

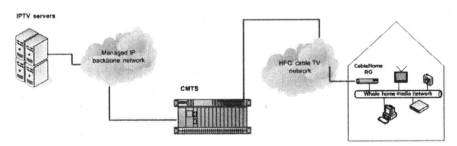

FIGURE 5.7 End-to-end CableHome reference network

lease and receive configuration information from the server. The provisioning of services on the RG and the discovery of in-home IP based devices is automatic.

5.4.5.2 Network Management   The installation of a network and the ability of the service provider to remotely manage that network go "hand in hand." Thus, CableLabs included a dedicated section to the management of home networks in its CableHome specification. The types of areas covered range from assigning IP addresses to diagnostics, and synchronization with servers located back at the headend. The CableHome RG architecture uses an embedded database called Management Information Base (MIB), which resides in the RG. The MIB stores information about the RG, devices, and PCs connected to the in-home network. The MIB stores information in a standardized format, which allows CableLabs to update whenever new features are required by the cable TV industry. The Simple Network Management Protocol (SNMP) is used to access the MIB and exchange messages back to the network management system at the cable headend. Note version 3 of the SNMP protocol is typically used to remotely manage a CableHome RG.

5.4.5.3 Quality of Service (QoS)   As more and more IP based services get delivered to people's homes, so will the traffic congestion on both broadband access and in-home networks increase exponentially. Increased traffic congestion leads to latency problems on the network, which in turn degrades the performance of the in-home network and affects the quality of the IPTV content transported around the house. Therefore, CableHome RGs incorporate support for different QoS features including the prioritization of traffic types to support real-time WHMN applications. To achieve this functionality, a queuing mechanism and a QoS policy server are used by the RG to prioritize the various types of packets flowing across the in-home network. The QoS server is resident within the gateway and allows both the IPTV network operator and the home user to manage and update the rules used in prioritizing data flow across the in-home networking system.

5.4.5.4 Securing the Network   Security has always been a major concern for telecom operators over the years. Nowadays with the proliferation of IP based technologies and broadband access networks, consumers with limited expertise in IT technologies are particularly prone to attacks from the public Internet. To minimize the risk of malicious attacks or viruses CableHome RGs includes strong protection mechanisms such as Network and Port Address Translation (NAPT), and parental control. The security systems supported by CableHome RGs focus on three particular areas:

- Secure RG remote management
- Prevention of unauthorized access from the public network to the home network
- Secure RG provisioning

The CableHome security architecture employed by CableHome RGs to protect in-home networks consists of a mix of technologies, including

A firewall—The firewall built into the CableHome RG implements a security policy between the in-home network and the IP broadband access network. It starts to operate automatically once the provisioning process has been completed. Its main function is to impede the ability of outsiders to gain access to WHMN services operating across the home network and is the first line of defense in protecting the in-home network against malicious attacks and traffic deemed as unwanted from external parties. A packet-filtering mechanism is generally used by the firewall to determine the types of streams allowed cross the firewall.

A key distribution center (KDC)—The KDC runs on a server located at the cable headend. It is responsible for managing and exchanging keys used in secure remote management of the RG.

A HTTPS server—The CableHome HTTPS sever is a Web server, that supports Transport Layer Security protocol (similar to the secure sockets layer (SSL) protocol). This component is used to establish a secure session for downloading RG configuration files.

Further security elements can also be used to increase the protection levels of CableHome RGs. Note that in 2006 CableLabs published another RG centric specification called "eRouter." The technology allows manufacturers to design gateway products without incorporating advanced features such as firewall capabilities included in the CableHome specification.

5.5 INTRODUCTION TO DIGITAL SET-TOP BOXES

The main purpose of a digital set-top box is to provide consumers with access to a variety of different types of digital entertainment content. The content varies from TV programming and movies to musical videos and sporting events. The content can be delivered to the set-top box over a variety of digital networking infrastructures, including

- Cable TV systems
- Satellite networks
- Terrestrial or wireless networks
- Telecommunication networks
- Mobile or cellular networks
- Wireless data hotspots

Digital set-top boxes also known as set-tops are best defined as types of computers that translate signals into a format that can be viewed on a television screen. They are complex electronic devices comprised of a number of different hardware and software components. A typical digital set-top box has a physical height of 2.5 in. and a width of 18 in. Interfaces are included on the back and front to support the flow of data into and out of the set-top box. These devices have advanced computational capabilities that add more value and convenience when networked. Set-top boxes have a number of attributes, including

- Portability—They are small enough to be transportable by an average person.
- Ease of use— They are simple enough to use and require no special training beyond an instruction manual.
- Affordable for most—They are inexpensive enough to be affordable by the average household.
- Easily controllable—Handheld remote controls or wireless keyboards are typically used to issue commands to the set-top box.
- Supports two-way data services—A back channel or return path is built into some set-top box models to facilitate communications to the network service provider.
- Support for interactive TV applications—They enable infotainment by providing services such as interactive TV and multiplayer gaming.
- Home networking capabilities—Many of the modern day set-top boxes provide interfaces that allow real-time communication with devices such as DVDs, digital cameras, and music servers.
- Different variations of set-tops—The type of set-top box varies with the type of network. So cable set-tops are required to access programming on a cable HFC network, terrestrial set-tops are needed to view content sent over a broadcast network, and so on.

For years the PC industry has predicted the demise of the set-top box. The set-top box has no intention of going away anytime soon for a number of reasons:

Large installation base—Set-top boxes have a huge installed base and help to generate large revenue streams for service providers and manufacturers across the globe.

Multifunctionality—Processing TV signals (standard and high definition), retrieving and storing VoD movies, interfacing with other digital consumer devices in the home are some of the functions possible on a set-top box. Most digital consumer devices are only dedicated to one or two functions.

Established industry—The set-top box industry has been in existence for years and software providers, manufacturers, standard bodies, and operators will continue to innovate and drive momentum in the years ahead.

Migration to broadcasting in digital format—The use of Internet technologies in combination with the rollout of digital TV broadcasting is having a positive impact on the worldwide sales of digital set-top boxes. Many countries around the world are expected to conclude the transition to digital TV over the next decade. Therefore, service providers are retrofitting subscriber's analog set-top boxes with new digital set-top boxes. Additionally, some of the more technologically developed countries are beginning to push second generation set-top boxes to support a range of new services.

Faster and cheaper set-top box components—According to Moore's law the processing power of semiconductors will continue to double approximately every 12–18 months. This coupled with technical advancements and greater business efficiencies means that every generation of set-top boxes are being built with more features and higher performance at lower costs.

It is because of the above reasons that digital set-top boxes will continue to maintain a stronghold on the delivery of entertainment services to households across the globe. The next section examines the historical evolution of the set-top box over the past 25 years.

5.6   THE EVOLUTION OF DIGITAL SET-TOP BOXES

The historical evolution of the set-top box over the past 25 years is graphically illustrated in Fig. 5.8.

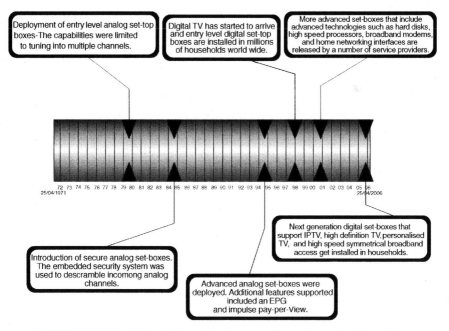

FIGURE 5.8   Evolution of the set-top box over the past quarter of a century

Today's set-top boxes are very different from the devices that were available in the past. In the early 1970s, most people only needed a standard television that they were able to purchase at their local store to watch TV. In the early 1980s, this simple model began to change. Cable and satellite providers required the consumer to connect their TVs to their networks to allow access to multichannel TV. Connecting to a TV service required the installation of an analog set-top box. These early day set-top boxes allowed pay TV subscribers to tune into over 20 different channels. Once a channel was received it then got converted to a format that was acceptable by standard VHF based televisions.

A couple of years later, service providers started to scramble their TV signals. This change in business model spurned the development and mass deployment of more advanced analog set-top boxes in the mid-to-late 1980s. These set-tops included the ability to descramble incoming TV signals but little else. A preloaded serial number assigned to the set-top box allowed equipment at the cable headend to enable access to premium programming content or tiers of programs.

In the early 1990s, the sophistication of set-top boxes continued to increase and service providers started to provide their customers with more advanced analog based set-top boxes that were able to run an EPG and impulse pay-per-view applications. The EPG was used by subscribers as an aid to select programs and channels. The bandwidth available in the vertical blanking interval (VBI) was used as a mechanism of keeping program schedule data updated. This chunk of bandwidth was also used by some service providers to download other types of information services such as news, sports, and weather forecasts to analog set-top boxes.

In the mid-to-late 1990s, the TV industry began showing an active interest in replacing analog set-top boxes with set-top boxes that would empower subscribers to access a range of digital TV services. These new boxes were very similar to their analog counterparts. The main function of these early stage digital set-top boxes was to receive and decode digital transmissions into a form suitable for display on analog television sets.

This new wave of digital set-top boxes coupled with upgraded cable network equipment increased the number of channels operators could offer, and pushed them one step closer to offering advanced services such as video on demand. In addition to receiving digital TV signals most, of these set-top boxes are also capable of receiving and processing analog signals.

Fast forwarding to the end of the 1990s and the start of the new millennium, set-top box manufacture's commenced the building and shipping of set-top boxes that used state-of-the-art technology to vastly increase channel capacity over existing networks, while providing significantly improved audio and video quality. The advanced user features and capabilities of these powerful set-top boxes provide support for a host of new services and an unparalleled level of flexibility and control.

At the time of writing this box the evolution of digital set-top boxes continues apace. The industry is more dynamic at the moment, and it is felt that more dramatic changes are on the horizon over the coming years.

5.7   CATEGORIES OF DIGITAL SET-TOP BOXES

The digital TV market comprises of a large population of set-top boxes working busily in homes across the globe. Table 5.7 segments the entire digital set-top box product space into a number of categories.

5.8   MAJOR TECHNOLOGICAL TRENDS FOR DIGITAL SET-TOP BOXES

With many countries finalizing standards and launching broadcast services, digital TV is becoming a force in the worldwide electronics industry. Set-top box manufacturers and semiconductor suppliers are continuing to ramp up production to meet the ongoing rise of digital TV and broadband subscriptions. As the digital set-top box market continues to expand, it is foreseen that a number of trends will emerge in the following areas:

Increased support for digital connections—As the choice of wireless and fixed wired based digital interfaces continues to grow, for the foreseeable future the set-top box will serve increasingly as a centralized device for managing the in-home distribution of digital content.

Increased storage capacities—The size of the hard disk drives used by set-top boxes will continue to rise to enable personal video recording and caching of data broadcasts. The larger disks will also be used by consumers to build in-home libraries of personal digital content. In addition to larger hard disks, many of the advanced set-top boxes are expected to include DVD recorders that could allow consumers to record IP based video content.

Increased support for HDTV—Manufacturers will continue to expand their use of high definition TV components to create brighter pictures for consumers.

Consolidation of silicon components—The trend in developing single-chip box solutions is expected to continue as pressure continues to reduce the unit price point of set-top boxes.

Migration to providing residential gateway capabilities—Consumers and service providers are increasingly demanding more functionality within the set-top box. Therefore, digital set-top boxes are evolving into digital home gateways capable of distributing digital media throughout the home. Traditional set-top box vendors are positioning their products as network hubs in the digital home of the future.

Support for advanced compression technologies—New compression standards such as H.264/AVC and VC-1 allow providers to deliver high quality digital content through their existing networking infrastructures. Therefore, set-top box vendors will continue to build support for these systems into their products.

TABLE 5.7   Categories of Set-top Boxes

Type of set-top box	Description
Dial-up digital set-top boxes	This category of set-top box allows subscribers to use the standard telephone network to access Internet content on their TV screens. They have a small hardware footprint. Applications that run on these set-tops include browser, e-mail, and instant messaging clients.
Entry level digital set-top boxes	Entry level set-top boxes are capable of providing traditional broadcast television that is complemented with a pay per view system and a very basic navigation tool. Characteristics of this type of box include low cost, limited quantities of memory, interface ports and processing power. These types of boxes are reserved for markets that demand exceptionally low prices and where interactivity over telephone networks is not an option.
Mid-range digital set-top boxes	Mid-range set-tops are the most popular model on offer today from TV operators. They normally include a return path or back channel, which provides communication with a server located at the head end. These types of boxes have double the processing power and storage capabilities of entry-level boxes. Mid-range set-tops enable service providers to deliver a range of different types of in-home entertainment products including high speed Internet access, and access to a sophisticated interactive television guide.
Next generation digital set-top boxes	This set-top box category bare close resemblance to a multimedia desktop computer and represent the future of digital TV viewing. They can contain more than ten times the processing power of a low-level broadcast TV set-top box. Enhanced storage capabilities of between *32MB and 64 MB* of flash memory (for code and data storage) in conjunction with a high speed return path, can be used to run a variety of advanced services such as video teleconferencing, whole home media distribution, IP telephony, DVD functionality, VoD, and high speed Internet TV services.

(Continued)

TABLE 5.7 (*Continued*)

Type of set-top box	Description
	Additionally, subscribers are able to use enhanced graphical capabilities within these types of boxes to receive high definition TV signals. Most next generation set-tops also incorporate digital video recording (DVR) functionality. High speed interface ports, which allow them to be used as a home gateway, are another feature. These set-top products represent the next wave of TV experiences beyond the personal computer.
IP digital set-top boxes	An IP set-top box is a digital device that provides an interface between an IP based broadband network and a television set or flat screen display. They are multi-purpose devices that allow network service providers to provide their subscribers with a whole range of revenue-generating IP based home entertainment services.

Increasing use of programmable high density integrated circuits—As standards and applications continue to evolve set-top manufactures are making wider use of chipsets that contain a high density of integrated circuits.

The use of multiple CPUs—The performance requirements of advanced set-top and IPTV applications have increased significantly in recent times. This has started to put pressure on single CPU based set-top boxes, and manufactures are now starting to examine the merits of including multiple CPUs in their hardware designs.

Seamless interoperability with cell phones—Next generation set-top boxes will support the movement of recorded digital video content from the internal hard disk directly onto a mobile phone. This level of interoperability between set-top boxes and mobile phones will help to enforce the whole concept of expanding the boundaries of traditional TV viewing by allowing consumers to watch TV anytime, anywhere.

So while the set-top world has historically been based on purely TV content, the next evolution of these devices promises to provide consumers with new and exciting ways of watching TV. Vast amounts of video and audio content will reside on set-top hard disks, while powerful processors and decoders will work at providing end users with a suite of rich digital TV and Internet based applications.

5.9 IP SET-TOP BOXES DEFINED

The set-top box represents a key piece of equipment in an end-to-end IPTV system. They are typically provided to subscribers by the supplier of the IPTV service and are charged a monthly rental fee. An IP set-top box connects directly to a television set and enables consumers to view IP based broadcast and video-on-demand programming content. The set-top box performs three primary tasks: receiving IP streams, decoding, and presenting a synchronized IPTV stream for display on the TV. The conversion and display of IP data and video requires a set-top box equipped with a sophisticated suite of hardware components. It is worth noting that a tuner is not required for an IP based set-top box. In contrast to RF based set-top boxes, an IP set-top box continuously communicates with IPTV servers at the headend. Installation is simple and involves plugging one cable into the TV set and another into the broadband router. Subscribers use a handheld remote control to gain access to the ever-increasing range of content and channels supported by IP set-top boxes. Some of the characteristics of IP set-top boxes include the following.

5.9.1 Presenting Synchronized Audio and Video Signals to a Display Screen

Similar to a cable or satellite set-top box, an IP based set-top box takes interactive, personalized digital audio and video content input from a broadband connection, decodes, and outputs to a television. They provide this functionality over existing and low cost Internet and IP based network infrastructures.

5.9.2 Support for Advanced WHMN Services

Entry level IP set-top boxes are used for decoding and rendering broadcast TV signals, while more advanced models are used as a home delivery platform for a wide range of WHMN services including, VoD, Web browsing, DVR functionality, home networking, and VoIP.

5.9.3 Processing Linear Multicast TV Signals

IP set-top boxes are capable of accessing multicast TV applications.

5.9.4 Handling Service Degradations

Lost video frames, network congestion, and dropped connections are all features of "IP broadband" access networks. IP set-top boxes are able compensate for some of these impairments and are able to smoothly support the degradation of an IPTV service.

5.9.5 Customizable User Interfaces

Many IP set-tops include software that allow service providers to add their own look and feel to the TV user interface.

5.10 TYPES OF IP SET-TOP BOXES

An overview of the various categories of IP set-top boxes are detailed below:

5.10.1 Multicast and Unicast IP Set-Tops

A block diagram representing the various hardware building blocks that make up a typical IP set-top box capable of supporting both multicast and unicast functionality is shown in Fig. 5.9.

The next sections will consider each of these modules in more detail.

5.10.1.1 Core Processor The underlying principles of all processors are the same. Fundamentally, they all take signals in the form of zeros and ones, manipulate them according to a set of instructions, and produce output in the form of zeros and ones. All of this happens in an electronic circuit called a transistor.

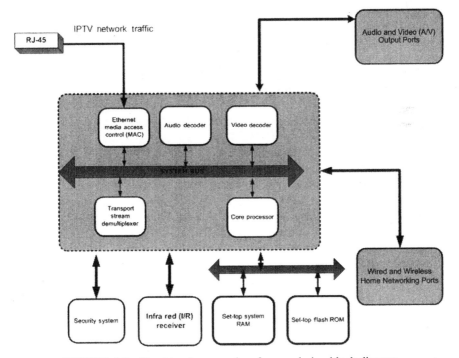

FIGURE 5.9 IP set-top box sample reference design block diagram

Today's IP set-top processors contain millions of transistors and are used to perform and execute instructions. The more electronic circuits a processor has, the faster it can process data. The actual speed of an IP set-top box processor is measured in megahertz (MHz), which means millions of cycles per second. The functions of an IP set-top box processor can include

- Initializing IP set-top box hardware components
- Reading IPTV packet streams
- Running the IP set-top real-time operating system
- Reading and writing data to memory
- Monitoring hardware components
- Executing the various services

Processors are available in different shapes, pin structures, architectures, and speeds.

Note that new advanced IPTV, HDTV, and IP-VoD applications can significantly increase the processing requirements of a single core IP set-top box processor. To solve this problem a number of IP manufacturers have turned to multicore processors. As the name suggests, a multicore processor integrates two or possibly more independent processors into a single silicon chip. Multicore processors allow for increases in the processing capabilities without increasing the heat dissipated by the chip.

5.10.1.2 Memory Just as a computer needs memory to function, an IP set-top box also requires memory to store and manipulate instructions issued by the subscriber or the operator. It acts, so to speak, as a staging post between the disk and the processor. The more data that is available in the memory, the faster the IP set-top box runs. Most elements within the set-top box will require memory to perform various tasks. Memory also allows the IP set-top box to simultaneously run multiple applications.

An IP set-top box uses two main categories of memory — RAM and ROM.

RAM Most functions performed by the IP set-top box will require access to random access memory (RAM). It is used as a temporary storage area for data flowing between the CPU and the various hardware components. RAM is available in several types. The most popular include

- *Video RAM* is used to process television signals. The amount of video RAM depends on the resolution of the TV signal.
- *Dynamic RAM (DRAM)* retains information stored in it only when the set-top box is powered up. The CPU uses DRAM to execute its software.
- A small amount of *nonvolatile RAM (NVRAM)* is required to store viewer preferences, PIN numbers and last viewed channels.

RAM is also used for processing graphics and running interactive IPTV applications.

ROM Read Only Memory (ROM) is nonvolatile, safeguarding in the absence of power the IP set-top boxes network settings, operating system, and many other items of important information. Once data has been written onto a ROM chip, the network operator or the IPTV end user cannot remove it. Most IP set-top boxes contain EEPROM and Flash ROM, which are variations of the basic ROM technology.

ABOUT EEPROM—ELECTRICALLY ERASABLE PROGRAMMABLE READ ONLY MEMORY   EEPROM is a special type of memory used in IP set-top boxes to store controls and boot up information. The data is permanently stored on the chip even when the subscriber powers off the IP set-top box. To remove this control information engineers need to expose the EEPROM chip to ultraviolet light and electrical charges. An IP set-top box will contain a small amount of EEPROM (usually Kbytes) and it has slower access rates than RAM. This memory type is commonly used to store the bootup software and any personalization parameters such as language types related to the IPTV end user.

ABOUT FLASH MEMORY   A flash memory chip is essentially a type of EEPROM. The main difference between the two is that certain flash memory chips can be erased and reprogrammed in blocks of data instead of one bit at a time. Flash memory components offer some very attractive features for storage of data. They are nonvolatile, so the data will be retained practically indefinitely without any power to the flash components. Flash memory consumes very little power and take up very little space. It uses solid-state technology and has no moving parts, so it can work in any living room conditions where mechanical disks might prove unsuitable in the long run. The flash memory market can be divided into two broad categories, based on its two dominant technologies, namely, NAND and NOR. Both technologies have unique features and are aimed at fulfilling different market needs. The "fast read" characteristics of a NOR based flash solution makes it suitable for storing small amounts of executable code. NAND based storage solutions, on the contrary, has been optimized for both code and data storage.

In addition to the silicon itself, a file system is required to interact with flash memory to provide the functionality of a mechanical hard drive on a solid-state silicon chip. Typically, a flash file management system is a software driver, which is used to make flash memory components emulate a disk drive. The flash memory is used to store software components such as the operating system (OS). So when the IP set-top box is powered up the OS is loaded into DRAM and executed.

5.10.1.3 A Transport Stream Demultiplexer   A highly simplified graphical view of the operations performed by a demultiplexer module that is built to process an MPEG-2 digital transport layer bit stream is depicted in Fig. 5.10.

The MPEG packets encapsulated in IP and UDP protocols arrive in the demultiplexer from the Ethernet port. Each packet comprises 184 bytes of video

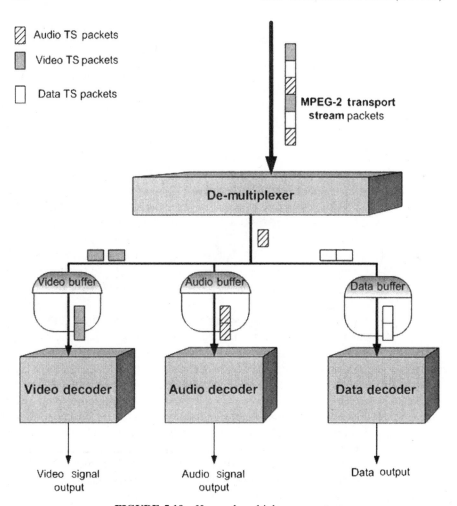

FIGURE 5.10   How a demultiplexer operates

payload and a 4 byte header. The most important field of the header is the 13 bits that define the PID. This is used to uniquely identify the content of the packet. Each IPTV channel will typically consist of at least three PID values:

(1) A PID for the video stream packets
(2) A PID for the audio stream packets
(3) A PID for additional text or data packets

The IP set-top box then consults a table of data, which provides pointers to the packets that are associated with the three PID types.

The transport demultiplexer uses this information to select and depacketize the audio, data and video packets of the program desired by the viewer. Once the demultiplexer has finished its work, the extracted packets are sent to the appropriate decoder chips. Note that Fig. 5.10 also includes a buffer for each of the media types. These buffers are required for storing bits used to build a complete picture frame and to ensure that the different media types are decoded in synchronization.

5.10.1.4 A Security System Another important functional module included with an IP set-top box is the security system. This subsystem provides service providers with control over what their IPTV subscribers watch and when. The primary purpose of this chip is to decrypt the incoming video and audio content.

5.10.1.5 The Video Decoding Module The video decoder is responsible for processing standard definition and high definition IPTV streams. It generally supports the decoding of the legacy MPEG-2 industry-standard compression algorithm in addition to the newer compression algorithms like MPEG-4 and VC-1. This function can be implemented either through a dedicated piece of hardware or else through programmable software. Executing the video decoding function in real time via software decompression algorithms requires a specialized chipset called a digital signal processor (DSP). Note that these chipsets are reprogrammable.

The use of DSPs inside IP set-top boxes has grown in recent years because of the flexibility that these chipsets offer to service providers. Consider an example of a service provider who uses MPEG-2 to compress IPTV content as it traverses the network. When the companies IPTV data center encoding equipment is migrated over to a next generation compression platform, the provider simply downloads the new algorithm to all IP set-top boxes in the field equipPed with DSP chipsets. For non-DSP based set-top boxes the devices will need to be completely replaced.

Alternatively, it is possible to integrate the video processing functionality directly into the main processor. In most instances a video signal is outputted to a digital-to-analog converter (DAC) Chip once processed by the decoder. The output from the DAC depends on the A/V interfaces supported by the IP set-top box.

5.10.1.6 An Audio Decoding Module Before entering a detailed description of the audio decoding module, it is important to understand how sound is recorded, stored, and subsequently encoded. Sound is pressure differences in air. When picked up by a microphone and fed through an amplifier, this is transformed into different voltage levels. The voltage is sampled a number of times per second. For humans to hear the sound the sampling rate needs to be between 20 and 20,000 times a second. According to the Nyquist's theorem the audio encoder needs to sample at least two times the highest sampling rate of the reproduction quality. Hence, CD audio quality is normally sampled at 44,100 times per second.

This sampling rate produces a 1.4 Mbps data rate that represents just 1 s of stereo music in CD quality. By using audio coding, it is possible to shrink down the original sound data from a CD by a factor of 12, without losing sound quality. So a 12 Mbyte sound file from a CD reduces to 1 Mbyte or so in compressed format. To achieve these types of results, audio encoders compress bit streams by eliminating those parts of the sound signal that are irrelevant to the human ear. Once parts of the signal that most people are unable to hear are removed the encoder applies standard lossless data compression techniques to the resultant signal. This technique does not work perfectly because the sensitivity of each person's hearing is different. But the sensitivity of human hearing does fall within a finite range, and researchers can determine a range that applies to the vast majority of people.

Once encoded and transported across the IPTV distribution network the audio encoding module receives packets of data from the front-end module, recovers the audio information to a level, and format suitable for consumption by the IPTV subscriber. This format is normally a stereo analog signal.

The method used by the set-top encoder to process digital audio signals depends on the compression format used by the service provider. There are a number of contenders for compressing digital audio content: MPEG, Dolby® Digital 5.1 surround sound, Dolby Digital Plus, and Microsoft® Windows Media™ Audio (WMA).

MPEG   The MPEG committee chose to recommend three compression methods and named them audio Layers I, II, and III. Each layer uses the same basic structure and includes the features of the layers below it. Higher layers offer progressively better sound quality at comparable bit rates and require increasingly complex encoding software. A brief description of each format is listed below:

- *MPEG Layer-I* is the lowest complexity and is specifically suitable for Digital Compact Cassette (DCC) applications. Using MPEG Layer I, one can achieve a typical data reduction of 4 : 1.

- *MPEG Layer-II* is directed toward "one to many" applications such as satellite broadcasting and writing streams to compact discs. Compared to Layer I, Layer II is able to apply the psychoacoustic threshold more efficiently. Using MPEG Layer-II, one can achieve a typical data reduction of 8 : 1.

- On the Internet, the most commonly used standard is the third "layer" of the MPEG-1 standard, often called "*MP3.*" It is the most powerful and complex member of the MPEG audio coding family and is widely used by IPTV applications. Using MPEG Layer III, one can achieve a typical data reduction of 12 : 1. Another feature of MP3 is its ability to carry five full audio channels and one low frequency enhancement.

- *MPEG AAC* is an audio encoding format defined in both the MPEG-2 and 4 standards. AAC is an advanced compression format that has a greater capacity to carry channels when compared to the MP3 format. In fact, AAC may be used to carry up to 10 times the number of channels carried by an MP3 stream. This is achieved because it compresses audio more efficiently when compared to older formats. High sampling frequencies are also supported by AAC, which helps to improve performance and quality of the audio signal. These various features allow IPTV service providers to effectively deliver surround sound services to end users.

Note that the various flavors of MPEG audio are compatible with each other. Layers I, II, and III are backward compatible. For instance, a Layer III decoder should also be able to decode a Layer I or II sound file, and a Layer II decoder should be able to decode a Layer I sound file. Note also that the bit rates for the various layers generally vary between 32 and 320 Kbps.

Dolby Digital 5.1 Surround Sound   The Dolby Digital 5.1 surround sound, also known as AC3, compression format is a surround sound technology that enables IP set-top boxes to provide consumers with the following high quality digital audio channels.

- Front center
- Center left
- Center right
- Rear left
- Rear right
- Low frequency effect (LFE)

The first five channels are typically reproduced in the home using speakers and the subwoofer is used to reproduce the LFE or bass channel. For IPTV subscribers that do not own a home surround system many IP set-tops also feature support for outputting two analog channels. The Dolby-AC3 is used extensively by standard and high definition digital TV networks in the United States and around the world. The interface supports data rates of approximately 448 Kbps.

Dolby Digital Plus   Dolby Digital Plus, also called E-AC-3 is an enhanced version of Dolby Digital and operates with next generation video compression formats. The technology is backward compatible with previous Dolby audio decoders, integrates tightly with various video decoding algorithms, while operating at lower bit rates. It delivers data as a single stream and supports 7.1 channels of audio. Specialized media interfaces are required to deliver that amount of channels to a home entertainment system. High Definition Multimedia Interface (HDMI) and IEEE 1394 are commonly used for this purpose. The DVB organization has recently added the Dolby Plus option into the organization's latest version of its IPTV specifications.

WMA and WMA Pro   Microsoft® Windows Media™ Audio (WMA) and WMA Pro codecs are popular audio coding systems, which form part of the Microsoft® Windows Media® series of technologies. They are designed to handle a variety of different types of audio and can operate in mono, stereo, and dual channel modes.

It should be noted that decoding an incoming audio signal is less demanding in terms of complexity when compared with video signal processing. Often, the CPU power required to decode the audio is minimal and is handled quite easily by most IP set-top box CPUs.

This is however changing with the advent of multiroom and multichannel audio systems. The shift upward in the requirements of audio complexity is having an effect on the types of audio systems used by IP set-top boxes to decode audio signals. Traditionally, the decoding of audio was handled by hardwired logic or a programmable DSP. Newer designs require a dedicated audio processor to run the audio decoding function.

Note, the audio encoder is also responsible for lip-syncing so that the actors and actresses voices match the movement of their lips in the video stream. This can be problematic in some implementations because the decoding of raw video content is more time consuming. As a result, audio decoder modules typically use an algorithm to delay the processing of audio packets in order to match the decoding rate of matching video packets.

5.10.1.7 A Graphics Accelerator   IP set-top boxes and PCs handle graphics in different ways. The limited hardware capabilities of IP set-top boxes reduce the sophistication levels of graphics that can run on a set-top box. The graphics functionality is typically built into the video decoder in most IP set-top box platforms. Thus, in addition to the video function the decoder also supports a graphic function, which is used to generate various user interface multimedia components.

Some manufacturers do however add an integrated graphics accelerator to their platforms to allow IPTV software developers create visually exciting applications for TV viewers and render a range of interactive TV file formats. The power of graphic accelerators continues to increase as TV providers attempt to differentiate themselves by offering exciting new applications such as 3D games to their subscriber base.

5.10.1.8 Interfaces and Connectors   IP based set-top boxes also have a number of interfaces that provide communication with external devices. These interfaces fall into six broad categories: input, analog video output, analog audio output, advanced digital output, home networking, and infrared interfaces.

Input Interfaces

  An Ethernet transceiver/controller—All IP digital set-top boxes are equipped with an Ethernet transceiver chip to provide interconnectivity with the broadband network. This generally supports networking speeds ranging from 10 to 100 Mbps.

  An infrared receiver—This is located at the front of the IP set-top box and accepts signals from a remote control.

RS-232 port—The RS-232 port may be used for diagnostic purposes.

Analog Video Output Interfaces An analog video signal contains three essential pieces of information:

(1) The *active video information* that includes details such as what images will be displayed, the different coloring schemes, and the levels of brightness associated with each part of the picture.
(2) The *Sync (synchronization)* information that identifies the location on screen of the active video information.
(3) The *blanking information* that is sent between video frames and is used to reposition the electron guns of a television display from the bottom of the previous frame to the top left hand side of the next frame.

Over the years a number of methods and associated interfaces have been designed to carry this information between electronic devices. In IP set-top box systems, the most common interfaces used to output analog video signals are

A *standard baseband RF output*—This interface provides an analog video output. The output generally uses an F-type connector, which allows an IPTV subscriber to use a standard coaxial cable to interconnect the IP set-top box with a TV set.

A *composite video output*—Radio Corporation of America (RCA) connectors are used to supply composite video signals to a television screen. In this format, the video signal information is transmitted serially across a single cable to the television display where it is assembled into a video picture. The NTSC and PAL television standards are both based on the composite video format.

A *super video (S-video) video output*—This third set-top analog output type provides superior signal quality when compared to composite video signals. The S-Video output port uses a physical 4-pin mini connector to send signals from the set-top box to external devices.

A *component video output*—Component video separates the signal into three separate colors:

- The "Y" video component represents the luminance or brightness of the video signal in terms of black and white.
- The "Pr" component represents the luminance or brightness of the video signal in terms of the red color.
- The "Pb" component represents the luminance or brightness of the video signal in terms of the blue color.

This separation of video color components provides for a very high quality output from the IP set-top box. This is the reason why DVDs are often encoded using this video format. At the IP set-top box output, three separate color coded cables connect to RCA style connectors and pass the information onwards to the TV display.

SCART Connectors Short for Syndicat des Constructeurs d'Appareile Radiorecepteurs et Televiseurs a (SCART) connector consists of 21 pin socket that is typically located on the rear of the IP set-top box. The main purpose of SCART connectors is to facilitate the transmission and receipt of three different types of video content, namely, composite, S-Video, and RGB.

Analog Audio Output Interfaces These interface types provide an analog audio output. RCA type connectors are generally used to allow the IP set-top box send audio to a peripheral device such as a CD player or a stereo tuner.

Advanced Digital Output Interfaces The more advanced IP set-top boxes will support interfaces such as

- *Digital audio*—This type of interface presents a digital audio signal to external music systems. The fact that the output remains in a digital state during transmission means that the end user can experience the full capabilities of a digital surround sound system. The connector used will depend on the file format of the compressed audio. The Sony/Philips Digital Interface (S/PDIF) protocol is an example of a format that allows the transmission of Dolby AC-3 compressed content from an IP set-top box directly into a multichannel amplifier or encoder.
- *IEEE 1394*— This interface provides a high performance digital output. It is starting to become a popular interface for interconnecting set-top boxes with other multimedia devices in the home and will be discussed further in Chapter 8.

Wired and Wireless Home Networking Ports IP set-top boxes typically include a number of different types of home networking interfaces. The number and type of interface technologies vary between IP set-top box implementations. Please refer to Chapter 8 for specific details on these different types of home networking interfaces.

Infrared (IR) Interface The infrared interface is normally connected directly with the media processor and allows two-way communication between the IP set-top box and either a remote control or a wireless keyboard. The communication protocol used over this link is called IrDa. An organization called the Infrared Data Association (IrDA) is responsible for defining the protocol. For the IP set-top box and a peripheral device to communicate via IrDA, they must have a direct line of sight. Once the commands from the remote control are received by this interface, they are sent to the processor for execution.

An xDSL Modem (Optional) The incorporation of a DSL modem into an IP set-top box removes the need to use separate devices in the home.

5.10.1.9 IP Set-Top Peripheral Devices A typical residential setting may possess many different pieces of peripheral devices that are able to interoperate with an IP set-top box. Typical IP set-top box peripheral equipment found in the home include a remote control, a wireless keyboard, and an IR blaster.

Remote Control A remote control is more than a device for channel zapping. It is the key component that enables IPTV subscribers to access a range of digital IPTV services.

The layout and design of remote controls varies between IPTV deployments. There are however a number of control keys, which are common to all remote controls (see Table 5.8).

The average number of buttons located on a remote control keypad is around 30, however, some remote controls are designed with a maximum of 48 different buttons.

The infrared interface that is embedded in the IP set-top box supports the reception of the American Standard Code for Information Interchange (ASCII) character set.

TABLE 5.8 Common Control Keys Used by IPTV Remote Controls

Key Description	IP Set-Top Action
On/Off	TV is instructed to go into power up or standby mode
Menu	Starts the installation menu for configuring the IP set-top box
EPG	Executes a navigational system that allows IPTV end users to select a specific multicast channel
Volume	Increases and decreases audio volume
Portal	Executes the IPTV portal application
Numeric block of keys	Used for entering PIN codes and selecting IPTV multicast channels
OK	Confirm the selection of IP VoD titles
Arrows—left, right, up, down	Facilitates channel surfing
Play, stop, pause, fast forward, and rewind	Used to issue commands while viewing an IP-VoD title
Subtitles	Used for switching subtitling on or off
Home	Allows *endusers* to navigate back to the start up screen
Help	Displays help information

Wireless Keyboards The IR interface is also capable of receiving information from a keyboard. IPTV service providers who are planning to offer interactive IPTV applications such as e-mail and chat will need to also supply their subscribers with an alphanumeric keyboard with an infrared transmitter. Wireless keyboards are ergonomically designed to fit comfortably on an IPTV subscribers lap while sitting on a couch and can operate at a distance of up to 30 ft away from the IP set-top box. Wireless keyboards are plug and play; in other words, the IPTV subscriber can begin using the keyboard without worrying about configuration settings and installing new device drivers. Some of the advanced wireless keyboards also come with an integrated mouse that allows a consumer to move a pointer around the television screen. Power to the keyboard is supplied from local batteries.

IR Blasters An IR blaster is a useful device for subscribers who would like to remotely control their television and VCR from a separate location within the home. It is basically a cable that plugs into one of the interfaces at the back of a set-top box that converts the signal transmitted along the wires into IR signals that remote controls can respond to.

5.10.1.10 IP Set-Top Front Panel The front panel unit consists of a numeric display made up of light emitting diodes (LEDs) that display selected channel, a power indicator, IR receiver notifications, or short messages.

5.10.1.11 IP Set-Top Physical Characteristics Multiple form factors can be used by IP set-top box manufacturers to house the functional modules described above. How all of these modules work together will depend on the needs of service providers and ultimately the consumer. It is also worth noting at this stage of the chapter that many of the functional building blocks described in Figure 5.9 are often absorbed into a single system-on-a-chip (SOC) module.

Advances in semiconductor technology over the years has enabled the development of smaller, more powerful, more flexible, less power hungry, and less expensive silicon. Such innovation has led to the emergence of system-on-a-chip technology. This technology integrates multiple functions such as microprocessing, logic, digital signal processing, and memory together on one chip, where, in the past, each function would be handled by a discrete chip (see Fig. 5.11).

Smaller sized chips that carry out all of the required functions of a device allow smaller product form factors, use less power, and this functionality integration bring costs down for manufacturers.

5.10.2 IP Set-Top DVRs

The IP set-top DVR category may be subdivided primarily into standalone and multiroom devices.

5.10.2.1 Stand-Alone IP Set-Top DVRs IP DVR set-top boxes serve as a media hub when connected to an in-home networking system and allow IPTV subscribers to

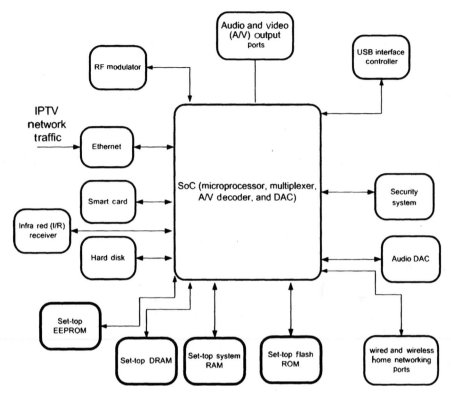

FIGURE 5.11 IP set-top box design using SoC technology

- Pause and rewind live programming
- Record shows and movies on the set-top hard drive
- Replay video content at any time

Most of the components at the core of an IP DVR set-top are almost identical to the parts used to build a standard IP multicast and unicast set-top box. The inclusion of a mechanical hard disk is the main differentiator of this type of IP set-top box. The main functions of the hard disk include

(1) Storing the IP set-top box's key components of the OS and software code.
(2) Storing system and user data (such as profiles, configuration details, the system registry, updateable system files).
(3) Storing IP-VoD content.

The following section provides a brief overview of mechanical hard disks when incorporated into an IP set-top box.

IP Set-Top Hard Disks   The basic technical architecture of a hard disk that is integrated with an IP DVR set-top comprises of magnetic coated circles of material that contain stored information. These circles are called platters and are stacked on top of each other to increase the storage capacity of the drive. Information is read from or stored to the hard disk using read and write heads.

The hard disk is a nonvolatile, bulk storage medium that acts as a repository for video content. The main benefit of this medium versus standard RAM systems is the sheer capacity of these devices and their ability to continue storing contents when the IP set-top box is switched off.

There are generally two types of internal bus interfaces used by drives integrated with IP set-top boxes—IDE and SATA.

Integrated drive electronics (IDE)—The IDE bus also known as ATA (Advanced Technology Attachment) connects the hard drive to the IP set-top motherboard. The IDE standard has been around for many years and has undergone several revisions over time. When integrated into a set-top box an IDE drive is capable of communicating in speeds in excess of a 100 Mbps.

Serial ATA (SATA)—ATA or IDE transfers data in parallel. SATA in contrast to the IDE standard uses a serial bus to communicate data. First generation SATA drives were released in 2001 and are capable of throughputs of 150 Mbps. Over the last couple of years, a number of IP DVR set-top manufactures have started to install drives that are based on an extension to the original SATA standard—SATA II. The newer SATA II drives offer higher rate interface speeds compared to their predecessors. The improved read/write times associated with SATA II help to improve the operational efficiencies of next generation IPTV applications.

While most DVR enabled IP set-top boxes use standard hard disks, some storage manufacturers have started to optimize their products for storing and streaming digital video streams. A drive optimized for DVR functionality inside an IP set-top box will generally include such features as

- *Advanced cooling capabilities*—Manufacturers can custom tune the cooling capabilities of IP set-top hard disks to improve overall power consumption of the unit.
- *Small form factors*—Hard disks with a form factor of only 3.5 in. are typically incorporated into IP set-top DVRs. The sizes of these drives will continue to decrease and 1.8 in. hard disk sizes are expected to become a standard feature in IP DVRs over the next couple of years.
- *Concurrent reading and writing of hard disk*—The simultaneous reading and writing of IPTV streams to disks allows consumers to time shift their viewing.
- *High capacities*—Video recording and IPTV streaming are very demanding of storage capacity. Therefore, IP set-top box hard disks come in a range of

sizes from 100 GB to 1 TB (TeraByte) capacity. Next generation technologies such as perpendicular magnetic recording (PMR), heat-assisted magnetic recording (HAMR), and holographic recording are expected to exponentially increase capacity of IP set-top box hard disks.

- *High RPMs (revolutions per minute)*—The RPM characteristic of a hard disk measures the rotational speed of the hard disk. Drives that are capable of operating at high RPMs (typically 7200 and above) allow for smooth recording of IPTV content and subsequent playback.

5.10.2.2 Multiroom IP Set-Top DVRs With these types of set-tops, IPTV subscribers can receive and watch streamed video content stored on their DVR set-top box in various rooms around the house. Each one of these rooms typically require an entry level IP set-top box to gain access to the video content. There are two approaches to implementing a multiroom system:

(1) A *centralized storage architecture* where only one IP set-top box that is connected to a home network includes a hard disk. Under this architecture the DVR functions are executed by the main or master IP DVR set-top box and the remaining non DVR capable digital set-top boxes assume the roles of access devices. The use of the centralized video storage architecture is a significant enhancement, however, IPTV subscribers cannot simultaneously access multiple programs from the DVR. In other words, the master IP DVR set-top box can only play one program at any particular instance in time.

(2) A *distributed storage architecture* where all IP set-top boxes connected to a home network include a hard disk.

5.10.3 Introduction to Hybrid IP Set-Top Boxes

A hybrid IP set-top box is defined as a device that can handle both the processing of IP broadband and digital terrestrial, satellite, or cable TV services. In other words, a single set-top box can be used to view linear TV content from one of the traditional providers side by side with on-demand video content from an IP based network. Figure 5.12 shows a block diagram of a hybrid IP set-top box sample design.

As shown hybrid IP set-top boxes incorporate a built-in tuner to provide access to traditional RF based terrestrial and satellite TV networks. The tuner receives the RF signal and sends to the demodulator, which converts the signal into baseband signals, which are then processed by the IP set-top box. Hybrid IP set-tops help in the gradual transition between traditional TV viewing and IP based video services.

The main benefit of hybrid IP set-top boxes come from their ability to access video content on an RF network, while reducing the need for large amounts of bandwidth capacity on the IP broadband network. Some manufacturers have even gone a step further by adding digital video recording capabilities to these hybrid set-top boxes.

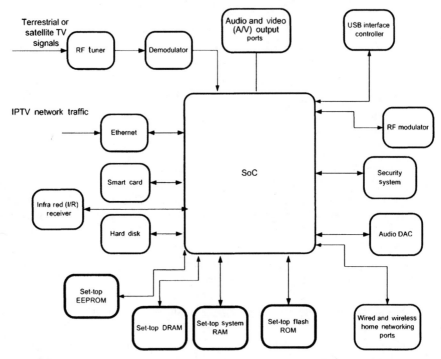

FIGURE 5.12   Hybrid IP set-top box sample reference design block diagram

5.10.4   Hybrid IP DTT Set-Tops

The front-end module, which is used to interface with the digital terrestrial TV (DTT) network, is the main differentiator of hybrid IP DTT set-top boxes. The two major components of a front-end module are the silicon tuner and the demodulator. The terrestrial signal is received via an F-type connector, which generally has an input impedance of the standard 75 ohms. The signal then gets passed onward to the tuner. The tuner will support at least one of three international industry standards for broadcasting digital terrestrial TV, namely, DVB-T, ATSC-T, and ISDB-T.

5.10.4.1   DVB-T   The incorporation of a digital video broadcast—Terrestrial (DVB-T) tuner into a digital set-top box enables the reception of over-the-air DTT signals. DVB-T is an international broadcasting standard, which was approved in 1995 and has since been taken over by the European Telecommunication Standard Institute and has become an official ETSI standard. As with the other DVB standards, MPEG-2 audio and video coding forms the payload of DVB-T.

5.10.4.2   ATSC-T   ATSC-T is also an international standard. It is used by North American Terrestrial Digital TV systems and is being promoted throughout the world by the ATSC organization. A tuner developed for the ATSC-T standard is used to recover the original video and audio information.

5.10.4.3 ISDB-T ISDB-T is an acronym for Integrated Services Digital Broadcasting-Terrestrial. It is a standard developed by a Japanese organization called ARIB and has a number of similarities with the DVB-T industry standard.

As with most if not all end-to-end digital TV solutions the ISDB-T uses the MPEG-2 compression standard to carry a variety of services over a network. Once the video content has been compressed, the ISDB-T standard supports the ability to split the channel bandwidth into 13 different and unique segments. This is a unique feature to ISDB-T and provides a broadcaster with the flexibility to mix and match the type of services included in a single channel.

5.10.4.4 Demodulators for Hybrid IP DTT Set-Tops The demodulator combines with the tuner chipset to provide a front-end subsystem that is capable of operating in demanding terrestrial environments. Demodulators found in modern day IP set-top boxes will typically support one of the following modulation formats.

VSB ATSC broadcasts use the VSB modulation format. Vestigial sideband (VSB) is primarily used in the United States for transmitting video signals over a terrestrial network. There are a couple of VSB variants available, namely, 8-VSB and 16-VSB.

- Through the use of eight amplitude levels 8-VSB is able to support data rates up to 19 Mbps in one single 6 MHz channel.
- 16-VSB is a newer version of the modulation scheme, which carries data at 16 different amplitude levels and is able to fit over 38 Mbps of digital data into the one channel.

VSB uses a single intermediate frequency (IF) carrier to transmit digital bits over the airwaves to the terrestrial set-top box. This method of transmission can make VSB prone to the multipath effects arising from signal obstructions while traversing the terrestrial network. One of the impacts of multipath is the late arrival of signals to the set-top box and is catered for in the design of the tuner and the demodulator.

COFDM Coded orthogonal Frequency Division Multiplexing (COFDM) is a modulation technique that is widely used by transmission networks in many European countries. It operates extremely well in conditions likely to be found in heavily built up areas where digital transmissions become distorted by line of sight obstacles such as buildings, bridges, and hills. In contrast to VSB, COFDM uses multiple carriers to send data from the IPTV data center to the terrestrial set-top box.

There are two formats which can be used—2K and 8K. 2K uses 1705 carriers for data whereas the 8K format uses 6817 carriers. The 2K modulation format is suitable for single transmitter operation and for small Single Frequency Networks with limited transmitter distances. The 8K modulation format on the other hand is typically used both for single transmitter operation and for small and large single frequency networks.

COFDM is resistant to multipath effects because it uses a large number of carriers each modulated at a relatively low symbol rate to spread the data to be transmitted evenly across the channel bandwidth.

5.10.5 Hybrid IP Satellite Set-Tops

Hybrid IP satellite set-top boxes are designed to assist satellite operators in complementing their existing RF based pay TV services with a new suite of SD and HD IPTV services.

The front end, which is used to interface with the network, is the main differentiator of hybrid IP satellite set-top boxes. The main function of the front-end module is to enable two-way communication between the set-top box and peripherals such as switches and Low Noise Block Converters (LNB) connected to the satellite dish. The two major components of a front-end module for a hybrid digital satellite box are the silicon tuner, and the demodulator.

5.10.5.1 Satellite Set-Top Tuners Satellite set-top tuners receive content from geostationary satellites, isolate a physical channel from a multiplex of channels and convert to baseband. Support is provided for by one of the following industry standard protocols.

DVB-S In Europe, digital satellite channels are broadcast using the agreed DVB-S standard. This is the satellite variant of the DVB group of standards.

The DVB-S protocol is a well established ETSI standard. It was originally adopted and ratified in the early 1990s and has been the engine behind the delivery of satellite TV services over the past decade. Its technical components include the MPEG-2 compression standard for transporting audio and video signals. The only adjustment or extension to the original standard came toward the end of the 1990s with the release of a standard called DVB-DSNG. The DSNG stands for Digital Satellite News Gathering, which enabled the transmission of MPEG-content over point-to-point or point-to-multipoint communication links. There are several million digital set-top boxes deployed in households worldwide that include tuners, which support the DVB-S protocol.

DVB-S2 DVB-S2 is the latest technique for transmission over a satellite network from the DVB. It is an open standard, which was developed by the DVB technical groups in 2004 and boosts performance advantages over the preceding DVB-S standard. The key applications that run over a DVB-S2 network include

- Standard broadcast video services.
- VoD services, including movies on demand and subscription VoD.

• Interactive services, including high speed broadband Internet access, games, and interactive advertising.

• High definition video content services.

From a technical perspective, the standard has largely benefited from the latest advancements in the techniques used for video content modulation and coding. DVB-S was only able to accommodate the transmission of MPEG-2 transport streams, which limits provider's ability to deploy sophisticated IP based applications. Therefore, the DVB designed the upgraded S2 format to support emerging compression formats such as MPEG-4 and Windows Media Format.

DVB-S2 equipment is capable of carrying up to 30% additional throughput of a DVB-S network. The deployment of a DVB-S2 system requires the installation of new modulators at the headend.

Accessing a DVB-S2 network requires manufacturers to incorporate a specific tuner into the hybrid IP set-top box. The design of this tuner needs to maintain backward compatibility with DVB-S networks to ensure that they operate with the existing installed base of legacy DVB-S set-tops during the migration period. Set-top box designs that incorporate dual standard capabilities provide consumers with an upgrade path to more sophisticated applications in the future.

ISDB-S   The DVB-S standard was used to broadcast satellite TV signals in Japan in the mid-1990s. The standard did not however satisfy the requirements of operators in the country who wanted to start deploying advanced digital TV technologies. Therefore, the industry standards group ARIB developed a new standard titled ISDB-S. Tuners that support ISDB-S are capable of receiving a bit stream of roughly 51 Mbit/s. This is nearly one and a half times the rate of a tuner that uses the DVB-S and is the main benefit of set-top boxes that support this standard.

DSS   DSS, also known as Digital Satellite System, is a proprietary TV transmission mechanism utilized by a U.S. based company called DirectTV.

A summary of the features of the four tuner types is shown in Table 5.9.

TABLE 5.9   Features of Hybrid IP Satellite Set-Top Boxes Tuners

	DVB-S	DVB-S2	ISDB-S	DSS
Coding of video content	MPEG-2 Video	MPEG-2, MPEG-4, and Windows WMV	MPEG-2 Video	Proprietary and MPEG-2 standards
Coding of audio content	MPEG-2 audio (AAC)	MPEG-2 audio (AAC)	MPEG-2 audio (AAC)	Musicam and AC3
Transmission bit rate (Mbps)	34	80	51	38

5.10.5.2 Demodulators for Digital Satellite Set-Top Boxes   The demodulator semiconductor is designed to integrate seamlessly with the tuner. It takes the output from the tuner and feeds into an analog-to-digital (A/D) converter, which is typically part of the demodulation chipset. The output from the A/D converter then gets demodulated and the MPEG-2 Transport Stream is recreated. Some demodulators found in modern day hybrid IP satellite set-top boxes will typically support one of the following important modulation formats:

- Binary Phase Shift Keying (BPSK)
- Quadrature-Phase-Shift-Keying (QPSK)
- 8-PSK

The following sections discuss these schemes in greater detail.

BPSK   This is a relatively simple mechanism of modulating the uniqueness of a data signal onto a satellite network. The modulated signal is created by combining a single bit transmission stream that consists of ones and zeros with a carrier.

QPSK   QPSK is a standard modulation scheme, which is in use for many years. It provides a means to deliver signals from the satellite TV provider to the satellite set-top box. The information encoding onto the network is achieved by phase shifting a waveform.

In addition to employing phase shifting, QPSK also transmits a second waveform. The amplitude of both waves are also modulated and are offset by 90% to each other. The net effect of transmitting data in this manner is the ability to transmit two bits of information for every hertz. This effectively doubles the bandwidth capacity of the network and makes efficient use of available radio frequency spectrum on the network.

8-PSK   This modulation scheme enables a carrier to transmit three bits of information instead of two, which is the case for QPSK. Such an approach further increases the rate at which data can be transmitted over a digital satellite network. This is a modulation technique in which the carrier can exist in one of eight different states. As such, each state can represent three bits—000 to 111. The output from the front-end module is then delivered into the core of the IP set-top box for further processing. The combination of the various chipsets described above results in a set-top box that can receive a wide range of digital TV services.

5.10.6 Hybrid IP Cable Set-Tops

Traditional cable set-top boxes are compatible with existing HFC networks and provide consumers with access to multichannel services. A hybrid IP cable set-top box goes a step further and is designed to receive standard RF based cable TV channels alongside digital IPTV services. A specialized front-end module is used to access broadcast TV channels available on the HFC network, whereas the embedded cable modem provides pay TV customers with access to IP-VoD and multicast IPTV services. These DOCSIS based set-top gateway products enable multiservice operators (MSO) to deliver video, data and voice streams, as well as other STB control features, over a single DOCSIS network.

5.10.6.1   Front-End Module   The main function of the front-end module is to enable two-way communication between the cable set-top box and the operator's headend equipment. The two major components of a front-end module are the silicon tuner and the demodulator.

Tuner   The tuners, which are built into hybrid IP cable set-top boxes, are capable of interoperating with different types of network standards. The standard varies from region to region. Table 5.10 shows the geographical spread of digital cable tuners across the globe.

The following paragraphs provide more detail on these standards.

- DVB-C is a European specification used for data transmission over cable TV networks. DVB-C was developed by DVB in the mid-1990s. The DVB-C has helped to focus European cable operators on a single set of standards to make cable modems and set-top boxes interoperable. The DVB-C specifications define both upstream (customer to provider) and downstream (provider to customer) communications.
- ISDB-C is an acronym for Integrated Services Digital Broadcasting—Cable. It is a standard developed by ARIB and Japan Cable Television Engineering Association (JCTEA). It uses MPEG-2 for compression and utilizes the QAM scheme for transferring digital data across a cable TV network.
- OpenCable is a digital TV standard created by an organization called CableLabs. OpenCable has two key components: a hardware specification and a software specification. The CableLabs hardware specification outlines the various flavors of silicon chips that are used to handle and process digital video and audio services. The software specification is OpenCable Applications Platform (OCAP) and provides a common software environment upon that interactive services may be deployed. The OpenCable

TABLE 5.10   Global Distribution of Digital Set-Top Box Tuners

Type of Tuner	Geographical Deployment
DVB-C	Europe, South America and other emerging markets around the world
ISDB-C	Japan
OpenCable	North America

networking infrastructure has the same core properties as DVB-C and ISDB-C.

A Demodulator   The demodulators found in a hybrid cable IP set-top box will support the QAM modulation scheme. QAM is a standard modulation scheme, which is in use for many years and provides a means to deliver signals from a cable operator's headend to hybrid IP cable set-top boxes in the field. QAM is a relatively simple technique for carrying digital information over a cable TV network at very high speeds. It uses two sinusoidal carriers, one exactly 90° out of phase with respect to the other to transmit data over a given physical channel.

It is important to note that some vendors provide a front end, which combines the QAM tuner and demodulator into a single silicon chip.

5.10.6.2   An Embedded Cable Modem   The inclusion of an embedded DOCSIS or EuroDOCSIS cable modem provides hybrid IP cable set-top boxes with the ability to process high speed IPTV broadband services. A cable modem is best defined as a client device that provides two-way communications (data, voice, and video) over the ordinary cable TV network cables. From a technical perspective, a cable modem is a digital modem that uses a coaxial cable connection. The cable modem communicates over the cable network via an IP connection to a device located in the headend called a CMTS over an OOB cable TV channel. The CMTS takes requests coming in from a group of hybrid IP cable set-top boxes and routes onward to the IPTV data center and vice versa. The requested IP video is either sent back over the OOB channel directly to the cable modem or distributed via the RF network and accessed using the front-end module of the cable set-top box.

The speed of the cable modem depends on traffic levels and the overall network architecture. This is a major drawback when using cable modems to access IPTV channels because the quality of the received signal may degrade as the number of people connected to the CMTS increases. The shared nature of cable TV broadband services is an issue for cable TV operators who are planning to offer IPTV services. This issue is addressed in version 3.0 of the DOCSIS standard. With regard to frequency ranges, cable TV operators define a portion of the frequency spectrum to carry the data. This frequency range can vary from network to network and from country to country.

5.10.7   IP Set-Top Boxes Feature Comparison

Table 5.11 compares some of the Key Features and characteristices of the various IP set-top box types used by IPTV Service providers across the glope. Clearly, IP Set-top boxes represent an exciting area of technology. For more information on the dynamics of this sector please visit www.tvmentors.com.

TABLE 5.11   Feature Comparison Between Different Types of IP Set-Top Boxes

IP Set-Top Box Type	Services Supported	Tuner Type	Transmission Standards Supported
Multicast and unicast IP set-top boxes	IP video multicast, IP-VoD and home networking connectivity	No RF tuner	IP suite of protocols
IP set-top DVRs	IP video multicast, IP-VoD, home networking connectivity, and local storage	No RF tuner	IP suite of protocols
Hybrid IP terrestrial set-top boxes	Terrestrial TV, IP-VoD, and home networking connectivity	COFDM and VSB	DVB-T, ATSC-T, ISDB-T, and IP
Hybrid IP satellite set-top boxes	Satellite TV, IP-VoD, and home networking connectivity	QPSK, BPSK and 8PSK	DVB-S, DVB-S2, ISDB-S, DSS, and IP
Hybrid IP cable set-top boxes	Cable TV, IP-VoD, and home networking connectivity	QAM	DVB-C, ISDB-C, OpenCable, and IP

5.11   OTHER EMERGING IPTV CONSUMER DEVICES

5.11.1   Gaming Consoles

A typical gaming console is made up of several key blocks that primarily feature proprietary hardware designs and software operating environments. There are however a number of resemblances between game consoles and IP set-top boxes. In fact, the components for game consoles and IP set-top boxes are based on a number of silicon components, which are common to both platforms. Common silicon chipsets range from decoders and video processors to connectors and security modules. There is however a number of unique differentiators.

5.11.1.1 Improved Graphic and Audio Capabilities   The design of a gaming console requires a step up in the quality of graphics. One of the main differences between a games console and a traditional IP set-top box lies in the inclusion of a subsystem for processing graphics. This subsystem is either built into the main system processor or included as part of a separate custom designed graphics processor. Its main function is to render very high quality gaming images and process three-dimensional (3D) components. In addition to supporting high quality graphics, games console platforms are also built to produce sophisticated audio outputs. This often requires the use of a dedicated audio processor to achieve the desired sound affects associated with some of the modern games.

5.11.1.2 Improved Storage Capabilities Advanced gaming consoles come equipped with a DVD unit for playing games. Additionally, most of the next generation gaming consoles include an in-built multigigabyte hard disk. This type of design speeds up play because the console can store games temporarily on its fast hard disk, instead of reading from a slow DVD.

5.11.1.3 Extensive I/O Capabilities Advanced set-top game console units are built with ports that support high speed connectivity to in-home networks.

5.11.1.4 High Powered Processor The processors used in game consoles will sometimes contain the same level of power as today's low end computers.

5.11.1.5 An I/O Port for a Games Controller Controller ports are sometimes added to support multiplayer gaming.

In addition to the above features, many of the major console manufacturers continue to extend the functionality of their products to include functionality that is typically associated with IP set-top boxes, namely, TV tuners and access to IPTV content.

5.11.2 Home Media Servers

Media servers are another type of IPTVCD that provides high speed access to IPTV services. Table 5.12 summarizes the functions of the core components used to build a home media server.

In addition to providing access to IPTV services, home media servers also permit digital content to be used throughout the home. This is achieved through the use of a number of sophisticated home networking interfaces or media center extenders.

SUMMARY

Residential Gateways are characterized by their ability to interconnect various types of digital consumer devices connected to an in-home distribution network with an IP broadband access network. In addition to connectivity RGs also support advanced functionality such as firewall protection, support for access to a wide range of IP services including IPTV.

Making residential gateways a reality requires the cooperation of a wide variety of companies, along with the development of new standards. Five groups are presently working on standards, which they see as enabling increased adoption and

TABLE 5.12   Inside a Home Media Server

Building Block	Functionality
Processor	High performance dual core processors are typically used to meet the demands of processing, storing, and managing WHMN applications and content.
Motherboard	Media server motherboards come with a variety of features ranging from in-built Gigabit networking to support for multiple hard disks.
Memory	Memory is a critical element of the media server and helps to ensure that the server operates at an optimum level.
Optical DVD drive	Media servers will typically include a next generation DVD drive, which is capable of recording both SD and HD IPTV content.
Hard disk(s)	Storage is a major unique selling point of media servers. As a result most media servers include one and in some cases two high capacity hard disks. The content stored on the media server hard disk(s) can be accessed by any other IPTVCD connected to the home network.
Graphics card	Some media centers build the graphics card functionality directly into the motherboard. More media servers used a dedicated graphics card, which reduces the burden on the processor.
Sound card	Some of the more advanced media servers on the market incorporate cards that produce surround sound.
TV Tuner card	The TV tuner enables media servers to access traditional RF digital TV signals in addition to IPTV multicast and on-demand channels. The TV tuner card includes a decoder and accepts input from a range of different connector types.
Casing	The casing needs to look esthetically pleasing because media servers are typically located in the living room.
Operating system	The media server OS provides home entertainment functionality, such as allowing end users to watch multicast IPTV programs, listen to music, and download IP-VoD titles from a remote server.
Keyboard and remote control	The keyboard and remote control are used to wirelessly issue commands via Bluetooth or InfraRed signals.

TABLE 5.13   Overview of IP Set-top Box Core Components

Building Block	Functionality
Processor	The processor provides the computing horsepower to run the various functions in the set-top box. It is needed to run demanding IP based software applications such as streaming IP video and online gaming. In addition to processing applications, the processor also needs to have enough power to support intense network activity. Processors with limited capacity can cause issues with the video stream including jitter and dropped frames.
Memory (RAM and flash ROM)	The amount of memory included in the IP set-top box is vendor dependent and can vary from 16 MB all the way up to over a 256 MB of capacity.
A video decoder	This hardware component is used to simultaneously decode multiple incoming IPTV content streams. The streams are normally compressed and error protected. This decoder will typically provide support for decoding high- and standard-definition video formats. An algorithm, which matches the algorithm used by the encoders at the IPTV data center is used by the decoder to decompress the IPTV packets.
An audio decoder	This element of the IP set-top box is responsible for decoding incoming audio signals.
A transport stream demultiplexer	The demultiplexer selects and depacketizes the audio and video packets of the IPTV channel desired by the viewer.
An Ethernet transceiver/controller	All IP digital set-top boxes are equipped with an Ethernet transceiver chip to provide interconnectivity with the broadband network. This generally supports networking speeds ranging from 10 to 100 Mbps.
A secure processor	Security is typically provided through a secure processor chipset and a DRM engine. The purpose of this component is to decrypt multicast and IP-VoD content.
Wired and wireless home networking ports	Home networking functionality is provided through the provision of high speed wired and wireless I/O ports.

(Continued)

TABLE 5.13   (*Continued*)

Building Block	Functionality
Audio and video (A/V) output ports	IP set-top boxes can output audio and video in a range of different formats.
Infrared receiver	This allows a consumer to use a remote control or a wireless keyboard to remotely control the functions on the set-top box.
A graphics accelerator	The graphics controller module has the ability to overlay and process graphics at very high resolutions. Advanced features of a graphics accelerator can include the ability to process high performance 3D content.
A hard disk	IP set-top box designs are capable of accommodating hard disk drives to support integrated personal video recorder (PVR) functionality. This allows IPTV subscribers to record, playback, fast-forward, and rewind live and scheduled programming. The amount of hours of TV recording permitted by the IP set-top box will depend on hard disk size.
An RF demodulator	This component is included in IP set-tops to allow service providers to offer access to free-to-air, satellites, and cable based digital broadcasts.

development of this technology—ISO/IEC, the DSL Forum, HGi, CableLabs, and the OSGi.

The IP set-top box is the device of choice for most telecom operators who want to add IPTV services to their product portfolios. An IP set-top box enables TVs and display panels to view IP based video content. Some of the more sophisticated IP set-top boxes include features such as time shifting of TV viewing, the ability to process high definition video content, and support for advanced home networking services.

Table 5.13 summarizes the functions of the core components used to build an IP set-top box.

Most of these modules can be integrated into a single chip. The advent of system-on-chip based IP set-top boxes makes it easier for service providers to deploy IPTV services. Note that the overall physical look and feel of a set-top box is vendor specific. The IP set-top box product range is segmented into a number of categories: multicast and unicast, DVRs, Hybrid IP Terrestrial, Hybrid IP Satellite and Hybrid IP Cable. Hybrid set-tops are a new category of device that can receive and decode standard terrestrial, satellite, or cable video content as well as enabling access to interactive IPTV services over a broadband network. Other types of

IPTVCDs include gaming consoles and home media servers. Game consoles are used by consumers for many forms of digital entertainment, including browsing the Internet and accessing IPTV services. Media servers provide a unique hardware and software platform that is ideal for accessing IPTV services.

6

IPTVCD SOFTWARE ARCHITECTURE

There are a wide variety of software components that run on a typical IPTVCD. These software components may be categorized into three primary categories, namely, drivers, an embedded real-time operating system (RTOS), and middleware.

The main function of integrated software drivers is to provide access to the hardware platform. The RTOS provides a variety of different functionalities ranging from parsing incoming streams to performing general configuration tasks. Another key element of the software architecture used by IPTVCDs is the middleware—a piece of software that acts as a communication bridge between the real-time OS and the interactive IPTV application. This chapter provides an introduction to the basics of IPTVCD software systems. Additionally, it discusses some of the industry initiatives and standard bodies that are working on defining middleware standards for the emerging IPTV marketplace.

6.1 WHAT MAKES AN IPTVCD TICK?

The architecture of an IPTVCD interactive TV system is built in layers, with each layer adding new capabilities. A high level overview of the architecture of a typical IPTVCD software system is shown in Fig. 6.1 and consists of three separate software layers:

Next Generation IPTV Services and Technologies, By Gerard O'Driscoll
Copyright © 2008 John Wiley & Sons, Inc.

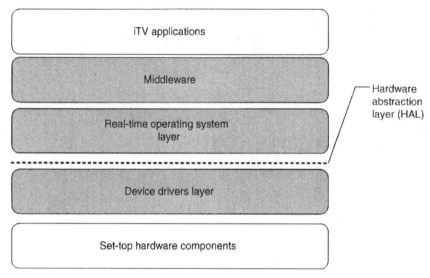

FIGURE 6.1   Generic IPTVCD software architecture

- Device drivers
- A real-time operating system
- A suite of middleware modules

The following sections give a brief overview of these layers when used in an IPTV networking environment.

6.1.1   Device Drivers

An IPTVCD RTOS requires pieces of software called drivers to control access and interact with the underlying hardware platform. Every hardware component in an IPTVCD whether it be an Ethernet port, a hard disk, or a decoding chipset must have a device driver. A driver is a program that translates commands from the real-time OS to a format that is recognizable by the hardware device. This negates the need for the real-time OS to interact directly with the underlying hardware devices. Every device contained within the IPTVCD has its own specialized set of commands, which is incorporated into the drivers program. Assembly language is typically used as the programming language for writing drivers. Therefore, if an IPTVCD manufacturer needs to add extra components to a particular design, then a new driver is written to allow the real-time OS control the new hardware. When the IPTVCD is powered up the RTOS installs and configures device drivers automatically.

6.1.2   An RTOS

The RTOS is the most important piece of software running on an IPTVCD. In most cases, the IPTVCD does not have the processing power of a multimedia computer and requires an operating system that uses a small hardware footprint, is fast, and is able to function in a real-time environment. An operating system designed for an IPTVCD includes a granular set of software components, which are used to perform a number of functions. At the heart of any IPTVCD OS is the kernel component, which is stored in ROM and occupies a very small memory footprint.

Typically, the kernel is responsible for managing memory resources, allocating tasks for execution by the processor in a specific order, and handling interrupts. An IPTVCD OS also supports multithreading and multitasking. These features allow an IPTVCD to execute different sections of a program and different programs simultaneously. Once the IPTVCD is powered up the kernel will be loaded first and remains in memory until the IPTVCD is powered down again.

The kernel is complemented by a number of additional components including a dynamic loader, a memory manager, an event manager, APIs, and a hardware abstraction layer (HAL). Table 6.1 explains these components in greater detail.

It is important to be aware that the components included in an RTOS can vary from product to product. The OS for an IPTVCD is often invisible to end users and as a result TV viewers are unable to interface directly with the IPTVCD OS.

TABLE 6.1 IPTVCD OS Components

IPTVCD OS Component	Description
Hardware abstraction layer	The HAL provides an interface between the IPTVCD hardware devices and system software. It shields the upper application layers from the physical hardware devices.
An event manager	As the name implies the event manager handles events that happen during the normal day-to-day operation of the IPTVCD. For instance, connecting the IPTVCD to a home network is considered to be an event and is processed by this IPTVCD OS component.
Loader	In addition to the kernel, a IPTVCD OS needs a loader to locate the OS executable files and load them into RAM when the IPTVCD is powered up.
Memory manager	This component ensures that programs run smoothly through the management of limited quantities of IPTVCD memory and ensures that applications do not run out of memory.
Advanced program interfaces	An IPTVCD OS needs to incorporate a set of APIs that allow software developers to write high level applications for the specific OS. An API is basically a set of building blocks used by programmers to write programs that are specific to a particular type of RTOS environment.

6.1.3   IPTVCD Middleware

Central to the IPTVCD software architecture is a connection layer that acts as a communications bridge between the IPTVCD OS and the interactive IPTV applications called "middleware." Middleware represents the logical abstraction of the middle and upper layers of the communication software stack used in IPTVCD software and communication systems. When compared to an IT environment, it equates to the presentation and session layers of the OSI (open systems interconnect) seven layer model. Middleware is used to isolate interactive IPTV application programs from the details of the underlying hardware and network components. Thus, these applications can operate transparently across a broadband network without having to be concerned about underlying network protocols. This considerably reduces the complexity of development because applications can be written to take advantage of a common platform. Consider the situation where a software developer is contracted by an IPTV service provider to write an application that uses a server's storage resources at the IPTV data center. To effectively deploy this distributed application, the developer will need to ensure that the interactive IPTV ap-plication interacts with the IP communication stack. Under the middleware approach, the developer is able to avail of the IP networking features without requiring a significant amount of knowledge on the architecture of the underly-ing protocols.

6.2   INTERACTIVE TV MIDDLEWARE STANDARDS

Adopting industry standards leads to competitive pricing from multiple suppliers and helps to reduce overall IPTV deployment costs. There is no common middleware standard defined for the overall global IPTV industry as of this writing. There are however a number of existing interactive TV middleware standards that are being adapted to provide IPTV functionality:

- MHP
- GEM
- OCAP
- ACAP
- ARIB B23

A description of these open middleware standards along with a brief overview of their related IPTV variants is provided in the following sections.

6.2.1   MHP

Before discussing the adaptation of Multimedia Home Platform (MHP) to sup-port IPTV it is first helpful to understand how MHP operates. The following sections give a brief overview of MHP when used in an end-to-end DVB digital TV system.

6.2.1.1 Introduction to MHP   Multimedia home platform is an open software framework that allows service providers to develop, deliver, and execute interactive and IP based TV applications. The standard is platform independent. Therefore, content developers can author applications or content once and be sure that their products will port across digital set-top boxes that use different underlying hardware resources and real-time operating systems. This is in stark contrast to the situation at the moment, which necessitates the rewriting of interactive TV (iTV) applications for a number of different proprietary systems. Usually the MHP middleware layer of software runs on top of an operating system. This enables easy portability of the middleware to different hardware architectures.

The main purpose of the MHP middleware layer is to manage the following functions:

- Scheduling CPU tasks and processes
- Device drivers
- Memory management
- Access to set-top box hardware resources
- Interaction between a remote control and the set-top box
- Presentation of iTV applications to consumers

MHP supports many kinds of applications, examples include

- EPGs
- Applications synchronized to TV content
- e-Commerce and secure transactions
- TV chat and e-mail
- Interactive commercials

6.2.1.2 Evolution of MHP   The European based digital TV consortium, the DVB group developed the MHP interactive TV specification. The key historical milestones of how MHP evolved since this group originally met are set forth in Fig. 6.2.

The development process commenced back in the mid-1990s when the MHP working group was born from a European project called UNITEL. The group comprised of manufacturers, broadcasters, and television operators whose aim was to raise awareness of the benefits associated with developing a common platform for accessing a wide range of digital multimedia services. In 1997, these activities were transferred to the DVB project that began defining a home

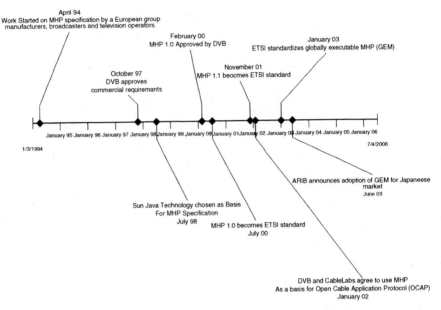

FIGURE 6.2   Evolution of the MHP specification

platform that would be the building block for interactive multimedia applications of the future. To aid this development, two working groups were set up by DVB. These were

- A *commercially oriented group called DVB-MHP*—The main responsibility of this group was to define user market requirements for enhanced TV, interactive TV, and Internet access in a dual broadcasting/Internet environment.
- A *technical group called DVB-TAM (technical aspects associated with MHP)*— This group has primarily been concentrating on defining technical specifications.

The DVB-MHP goal was simple: to develop a single, open, interactive TV development platform. This platform was to define a standard for a generic interface between interactive digital applications and the terminals on which they execute. Such terminals include integrated digital televisions, low and high end digital set-top boxes and multimedia PCs. While MHP is primarily targeted at these devices it also ties into home networking technologies. This standard became a reality when the first DVB-MHP specification was launched in 2000. Following further development requests from the industry specification version 1.1 was released early in 2002. The work did not finish for the DVB group and the organization proceeded to release a variant of the core standard in 2003 called Globally Executable MHP (GEM). While the MHP application environment is well tested and stable at this stage, the DVB will continue to evolve and develop the standard in line with market demand.

6.2.1.3 Technical Overview   The MHP middleware specification is a combination of different parts of many standardized technologies most of which are derivatives of Sun Microsystems Java technology. The DVB MHP group adopted the various standardized components, along with their own broadcasting standards and compiled a comprehensive specification. These various components can be classified into a number of major sections, namely, DVB-J, communication protocols, security, managing and locating MHP applications, and DVB-HTML.

DVB-J Technology Overview   The DVB-TAM group considered and evaluated a number of candidates for their MHP platform. Each middleware solution was examined for levels of openness, evolutionary capabilities, scalability, and flexibility. After a lot of debate and negotiations, the group decided to adopt Java technologies as the most suitable platform for receiving and presenting multimedia applications in a digital TV environment. The term DVB-J was applied to the chosen platform and it consists of two parts, namely, a Java Virtual Machine (JVM) and a number of APIs.

JVM   A virtual machine, also called an "engine" can be defined as a self-contained environment that behaves as if it is a separate set-top box. The virtual machine layer acts as the run-time environment for the MHP applications. So programmers can develop MHP applications and author content without worrying about the underlying hardware architecture. The JVM is a program that interprets Java bytecodes into machine code. This type of virtual system has two benefits:

(1) An MHP application will run in any JVM, regardless of the underlying hardware platform and real-time OS.
(2) Because the JVM has limited contacts with the real-time OS, there is little possibility of a Java program damaging other files or applications on the set-top box.

In addition to interpreting interactive MHP applications, the JVM is also used by the set-top box to verify the integrity of the code. A JVM instance is created once an interactive TV application is loaded and gets terminated once the application has been terminated by the MHP software.

APIs   The DVB-J platform uses APIs for the development of interactive TV applications. An API is an interface to the real-time operating system that allows an interactive TV application to avail of the resources and features of the underlying hardware platform. The Java language was not originally created for television and as a result it does not include some core functions needed by the DVB-J platform. Luckily, Java supports extensions to allow for such things, one of which is APIs that support the processing of digital broadcast of audio and video services. A list of Java APIs intrinsic to the operation of MHP is available in Table 6.2.

TABLE 6.2   List of MHP APIs

Java API	Description
Java.lang	This is one of the core packages providing some of the basic functionalities that are fundamental to the design of the Java programming language. The java.lang package contains a number of vital classes inherent to the Java programming language.
Java.util	The flow of information on any system usually conforms to a simple and defined structure. The MHP compliant set-top box is no different. Any implementation of its standard utilizes the java.util for the structuring of data.
Java.io	Another very important package for any MHP compliant set-top box is the java.io package. This package provides basic classes for the input and output of data streams, serialization as well as file handling and inter-thread communications. The java.io package can be used to control the input and output streams of Java programs. It is from the basic classes associated with the java.io package that instances are created and customized for a particular type of input or output.
Java.net	MHP and its Internet Profile have typified the convergence of digital television and the Internet. While it is not required that every implementation of the MHP standard provide access to the Internet the means by which access can be achieved is included through the java.net package. It is not necessary for the complete package to be implemented however two of its classes are required. These are the java.net.URL class and the java.net.InetAddress class.
javax.tv.service.selection	The javax.tv.service.selection API package identifies available services in the transport stream.
Streamed Media API	The combination of a digital broadcast of audio and video services with interactive applications requires a Streamed Media API. The main part of the DVB Streamed Media API is the Java Media Framework (JMF) developed by Sun Microsystems in 1998. This API allows audio and video content to be embedded in applications.
Java.beans	Of course with any implementation there is the need for creating graphical user interfaces, how these interfaces are constructed is aided by the inclusion of the java.beans API. A java bean allows a Java class to be a reusable software component. The interconnection of these beans is enabled by manipulation of the class's properties, events and methods.

TABLE 6.2 (*Continue*)

Java API	Description
Java.math	The manipulation of numbers and bits is provided by the java.math package; a useful function when decimal and integer numbers must be calculated.
Java.text	The java.text package handles all dates, numbers and general text on a DVB-J platform.

All of these APIs take into account the limitations associated with the relatively small footprint of set-top boxes.

Communication Protocols   Simply connecting a digital set-top box to a network does not guarantee successful communications between the subscribers' home and the service provider's headend system or data center. The device needs communication software to instruct how to transmit signals across the network. Broadcasting MHP applications requires the use of a suite of communication protocols. The communication protocols used by MHP are grouped into different layers that are stacked on top of each other. Each layer in the protocol stack specifies a particular function and uses the services of the protocol beneath it. This suite of protocols describes how an interactive TV application running on a backend server communicates with a software application running on an end-users digital set-top box. The four main protocols used by the MHP communications model are as follows:

MPEG-2—Like the other DVB variants, DVB-MHP is based around the MPEG-2 standard. This high quality video and audio compression system enables network operators to broadcast many more channels into customers' homes than they can via analog networks.

DSM-CC—Digital Storage Media-Command and Control (DSM-CC) is an open standard developed for the delivery of multimedia services over a broadband network. DSM-CC offers, among other things, a mechanism to broadcast objects (like files and directories) in a carousel-like fashion. It may be used for a wide variety of other purposes including the transmission of software applications to set-top boxes.

DVB object carousel—The DVB object carousel is a subset of the DSM-CC specification. Its main function within an MHP environment is to allow application developers to cyclically broadcast different types of software objects to a set-top box. The teletext service in the United Kingdom is a good example of how object carousels are used to efficiently deliver information pages to TV viewers. There are three different types of objects carried within a DVB carousel:

(1) *File*—File objects are data files that typically contain MHP executable programs.

(2) *Directory*—The set-top box uses directory objects to determine the location of specific file objects.

(3) *Stream*—A high level of interactivity can be achieved by using particular events in a TV program to trigger specific behaviors within the MHP application. These events, each with their unique ID and associated data, are broadcast within the object carousel as stream objects.

All the objects are located on a server at the network operators' headend. These objects are normally carried within the multiplexed transport stream in the same way as the other video, audio, and service information streams. A unique PID is assigned to each carousel of objects. The DVB SI service tables then reference this PID value.

DVB multiprotocol encapsulation—One of the major factors that affects the deployment of interactive TV services to the home is the availability of standardized protocols. Without such protocols, every service delivered to the home would require its own interface to receive it. Protocols that comply with open standards allow PCs and set-top boxes access to multiple services from multiple service providers. Multiprotocol encapsulation (MPE) is an open protocol that is commonly used on DVB compliant broadcast networks. It describes how IP packets are encapsulated and transported via the DVB MPEG2 transport stream.

Note MHP enabled devices can also support two-way communications over an IP based network.

MHP Security   MHP has a number of integral subsystems, the most important of which is the security subsystem. The function of this subsystem is exclusively to provide security for MHP applications using a range of security management services, including

Hashing—Hashing is the encryption of a string of characters into a usually shorter fixed-length value or key that represents the original string.

Digital signatures—A digital signature is a digest of information, encrypted using a private key.

Certificates—A certificate, in general, is a file that is used by MHP to verify the identities of end users and set-top boxes.

Note the subject of security is covered in Chapter 7.

Managing and Locating MHP Applications   A DVB-MHP application is basically a directory of different types of executable and data files. Interactive TV applications built for an MHP set-top box platform are also referred to as Xlets. There are generally three parts for running an MHP application on a digital set-top box:

(1) The set-top box needs to identify the location of the MHP service that was requested by the TV viewer.

(2) The set-top box needs to identify the data sources that are required by the application.

(3) The application needs to be loaded, started, paused, and stopped.

To meet the needs of the first and second parts, MHP has defined a specific table called the Application Information Table (AIT), which enables a set-top box to locate interactive TV applications on a broadband network. The AIT specifies how to embed an MHP application within an MPEG stream. It provides the set-top box with the following items of information:

- The exact location on the network of the application
- The location of the source data for the application
- The hardware resources required by the application
- The application type (e.g., DVB-J or DVB-HTML)

The remaining part associated with running an MHP application is dealt with by the application manager. It is the application manager's responsibility to manage the lifecycle of all interactive TV applications. This ensures the integrity (the look and feel) and interoperability of an application across various MHP implementations.

DVB-HTML   MHP also supports applications that are written in HTML. The use of HTML in MHP is called DVB-HTML and consists of a number of Internet standards. The main body of the DVB-HTML specification defines application formats that are based on the W3C recommendations. Note that the W3C has recommended such standards as XML, HTML, ECMAScript, CSS2, DOM2, and XHTML, which are core elements in today's Internet and now in MHP's DVB-HTML. Similar to DVB-J, the digital IPTVCD retrieves the DVB-HTML files from the incoming digital video stream.

6.2.1.2 MHP and IPTV Overview   In recent times the DVB organization has started to develop standards that allow the transmission of DVB services over IP broadband networks. One of the primary additions to the existing suite of DVB standards is an extension to the MHP middleware system called MHP-IPTV. The key technical and operational characteristics of MHP-IPTV include the following

- The extension allows interactive TV applications that run on top of the MHP protocol stack to use DVB video resources.

- It reuses many of the existing MHP APIs defined in Table 6.2. javax.tv. service.selection is an example of an MHP API used frequently by MHP-IPTV applications.

- New extensions have been developed for existing APIs that support the delivery of video content across an IP broadband network. For instance, the org.dvb.service API adds support for hybrid IP set-top boxes.

- MHP-IPTV provides interfaces to the various network protocols defined in the ETSI TS 102 034 standard. In particular, MHP-IPTV has developed an API called org.dvb.service.sds that provides access to the SDS protocol.

- An API called org.dvb.tvanytime has also been developed to allow access to Broadand Content Guide (BCG) meta-data.

Note that MHP-IPTV is a part of the MHP 1.2 specification, which has been approved by the DVB technical module in 2006 and was being prepared for submission to ETSI at the time of writing.

6.2.2 GEM

As more and more broadcasters in different countries started to offer digital TV services, a consensus emerged from the industry standard bodies of a need for a common worldwide middleware platform for interactive TV services. For this reason, a new standard—the Globally Executable MHP (GEM) specification—was developed. The work involved in developing the standard was carried out by the DVB organization. From a technical perspective GEM is based on MHP middleware technology and defines a set of mandatory core features. Owing to the fact that specifications for broadcasting digital TV varies around the world, the GEM specification has eliminated technical references related to DVB digital transmission technologies. Therefore, GEM acts a platform for building open standards. International standard organizations such as ATSC, CableLabs, and ARIB are all using GEM as a basis for building open middleware standards that take into account the needs of their local markets. Therefore, a content developer can now produce interactive TV content, which will run on GEM compliant digital set-top boxes anywhere in the world.

6.2.2.1 GEM and IPTV Overview   In addition to the development of MHP-IPTV, work is also ongoing with regard to producing an IPTV version of GEM. Details are light at the time of publication, however, the following high level features have been identified:

- It interoperates with networks that use proprietary IPTV systems.
- It reuses many of the MHP-IPTV APIs and technologies. The service selection package used to discover IPTV multicast services is one example of an API, which is common to both MHP-IPTV and GEM-IPTV.
- Provides support for RTSP, IGMP, and UDP protocols.

The modular and open approach of this specification has meant that the ACAP, OCAP, and ARIB B23, standards use GEM as a foundation for each technology. Additionally, current interactive TV software providers can also use GEM to build a product portfolio that complies with the various international standards. Although GEM is used as the core technology, there are some differences that make each standard appropriate to their own particular markets. The following sections will provide further details on these middleware technologies.

6.2.3 OCAP

OpenCable Applications Platform (OCAP) is the interactive TV middleware standard of choice for network service providers based in North America. The standard is developed by CableLabs. The latest version of the specification has evolved directly from the MHP and GEM specifications, however, it has removed all the elements that refer specifically to DVB transmission systems and adds elements that are appropriate for the OpenCable family of digital TV standards. OCAP 1.0 was released at the end of 2001 and OCAP 2.0 was released for industry review in 2002. Versions 1 and 2 both use the same core functionality, however, OCAP 2.0 includes support for new technologies such as HTML with the application known as OCAP-HTML. OCAP 2.0 requires additional memory and CPU resources beyond those required of OCAP 1.0. While OCAP reuses a number of MHP elements, there is however some slight differences. First, the standard is considered by some industry analysts to be slightly more complex when compared to MHP based systems. Second, the standard has as previously stated removed any parts that refer to DVB transmission standards. The API for DVB based service information is one obvious example. Third, CableLabs has added an application called "the monitor" to the software stack, which provides the cable operator with greater control over the set-top box.

6.2.3.1 Types of OCAP Interactive TV Applications   OCAP classifies iTV applications into three separate categories:

Bound—An iTV application that is classified as bound is only associated with the channel that is currently tuned on the set-top box. Therefore, if the viewer switches channel, then the application is terminated. So, for example, if a viewer has tuned to channel 500 and is playing a gaming application and the viewer decides to switch to channel 50. When this happens the gaming application releases all the resources that it is using and ceases to operate.

Unbound—In contrast to bound applications, unbound applications are available across multiple channels. So when a viewer switches channels, the application remains active. A stock information ticker is a practical example of an unbound application.

Native—Native or resident applications are defined as iTV applications that are written for specific set-top box hardware platforms.

6.2.3.2 Software Architecture for OCAP Enabled Set-Top Box   The key modules of the OCAP (version 1.0) middleware stack are shown in Fig. 6.3 and explained in the following sections.

The Cable Network Interface   As the name implies, this module provides an interface between the cable TV network and the digital set-top box. The protocols required for communication between client and server side applications are included in this module. As with MHP, the OCAP applications are typically placed in DSM-CC object carousels that are encoded into the OCAP transport stream.

The Execution Engine   The OCAP execution engine is based on the Java platform, which consists of a JVM and a number of reusable APIs. These APIs are organized into functional groups called "packages." OCAP has defined a number of packages that allow for the development of a wide range of interactive applications. Table 6.3 explains these API packages in greater detail.

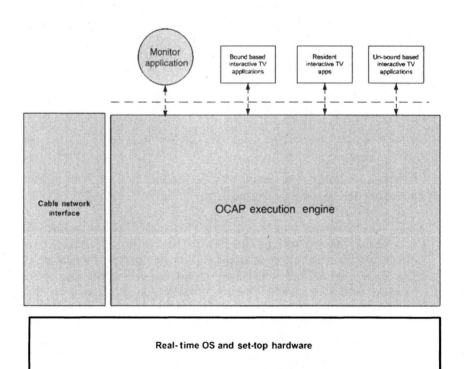

FIGURE 6.3   OCAP software stack

TABLE 6.3   Different Types of OCAP API Packages

Type of API Package	Description
Content presentation specific	The design of a user interface for interactive television differs greatly from designing interfaces for the World Wide Web. Therefore OCAP has added a number of extensions to the core Java platform to allow developers create applications that are specific to a TV environment.
Streaming media	The main purpose of the streamed media APIs is to provide TV viewers with a consistent interface for accessing various types of streaming digital content.
Data access	OCAP includes several new data access APIs, each of which has varying levels of support for accessing data residing on the set-top box and the network.
Service information	Service Information is embedded in the transport stream and provides a mechanism for set-top boxes to find out information about available interactive TV services. OCAP has defined service selection APIs that provide developers with an interface for selecting these services for presentation. As well as this the Service Information APIs enables OpenCable compliant set-top boxes to automatically tune to particular services.
Common infrastructure and security	The OCAP 1.0 Java platform includes Java packages that allow inter-application communication. Additionally the standard provides comprehensive support for plug-ins and advanced security standards.

In addition to reusing many of the APIs defined in the DVB-GEM specification, OCAP has also defined modules of related functionality that applies to a U.S. cable TV environment. Table 6.4 provides an overview of these modules.

The Monitor Application   The monitor application sits above the OCAP execution engine. It is a program that was specifically developed for cable TV systems that use the OCAP software environment. Its primary function is to monitor and track the lifecycle of OCAP iTV applications. To optimize performance, the OCAP monitor also centralizes basic network functions such as:

Application registration and validation—The monitor registers and validates interactive TV applications before they are allowed run on the set-top box.

Resource management—Managing the resources that are consumed by the OCAP interactive TV application is generally processed by the monitor application.

TABLE 6.4   OCAP Functional Modules

Functional Module	Description
Executive	This module is an integral part of the firmware and is responsible for launching the monitor application.
CableCard resources	This module is the link that allows a retail cable box to be portable across a variety of different cable system headends by standardizing the communication between individual addressable CableCARD modules and the connected set-top boxes.
The EAS module	This OCAP module is used to support the emergency alert system (EAS), which was established in 1994 to warn the public about emergency situations.
Closed captioning	This module handles closed caption subtitles.
Watch TV	This provides support for basic channel changing.
Service information	This parses and processes various types of SI data.
Download	This module is used to upgrade the set-top box software systems.
Copy protection	This enforces the copy protection system used by the OCAP set-top box.
Content advisory	This module deals with any parental rating information, which is available in the signal.

Handling Errors—The monitor deals with any errors that may occur within the OCAP middleware software stack.

Managing OCAP Applications—The sequence of steps by which an OCAP-J application is initialized, undergoes various state changes and is eventually destroyed is collectively known as the application lifecycle. The application manager resident in the set-top box is responsible for an OCAP applications lifecycle state change. The AIT or the Extended Application Information Table (XAIT) provides the application manager with the necessary information required to manage application lifecycles.

The descriptions so far of OCAP have focused on versions one and two of the specification. Since their publication CableLabs has continued to develop new versions of OCAP to support the emergence of next generation interactive TV applications. Recent OCAP versions include

OCAP Digital Video Recorder (OCAP-DVR)—Built as an extension to OCAP 2.0 this specification defines the requirements for a DVR software environment for DVR capable digital set-top boxes. Extending the functionality of the EPG and time shifting broadcast events are the two most popular features of this interface.

OCAP Front Panel Extension (OCAP-FPEXT)—Most if not all digital set-top boxes are designed with a front-panel display that includes multiple LEDs that are located adjacent to each other. This OCAP extension API is used to manage the functionality of this panel.

OCAP Home Networking Extension (OCAP-HNEXT1.0)—This extension of the OCAP standard was created to deal with home networks that are specifically connected to cable TV broadband networks. The types of re-quirements related to these types of home networking environments can range from protecting video programs to helping consumers access different types of digital entertainment content across the home network.

OnRamp to OCAP—This is best defined as a subset of the OCAP middleware standard, which was designed to run within the resource constraints of low end digital set-top boxes. Responsibility for developing OnRamp lies with an expert group formed by a number of U.S. cable television operators and equipment suppliers. Interactive TV applications written for OnRamp enabled set-top boxes are forward-compatible with advanced set-top hardware platforms that include support for the full OCAP specification.

The other primary extension, which is currently being defined at the time of writing this book, is the addition of "IP-tuning" capabilities to the OCAP environment.

6.2.3.2 OCAP and IPTV Owing to the fact that OCAP does not have a dependency on any particular network technology, OCAP remains a relevant middleware to the evolution of open IPTV middleware standards. At the time of writing CableLabs had not extended OCAP to provide direct support for IPTV, however, the organization has plans to incorporate IPTV enhancements from the MHP-GEM 1.2 specification into the next release of the technology.

6.2.4 ACAP

ACAP (Advanced Common Application Platform) is an open interactive TV middleware standard used by digital set-top boxes that receive signals from U.S. based terrestrial broadcast and cable TV networks. The standard was formally approved by the ATSC in 2005. A timeline of how the standard came into existence is shown in Fig. 6.4.

The ATSC's technology specialist group (T3/S2) is responsible for developing the ACAP standard.

The technical architecture of the ACAP software has evolved from a harmonization process between ATSC's Digital Television Application Software Environment (DASE) and CableLabs OCAP specifications. It uses the application model and Java API found in the GEM specification but adds a number of technical components that are specific to the U.S. terrestrial TV marketplace. The processing of closed captioning information is one such example. ACAP enabled set-top boxes support two different types of applications, namely, ACAP-J and ACAP-X.

6.2.4.1 ACAP-J ACAP –J is based on DVB-J, which consists of a Java virtual machine and a number of reusable libraries of Java programs.

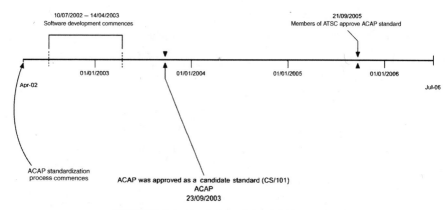

FIGURE 6.4   Evolution of the ACAP standard

6.2.4.2 ACAP-X   ACAP has been designed to embrace a variety of open Web standards and features its own XHTML platform. The platform is known as ACAP-X and supports a number of different media types ranging from JPEG images to XML based Web pages. The methods used for transporting, launching ACAP-X applications are more or less the same as those used in the GEM specification and a number of different networking protocols provide interconnectivity between an ACAP-X client set-top box and an Interactive TV application server. The ACAP security model is fully conformant to the GEM security model, however, it does include details of an extension to this model, which take into account the security risks associated with broadcasting content over a terrestrial network.

6.2.4.3 ACAP and IPTV   As described above, ACAP draws heavily from the GEM specification. As a result, the IPTV "flavor" of ACAP is constructed using GEM combined with IPTV specific APIs. Note that the use of ACAP as a middleware for deploying IPTV networks is in its early stages at the time of this writing.

6.2.5   ARIB B23

ARIB, the Japanese TV standards body, has also adopted GEM as a basis for Japanese iTV standards. The Application Execution Engine Platform for Digital Broadcasting (ARIB-AE) Standard, also known as B23 was officially approved by ARIB in 2003. In addition to B-23, Japaneese TV serice providers also use Broadcast Markup Language (BML) for creating interactive TV applications. BML is an XML-based standard.

6.2.5.1 ARIB B23 and IPTV   Similar to other international standards ARIB B23 is the most likely candidate of becoming the underlying platform for the delivery of IPTV services for the Japaneese marketplace.

6.3 PROPRIETARY MIDDLEWARE SOLUTIONS

Even though the market for standardized platforms grows on a global basis the vast majority of middleware software running on IPTVCDs at the time of writing is provided by a number of Interactive TV software providers. Each of these environments has their own capabilities and deployment configurations vary between IPTV networking infrastructures. A more comprehensive and up-to-date analysis of the IPTVCD middleware marketplace is available at www.tvmentors.com.

SUMMARY

Software running on an IPTVCD enables the smooth operation of the device. It hides the complexities of the system elements from the end user. A typical IPTVCD software environment includes drivers, RTOS, and a middleware layer.

Drivers—This component supports a range of media types and networking protocols.

RTOS—The RTOS is a key component of the IPTVCD software platform. Operating systems that run on IP set-top boxes and RG platforms are able to meet the strong demands associated with delivering IPTV services to consumers. An RTOS is able to operate within a small memory space, is extremely reliable, and boots very quickly during device startup. In addition to processing and de-coding television signals, an IPTVCD RTOS must also regulate the allocation of resources, such as memory and processor utilization.

Middleware—Middleware, in an IPTV context, is a type of software that sits on an IP set-top box and acts as a conversion or translation layer. Middleware can be used with a server to off-load some processing and storage tasks from the IP set-top box to the backend server. In addition, middleware plays a key role in ensuring that IPTV services are processed correctly.

The current business model for IPTV services comprises of a subscriber buying or leasing an IPTVCD running a proprietary middleware software system. With the move toward a retail model, the importance of open standards becomes key in ensuring that their IPTVCD investment is safe. The demand for open standards within the world of IPTV iTV middleware systems has resulted in a series of plans to extend the functionality of the following standards to support the delivery of interactive IPTV services.

- MHP is a single, open, interactive television development platform based on the Java platform, which consists of a JVM and a number of reusable libraries of Java programs. A key benefit of MHP is that it provides developers with the ability to easily build applications that are based on open standards. MHP-IPTV is an extension to the MHP middleware system.

- GEM is a standardized interactive TV platform based on MHP technologies.
- OCAP is an open middleware software platform that was designed for the U.S. cable industry. It allows operators to deliver sophisticated interactive TV services over their HFC based networks. OCAP incorporates a large set of APIs that are used by the development community to write iTV applications.
- ACAP is a standard that integrates two interactive TV middleware platforms, namely, OCAP and ATSC's own DASE software. The ACAP standard follows the technologies used in the GEM specification quite closely.

While the work continues on the definition of a worldwide interoperable standard for the deployment of interactive IPTV applications, a number of software vendors have launched interim solutions.

7

IPTV CONDITIONAL ACCESS AND DRM SYSTEMS

To protect unauthorized access of IPTV services and content, telecom operators need to ensure that a security system is installed that will provide the identity of each subscriber that accesses the IP based content. In addition to authentication, the IPTV security system also needs to protect content at all stages of delivery—from the broadband access point right through the last mile to the IPTVCD. In addition to protecting on-demand and multicast TV content assets, an IPTV security system needs to also protect end-users private data and network infrastructure components. There are a couple of different approaches that can be used by service providers to prevent theft. This chapter provides a technical overview of commonly deployed traditional pay TV and IPTV security systems.

7.1 INTRODUCTION TO IPTV SECURITY

Security is the number one priority for IPTV service providers because video content producers are reluctant to grant license rights to distribute premium content over digital networks unless there is a powerful mechanism in place, which will secure that content. The objective of a security system is to ensure that only authorized paying subscribers can access broadcast IPTV channels and VoD content. The entire system from content production to the in-home network needs to be secured.

Next Generation IPTV Services and Technologies, By Gerard O'Driscoll
Copyright © 2008 John Wiley & Sons, Inc.

There are a number of IPTV content protection schemes available on the market. These protection schemes fall into two broad categories: CA and DRM environments. The CA system is primarily responsible for ensuring unauthorized access of the IPTV service, while the DRM system enforces content owner's business models and rights.

7.2 DEFINING IPTV CA SECURITY SYSTEMS

For network operators to take maximum advantage of the IPTV revolution, they need a CA system to prevent theft of their services. A CA system is best described as a virtual gateway that secures the delivery of multichannel pay TV and video on demand services to IPTV subscribers.

In contrast to a traditional pay TV network, an IPTV networking infrastructure uses point-to-point connections to each IPTVCD. This basically means that only channels that are been watched by the IPTV subscriber are transmitted over the networking infrastructure. This method of communication contrasts with cable, satellite, or terrestrial TV networks where multiple channels are broadcasted across their networks. So from a security perspective it should in theory be more difficult to steal from a network with content that gets transmitted intermittently as required versus a network that has a continuous flow of digital services such as broadcast TV networks. Therefore, the security system deployed on an IPTV network is different to those used in cable, satellite, or terrestrial broadcast based TV systems. The main goal of any IPTV CA system is to control subscribers' access to IPTV pay services and secure the operators revenue streams. Consequently, only customers that have a valid contract or paid subscription with the network operator can access a particular IPTV service. There are three alternative approaches to IPTV conditional access systems.

(1) Hardware
(2) Software
(3) Hybrid hardware and software solutions

7.2.1   Hardware Centric CA Systems

The general architecture of a hardware centric CA system is based on a client-server networking infrastructure. A simplified block diagram depicting the interoperability of the main headend components is depicted in Fig. 7.1 and explained below.

7.2.1.1   Customer Authorization System (CAS)   The CAS component typically includes a database that is capable of storing IP-VoD and multicast TV product information, identification numbers of smart cards, subscriber profile details, and schedule data. It interfaces with the OBSS, which provides the support required to accurately manage the IPTV business model. The main goal of the OBSS system is to ensure that subscribers view exactly what they pay for.

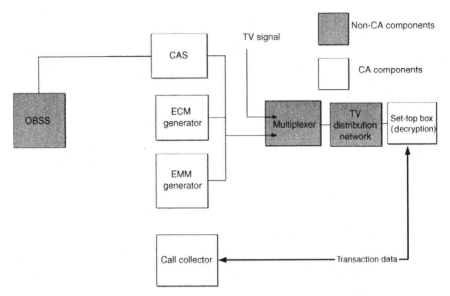

FIGURE 7.1   Key components of an end-to-end smart card based CA system

Subscribers can only access digital television channels or on-demand video assets if they purchase the relevant authorization rights. Once the rights are purchased, the OBSS communicates this information to the CA system. This information is fed into the CAS system and a message called an Entitlement Management Message (EMM) is sent over the distribution network to a smart card inserted into a digital set-top box. An EMM allows a customer to watch a particular TV broadcast channel or special event. It can contain the following items of information:

- Authorization keys to view digital IPTV services
- Credit for future purchases
- Service cancellations and renewals
- Other subscriber data such as address and billing details

The CA system also generates EMMs when there is a change to the customer's subscription details. For example, when a customer subscribes to a new on-demand service, an EMM is sent to authorize access to the service. Note that in an IPTV networking environment EMMs are sent separate to the IPTV bitstream and often use a TCP communication session for delivery to the destination IPTVCD.

7.2.1.2 Encryption   In traditional RF based TV systems, network operators used a technique called scrambling to protect their content. The basis of a scrambling system is the method by which it renders the picture unwatchable. It uses an algorithm to transform the video signal into an unreadable format. The set-top box uses the corresponding descrambling algorithm to restore the signal to its viewable form. There are two basic types of scrambling system: dumb and addressable. The dumb system does not have any over-the-air addressing. As a result, the service provider cannot turn a subscriber's set-top box off. This type of system has become outdated and offers minimal security. As a result, it is not used for IPTV applications. Restricting access to modern digital IPTV services is accomplished using encryption mechanisms.

In contrast to scrambling, encryption transforms each data element using a sophisticated cryptographic system. The CA system uses an encryption key, also referred to as a control word in conjunction with an algorithm to encrypt and decrypt information. Note that encryption is only applied to the video payload, while the packet header information is left in the clear. This allows digital set-top boxes to parse the incoming stream and identify packets associated with the different elementary streams.

The keys used in IPTV CA systems are extremely hard to break and provide a high level of security. Furthermore, different keys are often used for different channels. The keys are sent via the IP stream to an RF based digital set-top box or an IPTVCD in an encrypted form as part of an entitlement control message (ECM). ECMs are generated in the IPTV data center using specialized hardware devices and are changed at regular intervals to maximize security levels on the system. In contrast to EMMs who are specific to the subscriber, ECMs are specific to individual multicast and broadcast channels.

The application of encryption at the IPTV headend differs according to the type of content:

Multicast IPTV—Content encryption is done in real time as the stream gets multicasted across the distribution network. In parallel or in some cases prior to the commencement of streaming the decryption keys are sent to the IPTVCD to authorize viewing of all broadcast channels associated with a particular product offering.

IP-VoD content—With this type of service, network operators have an option to either encrypt the VoD asset in real time or preencrypt prior to unicasting. During the unicast session security keys are generated and sent along with the IP-VoD stream to the client device. A different set of keys is generated for each unicast of IP-VoD content. This contrasts with the approach of multicast TV systems where the keys are able to decrypt a number of different IPTV assets, namely multicast channels.

Note on an IPTV network the ECMs are sent as part of the UDP bitstream to the IPTVCD.

7.2.1.3 A Return Path   A CA system makes extensive use of the bidirectional capabilities of IP broadband networks to gather transactional data from IPTVCDs. The specific information retrieved will differ from CA to CA; however, most systems will support the collection of the following key pieces of data:

- Identification codes for the smart card and the IPTVCD
- IP-VoD assets ordered
- Credit amount

The frequency of how often this information is collected differs between systems. In traditional pay TV environments, the telephone network is often used as the return path, which means that digital set-top boxes are interrogated once a day or once a week. IP broadband networks provide for real-time interrogation of IPTVCDs, which is a useful feature for service providers.

7.2.1.4   Client IPTV CA System   Decryption is the reverse process of encryption and is used to convert the digital TV signal back to its original format. Once the signal is encrypted, it can only be decrypted by means of a client IPTV CA system. The client CA components comprise of a deencryption chip, a secure processor, and some appropriate hardware drivers.

- The deencryption chip is responsible for holding the algorithm section of the CA.
- The secure processor is a silicon chip, which contains the necessary keys needed to decrypt various types of digital TV services.

A simplified block diagram depicting the interoperability of the client IPTV CA components is depicted in Fig. 7.2.

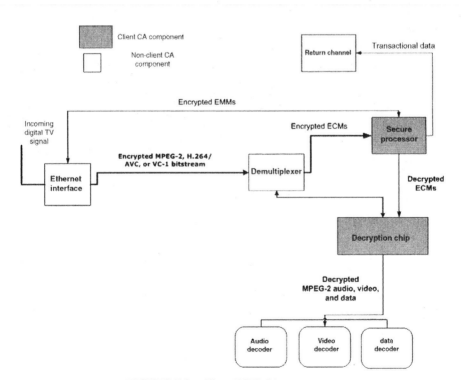

FIGURE 7.2   Client IPTV CA components

Digital signals from the access network are typically formatted into streams of IP packets. This stream arrives at the IPTVCD and content is extracted from the stream using the demultiplexer and sent via the decryption chip to the appropriate decoders. The encrypted ECM, which is sent along with each video stream, is also extracted and sent to the secure processor for decryption. The decryption of the ECM can only take place when authorized to do so. The authorization to decrypt the ECM comes from an EMM, which is also processed by the secure processor. The ECM contains a key, which gets regenerated by the processor and sent onward to the deencryption chip. The key is then used to generate a clear IPTV video stream. From here, the standard decoding of the compressed IPTV packets takes place. The secure processor also uses the bidirectional functionality of the IP broadband network to provide the headend CA system with transactional data. The security scheme just described is very general and does vary between different types of commercial CA systems. Note that the client set-top CA system can be embedded within the box or implemented using a removable smart card.

7.2.1.5 A Smart Card   An IPTV CA system employs a two-tier security system to protect against content theft. The first tier involves the encryption of the content, while the second tier deals with authenticating the IPTVCD. Smart cards are sometimes used to provide authentication services to the IPTV CA system. A smart card is a removable security device, which can readily be replaced if needed. It is typically owned by the service provider and is used by IPTV CA systems to securely identify subscribers of IP multicast and IP-VoD services. A typical smart card is made of plastic and consists of a microprocessor, software, and a quantity of memory.

A microprocessor—The embedded microprocessor allows the card to make computations and decisions and manipulate data; in other words, this is the "smart" part of the card. The sophistication and capabilities of smart cards are continually growing to meet the demands of sophisticated IPTV and HD based applications. This chip is unique to every vendor's CA system.

Software—The software generally consists of an operating system, which is customized specifically for CA systems.

Memory—There are a couple of memory types used by set-top smart cards — RAM for temporary data storage, which is only used when power is applied (usually when the card is in contact with the reader), and ROM that stores fixed data and the operating system. The use of nonvolatile memory such as EEPROM or Flash memory is ideal for storing data such as an account Personal Identification Number (PIN), encryption keys, or entitle-ment data for subscription and PPV services, the type of data that must remain stored once the power is removed.

A physical smart card is very similar in appearance to an ordinary credit card but with the addition of a number of different types of embedded integrated circuits.

The architecture of smart cards generally complies with an international standard called ISO/IEC 7816. To minimize security risks some CA systems service providers create a link between the smart card and the IPTVCD. This process is called pairing and occurs when the smart card is initially authorized on the network. At this stage the hardware ID of the IPTVCD is written to the smart card memory. Once this is completed the rights to view subscription, PPV, or VoD content are only granted when the card is inserted into the paired IPTVCD. One of the major benefits of pairing becomes evident when the card is compromised. In such cases the security of the CA system is restored by removing the old card and replacing with a new one, which was received from the service provider. Note that due to the sophistication of smart card chipsets it is not easy to corrupt or extract data from a smart card.

7.2.1.6 Removable Security Modules   A removable security module is a stand-alone piece of hardware that performs the functions of the embedded CA system, namely, the decryption of digital content. The module itself is quite compact, about the size of a standard Personal Computer Memory Card International Association (PCMCIA) card, which is typically used in laptops. It contains the hardware chipsets and related software required to decrypt video content. By making the hardware module a separable unit from the digital set-top, the boxes become inter-changeable systems that can be sold through the retail channel. In other words, removable security modules allow consumers to choose and purchase a digital set-top box independent of the service provider. So for instance, if I decide to relocate to another part of Ireland served by a different TV service provider, I can simply return the removable security module back to my current provider, take the set-top box with me, reconnect it upon moving to the new location, and obtain a new security module from my new service provider.

When a digital signal arrives from the network, it gets passed through the removable security module where encrypted information gets processed. Removable security modules will typically comply with one or more major international standards, namely, DVB-CI and CableCARD.

DVB-CI   The Common Interface (CI) is a standardized interface that has been defined by the DVB project between the set-top box and a separate hardware module. Its function is to accept the removable security module. The DVB-CI standard has been particularly popular amongst European operators and has also been selected by China.

CableCARD   CableCARD is the U.S. equivalent of DVB's common interface. Formerly known as a Point of Deployment (PoD) module, CableCARD is OpenCable's standard for removable security. CableCARD is a plug-in hardware device that decodes encrypted or scrambled content. The deployment of these modules is in response to a mandate from the U.S. based FCC, which states that all providers must make CableCARD plug-in security modules available to their subscribers. The first revision of the technology, known as CableCARD 1.0, was approved by the ITU in 1998 and has been through a number of revisions over the past 9 years. The key technical and operational characteristics of CableCARD 1.0 are as follows:

- It operates with digital set-top boxes and integrated TVs, which comply with OpenCable standards and specifications.
- It supports one-way operation. The telephone modem is used by the CableCARD for PPV and IPPV purchases.
- It supports the decryption of protected MPEG video streams.
- It uses PCMCIA type II PC card modules to provide connectivity with the chosen provider. The CableCARD 1.0 hardware module gets inserted into a digital cable set-top box and is typically assigned a unique ID number.

Lack of support for an OOB two-way communication channel in the first version of CableCARD combined with some technical glitches meant version one of the standard was unable to support interactive TV applications. Therefore, CableLabs went back to the drawing board and produced an updated specification called CableCARD 2.0. The key characteristics of the CableCARD 2.0 standard are as follows:

- It supports two-way operation. So consumers can purchase an OpenCable compliant set-top box in their local store, receive a CableCARD 2.0 removable security module from their service provider and access a range of VoD, interactive, IPTV, and linear digital TV services.
- It supports multiple tuners, which is important for DVR type applications.
- It can be fitted to a computer that uses the Windows Media Center operating system. This allows consumers to view cable TV video content on their home PCs.

To complement the CableCARD standard CableLabs has also developed a two-way multistream version of the CableCARD (also known as the "M-CARD"). Some of the key features of this M-Card interface include

- Support for the simultaneous decryption of multiple video streams.
- Allows pay TV subscribers to view one channel, while decrypting and recording a separate digital TV channel.
- It operates in two modes, namely, single stream and multistream. The single stream mode is typically used for unidirectional digital TV screens or set-top boxes, whereas the multistream mode is used by two-way access devices.

In addition to DVB-CI and CableCARD the Japanese have also standardized an interface as part of the ISDB digital TV specifications called B-CAS. The B-CAS interface is typically integrated into digital set-top boxes and televisions. The interface accepts B-CAS smart cards that are used to decrypt video content. Each

B-CAS card contains a unique key, which is used for decryption purposes. There are two types of B-CAS cards available to Japanese digital TV subscribers, namely, red and blue. Both card types are used to access different types of digital TV signals.

Note that removable security modules have experienced low penetration rates in the traditional pay TV security sector at the time of writing. As such the impact on the IPTV security sector of these modules is expected to be limited over the next number of years.

Hardware centric CA systems are a mature and proven solution for protecting digital content. One of the main drawbacks of these systems is the difficulty experienced when a serious security breach occurs. Because the system is implemented using physical circuitry, the logical approach used in restoring high security levels is to replace the various CA hardware components. This can be a very expensive means of correcting a security flaw. Although such flaws and breaches are rare due to the level of sophistication employed by modern day CA systems, it is a risk that is ever present.

7.2.2 Software Centric IPTV CA Solutions

Employing a software centric CA solution is another mechanism for securing digital content. Also known as smart card less CA mechanisms these types of systems have the following characteristics:

No smart card reader required—One of the most obvious benefits of these systems is the fact that IPTV subscribers do not require smart cards and readers built into their IPTVCDs to view broadcast and on-demand IP content.

Fast recovery from security breaches—Once a breach occurs a software patch or fix can be applied to the system within seconds to minimize affects of the security breach.

Available for a wide variety of IPTV network types—These systems can be deployed over IP broadband networks as well as traditional satellite, terrestrial, and cable pay TV networks.

Support well established encryption standards—A software centric CA system is based on proven and mature Internet encryption algorithms that are able to maintain the security of IPTV connections.

Extensive use of digital certificates—These software driven CA systems are designed to use digital certificates to identify each element connected to the IPTV security system.

Works in parallel to hardware centric CA systems—Both hardware and software centric CA systems can operate together in a hybrid IP and broadcast networking environment.

Always on connection—A real-time networking environment is required to support software-centric CA systems. Therefore, an always-on and two-way

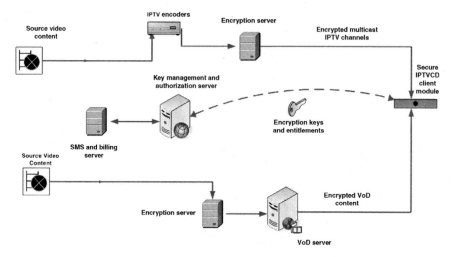

FIGURE 7.3 Network architecture of a software centric CA system

broadband connection is required between the IPTV data center and the IPTVCDs deployed in subscriber's homes. The following technical components should appear in a fully rounded end-to-end software centric IPTV CA system:

- An encryption server
- A key management and authorization server (KMAS)
- A secure IPTVCD client module

The role of each component is depicted in Fig. 7.3 and discussed in the following sections.

7.2.2.1 Encryption Server   The most common method of protecting sensitive information on a broadband digital TV network is encryption. When information is encrypted, it is altered so that it appears as meaningless garble to anyone other than the intended recipient of the information. Decryption is the reverse of encryption where the IPTV data in its unreadable appearance is again put back into its original form. A piece of hardware called an encryption server is responsible for securely encrypting digital content, prior to entering the distribution network. These servers employ encryption algorithms to make video content unreadable to persons who do not have authorization to view the content. Different encryption algorithms are used to provide different levels of security. Some of the most common types of encryption algorithms used by IPTV CA systems are available in Table 7.1.

In addition to using the above algorithms, some commercially available encryption servers use the IPSec suite of protocols as a means of extending authentication and encryption capabilities.

TABLE 7.1   Commonly Used CA Algorithms

IPTV CA Algorithm	Description
RC4 (RCA)	This is a cryptographic software system supplied by a company called RSA Security Inc.
Data Encryption Standard (DES) Algorithm	DES is an encryption algorithm used to protect IPTV content. DES takes in a continuous stream of bits in the clear and outputs an encrypted stream of bits. There are different variants of DES. The most secure flavor is the Triple DES, in which three separate keys (168 bits) are used to perform a cryptographic operation.
Advanced Encryption Standard (AES)	AES is an encryption standard that has been adopted by the US government. AES uses 128-bit keys to encrypt video content and is widely deployed by the IPTV industry sector.
Common Scrambling Algorithm	The DVB has defined a Common Scrambling Algorithm (CSA) for transmitting digital television. The specification and licensing rights to the CSA and the Common Descrambling Algorithm are distributed under arrangements with the European Telecommunications Standards Institute.

Designed by the IETF, IPSec is a standard part of IPv6 that is used to secure and encrypt communications between two IP devices. The standard also provides an authentication framework that utilizes a key management service. From a technical perspective IPSec consists of three different protocols (Table 7.2).

The content encrypted at an IPTV data center falls into two broad categories: linear broadcast channels and VoD assets. Both of these categories are detailed in other parts of this book.

TABLE 7.2   IPSec Protocol Descriptions

Protocol	Description
Encapsulating Security Payload (ESP)	This cryptographic protocol uses one of the algorithms described in Table 7.1 to take care of encrypting and authenticating IPTV content.
Authentication Header (AH)	AH is used to authenticate packets.
Internet Key Exchange (IKE)	IKE is required to setup, negotiate, and manage the exchange of security keys.

7.2.2.2 Key Management and Authorization Server (KMAS)   The main function of the KMAS is to

- generate, store in a database, and manage the encryption keys, which are used to encrypt linear broadcast channels or VoD content;
- securely distribute and deliver keys to authorized IPTV clients;
- provide transactional data to the OBSS.

In addition to the above functions, the KMAS is also responsible for changing security keys on a frequent basis. The hardware architecture of a typical KMAS is made up of a number of secure application specific integrated circuits (ASICs). In addition to supporting the core KMAS functionalities ASICs also offer strong protection for private keys.

7.2.2.3 A Secure IPTVCD Client Module   The IPTVCD client software component is responsible for decrypting the incoming encrypted video content in real time and providing subscribers with access to content they paid for. It is integrated with the IPTVCD and is easily upgraded or replaced through a software download from the service provider's data center.

7.2.2.4 Software Centric CAs and PKI   It is important for telecom operators to establish the identity of a subscriber to an IPTV service prior to delivering services over the network. Traditionally, one-way pay TV networks used a security mechanism called "shared secrets" to validate the identity of both the sending and receiving devices. Under this approach, "secret keys" are embedded into a chipset during the manufacture of a receiving device such as a digital set-top box. Alternatively, the shared key is embedded onto a smart card. During broadcasting the security servers use the shared key(s) to encrypt content while the channels are transmitted across the network. The receiving digital set-top box uses the shared key stored on the dedicated on-board or smart card chipsets to decrypt the various digital TV channels. In general, the shared secrets approach has been and continues to be a relatively secure method for protecting access to digital TV channels over one-way broadcast type networks.

With the advent of two-way IP broadband networks an enhanced security architecture for authenticating content senders and receivers has emerged—Public Key Infrastructure (PKI). Under the PKI approach participants of a communication session over an IPTV network are able to exchange content and control information in the confidence that it will not be tampered with or illegally accessed during transmission.

Software centric CA systems make extensive use of PKI technologies. Technically, PKI refers to a framework that uses a pair of mathematically related keys to encrypt and decrypt digital content. Each of these pairs consist of a private and a public key. The private key, which is known only to the sender, is used to transform the IPTV content into a seemingly unintelligible form. The receiving device must have the corresponding public key in order to return the video content to its original form. Note also that under the PKI security framework each IPTVCD has its own unique private key. Therefore, only a single or a very small number of IPTVCDs are affected during a security breach. This is one of the differentiators of PKI based systems when compared to traditional systems where a compromised key can affect the security of all devices connected to the networking infrastructure. In addition to encrypting and decrypting content, the PKI system also uses digital signatures and certificates to authenticate the identity of an IPTV client or the backend server. A digital signature is generally used to verify the identities of digital devices. A digital signature is the functional equivalent of a paper signature. The signatures are created and verified by means of a technique called cryptography, a mathematical system that is responsible for transforming data and applications into unreadable formats. The term "digital signature" describes the use of an algorithm and the sender's private key to create a small summary of the source data called a "hash" or "digest." One of the key advantages of digital signatures is that they enjoy protection in law and cannot be duplicated. In the context of an IPTV environment, this hash value accompanies the data during transmission across the network. Once the hash file and data reaches the IPTVCD the signature is verified by using an algorithm to match the received hash value with the original value. If they match, then the attached security message is processed. If there is a discrepancy, then the message has been altered or "spoofed" by another source. In such a case the data is discarded and the CA system is alerted of a security breach.

Although signatures are adequate in many cases for authentication, IPTV providers are demanding that digital certificates be included in IPTV software centric CA systems as an extra layer of security. Digital certificates are electronic credentials that authenticate IPTV client and server devices. A certificate is a file attached to an electronic message that guarantees the identity of a particular IP device.

Table 7.3 shows some of the information contained in a basic certificate used by IPTV clients and servers.

TABLE 7.3   IPTV Digital Certificate Attributes

Attribute	Description
Version	The version number of the certificate.
IPTV end users details	The name and address of the end user.
Keys	A public key that matches the end users private key. In addition, the certificate may also contain the CAs public key.
Digital signatures	Signatures for both the CA responsible for issuing the certificate and the IPTVCD are typically included in the certificate.
Validity period	Certificates are given a set life span when issued. In the case of IPTV systems the certificate is typically valid for a couple of years. Once the certificate has expired a new certificate must be issued.

Certificates can hold other items of information, such as an e-mail address or a telephone number. Once a security system of this nature is implemented on a broadband network, the chances of a hacker intercepting details about the subscriber are almost nonexistent. One of the most popular standards used by IPTV systems to specify the contents of a certificate is X.509, a standard published by the ITU.

Organizations called certification authorities (CAs) also known, as trusted third parties are responsible for generating and issuing valid digital certificates. PKI deployments will often use a hierarchy of certificate authorities to establish a subscriber's identity. So in the example of an IPTV environment, a root CA provider, which could be the vendor of the security system issues a certificate to the service provider, who in turn applies its own digital signature, which subsequently gets forwarded onward to the IPTV subscriber. The streaming of the IPTV content starts once all parties have been validated. CAs also use certificate revocation lists. These lists detail certificates that are no longer valid. A certificate may be declared invalid if it is stolen.

7.2.2.5 Overview of How an IPTV Software Centric CA Solution Works? An example of how a typical software centric CA solution operates is depicted in Fig. 7.4 and explained in the following sections.

(1) Once the IP set-top box is powered up, it scans the network for a server that provides decryption keys and authentication services. The IP address parameter of the key management and authorization server is typically preconfigured on the IPTVCD.

(2) After locating the server, the secure IP set-top box software client informs the data center of its presence on the broadband network and requests a key to decrypt incoming video content.

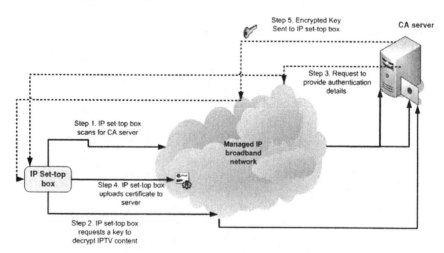

FIGURE 7.4 An IPTV software centric solution

(3) Once this request is received, the server will prompt the IP set-top box software to provide authentication details.

(4) The IP set-top box software will then upload a digital certificate to the server.

(5) Once received and authenticated by the server, the key is retrieved from the database, encrypted and returned back to the IP set-top box client software. The key is then used to decrypt the video content before passing onwards to the decoder chip for further processing.

7.2.3   Hybrid Hardware and Software Solutions

The hybrid approach embeds the decryption capabilities directly onto the IP set-top video box and does not require a smart card. So the decryption of the content takes place inside either the video processor or a special security processor. The decryption of incoming signals involves the set-top processor communicating directly over the two-way IP based networks with CA servers located at the headend.

7.3   CA INDUSTRY INITIATIVES

In addition to the proprietary IPTV security solutions in the marketplace, there are a couple of industry groups developing technical specifications and initiatives that are used to fight piracy within the digital TV and IPTV industry sectors.

7.3.1   Securing DVB Digital Channels

When the DVB standards were originally developed, it was recognized that provision must be made for CA systems to be used when there is need to control access to a broadcast service. Recognizing this fact, DVB opted for the development of two new protocols: SimulCrypt and MultiCrypt.

7.3.1.1   SimulCrypt   A SimulCrypt based system allows two CA systems to operate within a particular geographical area in parallel. The system uses multiple set-top boxes, each using a different CA system, to authorize the programs for display. The general architecture of a SimulCrypt access system is based on the transmission of ECMs and EMMs required by each CA system across the TV network. Each individual set-top box recognizes and uses the appropriate ECM and EMM needed for authorization while ignoring entitlement messages that are destined for other set-top boxes. In SimulCrypt based systems, scrambling and encryption are separate processes.

7.3.1.2   MultiCrypt   The MultiCrypt concept allows the use of multiple CA systems in the same digital set-top box.

7.3.2 European Antipiracy Association AEPOC

AEPOC is a European based organization, which was established in the late 1990s and consists of 35 leading digital television and telecommunication companies including broadcasters, CA providers, transmission infrastructure vendors, and manufacturers of related hardware. The organization works with the broadband industry sector to fight piracy of IP based content. For more information visit www. aepoc.org.

7.3.3 National Cable and Television Association (NCTA)

The NCTA, a trade association representing the U.S. cable industry sector has made a commitment to the FCC to roll out a software centric security system called downloadable conditional access (DCAS) that allows cable TV providers to download CA software of their choice directly to digital set-top boxes. Some of the key milestones identified by a recent NCTA submission to the FCC are shown in Fig. 7.5.

Using the software centric approach is seen as an alternative to deploying CableCARD plug-in security modules. By combining the cable industry's next generation DOCSIS 3.0 technology platform with DCAS, cable operators are able to use this software centric security environment to protect content traversing both RF and IP cable networking infrastructures. Some DCAS technical and general characteristics include

(1) The removal of the need for expensive hardware elements such as physical smart cards.

(2) Interoperability with existing and legacy hardware centric CA systems.

(3) Support provided for the portability of digital set-top boxes. Thus, when a cable set-top box or a hybrid IP model is plugged into any cable TV network

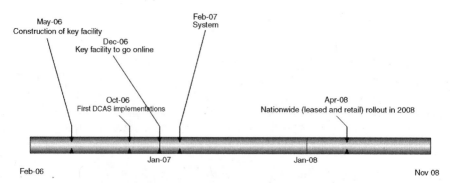

FIGURE 7.5 Evolution of DCAS initiative

with a DCAS security system the chip will automatically download the new CA system onto the internal silicon chip.

(4) The system employs three key components, namely,

- *A secure microprocessor*—The secure microprocessor includes a unique hardware characteristic, which is used to establish trust between the server and the IP set-top box.

- *A secure software based bootloader*—When a DCAS enabled digital set-top box or some other type of digital device is first turned on it does not have a CA system in its secure microprocessor. The real-time operating system used by the set-top box is often unable to deal with complex tasks such as downloading a CA system. Therefore, a small and specialized software program called a bootstrap loader or boot loader is used to download the client element of the CA system from the IPTV service provider's data center or the cable headend. The bootstrap program itself consists of a very small set of instructions and its sole purpose is to load DCAS systems from the headend or IPTV data center servers. Once the CA has been loaded and executed the set-top box is then able to process the various CA commands and instructions, which are issued by the backend server.

- *A keying facility*—The keying facility is responsible for developing and installing security keys in the microprocessor. Additionally, this facility also manages and tracks all the microprocessors.

(5) Protection for high value video services and provides support for two-way interactive IPTV services.

(6) Easy and straightforward replacement of the security infrastructure if a breach occurs.

(7) Designed to be integrated into retail devices such as hybrid IP set-top boxes, game consoles, and portable media players,

There are other industry groups involved with promoting IPTV security standards, however, they are primarily focused on digital rights management systems.

7.4 INTRODUCTION TO NEXT GENERATION DRM SOLUTIONS

Although the CA system protects against the illegal use of IPTV services, it does not protect against the piracy and theft of premium video content. Therefore, a CA system needs to be complemented with a Digital Rights Management environment that controls how consumers use and distribute content to other digital devices in the home. In other words, a layered approach has to be adopted to secure premium content sent over an IPTV network. Such an approach will ensure that content delivered to a subscribers home will not be compromised, which in turn will help to strengthen working relationships between content suppliers and network operators.

The sophistication of the DRM system implemented by a service provider will depend on a couple of factors:

(1) *The types of IPTV services*—There is a debate ongoing within the IPTV industry on the need to apply a sophisticated DRM system to live broadcast channels. In the majority of cases the material is consumed immediately, negating the need for DRM technologies to protect multicast TV channels. The consumption habit of VoD content differs and does necessitate the use of a DRM system.

(2) *The hardware features of the IP set-top box*—Hard disk enabled set-top boxes used to store video content are more conducive to the piracy of digital content when compared to "diskless" IP set-top boxes that simply ingest and process the incoming IPTV stream in real time.

IPTV service providers need to take both of these factors into account when deciding on the level of security features that are required for their DRM systems.

7.4.1 DRM Defined

Transmitting video over an IP based network raises obvious questions on establishing the right to content and protecting that right. To meet the challenge of securing digital content the IPTV industry sector has developed end-to-end DRM technologies to protect theft of content once it is received in the home. In addition to protecting storage of digital content a DRM system also prevents piracy and unauthorized sharing when routed over an in-home distribution network. DRM systems can also be configured to enforce other types of content control regulations such as rental time periods and revocation of rights to a video asset. The strength of the DRM system deployed will vary between service providers and associated security threat levels.

In technical terms, a DRM system is classified as a mechanism of managing and monitoring access rights to copy protected content. The types of content can range from VoD movies to high definition recorded subscription services.

7.4.2 DRM and the Content Protection Value Chain

The IPTV content reception chain is illustrated in Fig. 7.6 and consists of three separate segments.

7.4.2.1 Content Providers   This segment of companies are responsible for creating the original content. The types of companies included in this segment include movie studios, broadcasters, TV production companies, and record companies. The creation of this content can involve a whole range of processing steps from filming and special effects to voice-overs and editing.

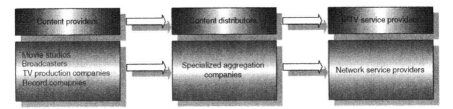

FIGURE 7.6 IPTV content reception chain

The types of content produced by this segment typically vary from sporting events and TV programming to music concerts and various types of movies. Finished video content is generally speaking a valuable asset and ensuring that this content is securely controlled is the first priority for providers. Therefore, a number of business and access rules are applied to each finished asset to define the costs and security requirements of distributing this content. The types of rules can include

- Restrictions on when the IPTV operator can broadcast or unicast the video asset
- The terms of payment for the asset (a fixed fee or a license charge per viewing)

These rules are contractually agreed between the parties and enforced through the DRM system.

7.4.2.2 Content Distributors Rather than dealing directly and engaging with time-consuming licensing negotiations with the content producers, some IPTV network operators are starting to turn to companies that specialize in aggregating and licensing content libraries.

These aggregation companies are effectively brokers that act as single point of contact for the IPTV service provider. They prepare the content assets for delivery to the IPTV service provider. Preparation of the content typically can include

- Encrypting the content
- Encoding the video assets
- Applying additional DRM viewing rules
- Adding meta-data such as censorship and subtitle information

A single point of contact and easier management of the source VoD and broadcast channel content are the main benefits for telecom operators who use content distributors. Once prepared the IPTV content is then ready for licensing to IPTV service providers. A variety of different network media may be used to deliver content to the IPTV service provider's data center.

7.4.2.3 IPTV Service Providers The last link in the content reception system is the distribution of IPTV content to consumers. Broadband service providers, satellite companies, wireless and cable TV companies are examples of IPTV service providers. Their main functions include

- Managing and growing a base of IPTV subscribers.
- Negotiating licensing deals with various content producers and aggregators across the world.
- Receiving and processing incoming video content.
- Managing a high speed IP based network.
- Delivering video assets to IPTV households.
- Ensuring the various business rules are enforced through a combined CA and DRM security system.

Once a contractual relationship has been established with the content supplier the electronic components that make up the IPTV data center process the incoming signals and route onward to the IPTV distribution network. Note that the content, which is received by the service providers can be prerecorded or come from a live feed.

7.4.3 DRM Software and Hardware Architecture

A simplified block diagram depicting the interoperability of the various hardware and software modules used by an end-to-end DRM system is depicted in Fig. 7.7.
The key features of the diagram are discussed in following sections.

7.4.3.1 Digital Watermarks It is possible to identify IPTV content with its owner by marking the material with a definite pattern of bits known as digital

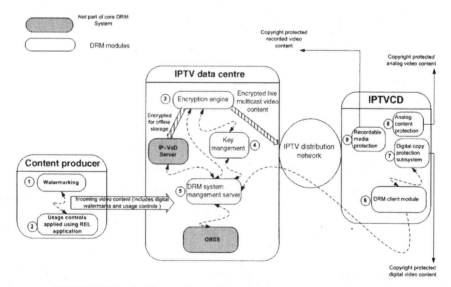

FIGURE 7.7   Architectural overview of end-to-end IPTV DRM system

watermarks. The technique is also known as data embedding and information hiding. These bits are embedded in the digital video or audio clip and are scattered throughout the file. Unlike paper watermarks bits are totally invisible or inaudible, making it difficult for anyone to identify or tamper with them. The type of information stored in a watermark varies between implementations. The name of the content creator, its distributor, and the licensing models are typically included in watermarks. In most cases digital watermarks are made resilient enough to withstand any changes that may be made to the file, such as editing, compression, decompression, encryption, and changing into analog format. In some configurations watermarks are "lightly" applied to the video content and are destroyed in the event that the material is changed, thus highlighting the fact that tampering has taken place.

Watermarks enable copy protection by either restricting access to the material that has a copyright or by hindering the process of copying itself. It also enables IPTV telecom operators to detect the illegal redistribution and piracy of video content. Watermarks remain traceable throughout the IPTV distribution process. The special nature of digital watermarks makes it essential to view them with the help of special software detection programs. In this example the watermarking has been applied at source.

7.4.3.2 A Rights Expression Language (REL) Application
REL is a DRM technology that allows content owners to define usage rights for IPTV content assets. The goal of any REL is to express the preferences of content owners in terms of copyright, restrictions, and usage control. A typical REL consists of a number of standard elements, which are listed and briefly described in Table 7.4.

The various elements defined in Table 7.4 are formatted into a file using the XML syntax. Of the various RELs available on the market as of this writing the MPEG-REL, which is included in part 5 of the MPEG-21 standard, is the most relevant with regard to IPTV deployments. Standardized in 2006, MPEG-REL is derived directly from the XrML 2.0 rights language. For more information on this language visit www.xrml.org.

7.4.3.3 Encryption Engine
DRM uses customized encryption in order to restrict viewing of content and ensure that only authorized and paying customers are able to access IPTV content. The text used for encryption is called cipher text. Different encryption algorithms and keys are used to provide different levels of security. The keys are used to lock and unlock content and the rights associated with that particular piece of content. The strength of the protection can be read from the number of possible keys and the size of the key. Many of the DRM systems currently on the market use the AES algorithm standardized by the U.S. National Institute of Standards and Technology. As shown in the diagram, the encryption engine deployed at the IPTV data center typically takes care of this function. The main function of this module is to ingest the incoming video content from a content aggregator or source provider and use the keys obtained from the key management module to encrypt both real-time and offline video content. The encrypted real-time video content is outputted directly onto the live network, whereas the offline

TABLE 7.4 Core REL Elements

Element	Description
Resource	In the context of an IPTV environment a resource defines either an IP-VoD asset or a multicast channel.
Agent	The definition of this term varies between RELs. In most cases however agent refers to the company or person who holds the rights to the content.
Rights	The terminology used to define this element varies between RELs. However, the purpose of this REL element remains consistent, in the sense that rights determine a set of allowed actions over the IPTV resource. Typical rights applied to IPTV assets include • Ability to decrypt and view the resource content. • Ability to store locally on the IPTVCD. • Ability to share the content with mobile IPTVCDs. • Ability to make copies of the IPTV resource.
Constraints	As the name implies the constraints element applies restrictions to IPTV resources. The number of copies that are allowable is one example of a constraint that is often applied to IPTV resources.
Conditions	This element defines a set of conditions that need to be met before the end-user can exercise any of the rights associated with the IPTV resource. Payment is probably the number one condition applied to most resources.

content is sent to the VoD server for storage. The encrypted content stored on the server is subsequently played out to the IPTV end user, when an instruction is received from the OBSS.

7.4.3.4 Key Management   The main function of this module is to generate keys for the encryption module. This module in particular is located in a secure area within the IPTV data center, which is accessible to a limited number of employees.

7.4.3.5 DRM System Management Server   The management server sits at the core of the end-to-end DRM system. It interfaces with the OBSS and receives details on viewing rights for individual IPTV subscribers. The DRM management server ensures that these access rights are enforced. In addition to enforcing protection rules for the content owner and making sure that end users get what they paid for, the DRM management server is also responsible for collecting statistics on the number of views that each piece of IPTV content receives. Such information is required to ensure that the service providers pay appropriate royalty fees to content owners. This information may also be used for billing purposes.

7.4.3.6 A DRM Client Module   Most if not all IPTVCDs will include an Ethernet port, which can cause difficulties when trying to protect against unauthorized playback and illegal duplication of content. Therefore, to ensure safe and authorized usage of digital video content, IPTVCDs can include an advanced DRM software subsystem. This module protects IPTV content from unauthorized playback or duplication and registers the digital rights of a content owner by noting the identity information. It also manages the business rules that are assigned to an IPTV broadcast channel or VoD asset. Examples of business rules include

- A timeframe for viewing the content
- Types of devices that are permitted to play the content
- Costs associated with copying the IPTV content

It is worth noting that DRM systems allow IPTV telecom operators to remotely modify these business rules. Once the business rules are processed the DRM client module proceeds to decrypt the file.

7.4.3.7 A Digital Program Copy Protection Subsystem DRM limits the kind of actions that the receiver of IP content can perform with regard to that content. This enables owners of high value IPTV services to prevent unauthorized distribution of their content.

The widespread deployment of set-top boxes that support high capacity digital output interfaces raises concerns about the whole area of protecting the copyright of digital audio and video content as it leaves the IPTVCD. Table 7.5 explains the main digital output interface types used to transport IPTV signals around a house.

Without some element of protection it would be very easy for pirates to use these various types of digital ports on an IPTVCD to make perfect copies of video and

TABLE 7.5   Types of Digital Output Interfaces

Connection Type	Description
USB	USB is a widely used industry standard output port that supports a full range of popular PC peripherals. Version two of the technology allows IPTVCDs connect to a home network at very high speeds. These rates are capable of carrying multiple streams of IPTV content.
IEEE 1394	IEEE 1394 has become a mainstream consumer electronics interface and is found in a wide variety of home entertainment devices including MP3 players, digital video cameras, digital televisions, set-top boxes, digital projectors, DVRs, and DVD players.
DVI	Short for Digital Visual Interface, DVI is defined as a high speed serial interface that provides digital connectivity between devices such as set-top boxes and digital displays. It is designed to transfer uncompressed real-time digital video and supports the resolution settings of high definition television.
HDMI	IPTVCD manufacturers have started to add High Definition Multimedia Interface (HDMI) outputs to their product ranges in recent years. The technology behind HDMI is quite similar to DVI and preserves the picture quality when it gets sent from the IPTVCD to the digital display.

audio content. Therefore, to minimize the risk of such occurrences a digital program copy protection subsystem is typically added to the IPTVCD as part of an end-to-end DRM system. The function of the digital program copy protection subsystem is to secure digital programs that are outputted in digital format. There are two major copy protection technologies used to secure content traversing the digital connections.

DTCP and DTCP-IP Link Protection Technology   Digital Transmission Content Protection (DTCP) is a protocol that prevents someone from illegally recording content at all or simply limits the number of copies allowable by the content owner. The security standard is also known as 5C and is maintained, developed, and promoted by Intel and four other companies, namely, Hitachi, Matsushita (Panasonic), Sony, and Toshiba. Additionally, Microsoft has also pledged support for DTCP copy protection technologies. DTCP uses a cryptographic protocol to protect illegal copying, intercepting, and misuse of IPTV content as it traverses an in-home network. Similar to other content protection systems DTCP employs a layered approach to securing digital content. The elements of the DTCP technical architecture are logically portioned into five separate layers, namely, authentication, key management, encryption, copy control, and system renewability. The following sections give a brief overview of these five layers when used in an IPTV infrastructure.

AUTHENTICATION OF IPTVCDS   5C defines two variants of authentication, full and restricted. Full authentication applies to all types of content, whereas the restricted version facilitates the protection of copy-one-generation and no-more-copies content specifically.

KEY MANAGEMENT   Three different key types are used by DTCP to enable IPTVCDs support various exchanges between source and destination devices, namely, authentication, exchange, and content.

ENCRYPTION OF CONTENT   An advanced cipher or algorithm is used to encrypt the IPTV content or WHMN services that are exchanged between DTCP compliant devices. DTCP specifies the use of Hitachi C6 cipher and allows for inclusion of optional ciphers as part of the implementation.

COPY CONTROL INFORMATION (CCI)   This is the rights management element of DTCP and comprises a number of bits, which are typically carried as part of the IPTV stream. Table 7.6 summarizes the various states that CCI represents:

SYSTEM RENEWABILITY   This layer deals with generating and delivering system renewability messages (SRMs). SRMs contain lists of devices that have become compromised. These lists are maintained and generated by the organization responsible for licensing the intellectual property and cryptographic materials used to implement DTCP—the Digital Transmission Licensing Administrator (DTLA).

TABLE 7.6 Different CCI States Used by a DTCP Enabled IPTVCD

CCI Bit Value	State	Description
11	Copy never	This imposes the highest restriction level of all four CCI states and typically applies to premium content types such as DVD movies.
10	Copy one generation	This refers to the copying of content from a prerecorded media.
01	No more copies	This state indicates that no more copies are allowable.
00	Copy freely	No authentication or encryption is required and the content is allowed to be copied freely.

There are a couple of mechanisms for delivering updated SRMs to DTCP enabled devices connected to in-home networks, namely, broadband networks or via new devices that are purchased and connected into the digital home ecosystem. Combining the five layers described above creates a secure environment for transferring IPTV content.

Consider the simple example in Fig. 7.8. In this case, a portable media player (PMP) has requested to download an IP-VoD asset from a DVR. As illustrated the steps associated with using DTCP in this particular exchange of data involve the following:

- *Step 1. Content request*—The PMP makes a request for the copy protected IP-VoD asset.
- *Step 2. Stream initiated*—Streaming of encrypted content commences. This stream is tagged with a CCI identifier that defines the rights associated with the stream.

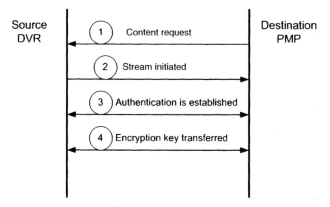

FIGURE 7.8 Example of DTCP in operation

- *Step 3. Authentication is established*—Depending on the CCI tags, the authentication may need to be executed.
- *Step 4. Encryption key transferred*—Keys are exchanged and the IP-VoD asset is decrypted.

Note that DTCP typically operates over USB and IEEE 1394 interfaces. An upgrade to the specification called DTCP-IP was recently developed to protect IPTV content over Ethernet and wireless LAN networking connections. Further details on DTCP are available at www.dtcp.com.

HDCP An encryption system developed by Intel called the High-Bandwidth Digital Content Protection (HDCP) standard is typically built into most high definition capable IPTVCDs. It is included as part of the hardware platform and encrypts high definition IPTV content before sending it to a display device that include either a DVI or a HDMI digital interface. The HDCP compliant digital device will then decrypt the content and display on screen. This ensures that content is not intercepted as it travels from the IPTVCD over to the display panel. The whole process of encrypting and decrypting the data is based on a set of unique secret 40 or 56 bit keys combined with a nonsecret identifier called a key selection vector (KSV). These keys are obtained from an organization, which acts as a CA called the Digital Content Protection, LLC. In addition to encrypting and decrypting digital video content, HDCP includes an authentication mechanism to verify that the client device, typically a flat screen display, is licensed to receive the content. So if an illegal recording device attempts to copy video content, the IPTVCD will not permit the transmission of the encrypted content. As shown in Fig. 7.9 the software for HDCP is typically built into the decoder chip.

Note that a HDCP-equipped IPTVCD is able to determine whether or not the display device it is connected to also has HDCP functionality. If a digital display, which does not support HDCP connects to the HDMI or DVI interface then the IPTVCD in some cases will lower the image quality to protect the HD content.

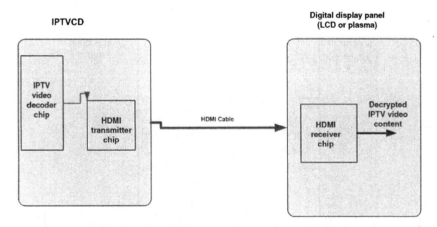

FIGURE 7.9 Implementation of HDCP on an IPTVCD

With regard to dealing with security compromises HDCP employs a technique similar to the approach adopted by DTCP. Once the Digital Content Protection, LLC determines that a HDCP enabled IPTVCD has been compromised then the corresponding KSV is added to a revocation list. Once the list is updated it is encapsulated inside an SRM and delivered to the HDCP enabled IPTVCD. This list is subsequently referenced during the authentication process to ensure that all HDCP receiver devices are authorized to receive the copyrighted IPTV content.

FCC's Broadcast Flag Mandate   In addition to the above IPTVCD interface copy protection technologies, the FCC in the United States proposed that manufacturers of products for the digital TV industry sector should be mandated to include a DRM technology called broadcast flag. The proposal involves the insertion of a single bit flag into the digital bit stream. The IPTVCD would identify the flag in the stream and execute appropriate security measures. At the time of writing a legal challenge was taken against the proposal and the court ruled against the FCC. Thus, the FCC broadcast flag mandate is not of immediate concern to IPTV providers.

7.4.3.8   An Analog Content Protection System   In addition to applying copyright protection technologies to digital ports, it is also important to use a security mechanism at analog interfaces to ensure that content copyright is not compromised. At these interfaces digital IPTV content is outputted in analog format, captured by a connecting device, and reconverted back into digital format. Under this process, the rights that were originally associated with the IPTV content are lost. As a result, IPTVCD manufacturers can also implement copyright protection technologies at analog interfaces. There are three dominant security technologies used to protect analog interfaces:

- CGMS-A
- Watermarking
- Macrovision

We explore these technologies in the following subsections.

CGMS-A   A mechanism called Content Generation Management System for Analog (CGMS-A) is one of the more commonly used systems for preventing the unauthorized redistribution of content via IPTVCD analog output interfaces. CGMS-A uses two copy control information bits, which are carried in the vertical blanking interval portion of the video picture. These bits are added to the video picture at source and are used to define the rights associated with the video picture:

(1) Copy never
(2) Copy one generation
(3) Copy no more
(4) Copy control not asserted

Note that CGMS-A has been adopted by the CEA for inclusion in some of its specifications. Full details of the CGMS-A technology platform are published in the IEC 61880 standard.

Watermarking Watermarking can also be applied to analog based signals. In fact, the technology has been used for many years to inhibit unauthorized copying of movies on VCRs and DVD recorders. The fundamental of how analog watermarking operates is similar to the techniques used for watermarking digital content.

Macrovision Macrovision is a video copy prevention scheme that inserts pulses into the VBI portion of the analog signal. These signals do not affect the picture quality; however, they prevent recording of the content.

7.4.3.9 Copyright Protection for Recordable Media Some of the advanced IPTVCDs include DVD units that allow end users to record IPTV content. It is generally accepted by the industry that recording for personal use is a reasonable activity. However, in an effort to limit the copying capabilities of individual end users, the industry uses two removable media copy protection mechanisms, namely, CPRM and VCPS.

CPRM Content Protection for Recordable Media (CPRM) is used to copy protect devices such as DVD-Rs and memory flash cards. Under this DRM mechanism the IPTV end user is limited to making a single copy of a particular IPTV asset for personal use. As with other security schemes CPRM makes extensive use of encryption and secret keys. Revocation lists of compromised IPTVCDs are also supported.

VCPS Video Content Protection System (VCPS) is used to encrypt video recordings on DVD+R and DVD+RW disks. The technology uses a 128-bit AES cipher when encrypting MPEG-2 content streamed from an IPTVCD directly to a DVD disk.

7.4.4 Understanding the DRM Process

A DRM system can be applied to both VoD content and live streaming IPTV channels:

> *Applying DRM to VoD Assets*—The delivery of on-demand content involves sending encrypted content from a video server to an IPTVCD for subsequent viewing. When content is stored on an IPTVCD, the risk of piracy is quite high because it can be forwarded onward to other digital devices. The basic steps involved in using DRM to protect an IP-VoD asset are
>
> • *Step 1. Registration of content owners rights*—Once the video asset has been encoded and prepared for transmission over the IPTV network, it is sent to the DRM subsystem. This subsystem consists of a software

application that registers details of the owner and the rights assigned to that particular video asset. In addition, the software applies additional information to the video asset such as licensing details.

- *Step 2. Post the protected IPTV asset on the VoD server*—Once protected the asset is posted to the IP-VoD server and made available for download over the IP based network.

- *Step 3. Playing the protected IPTV content*—To play the protected IP VoD asset the IPTVCD must include a DRM client piece of software that is capable of processing the file. This generally involves executing the licensing rights and playing the content on the subscribers TV.

Applying DRM to real-time IPTV—Traditionally, DRM was only applied to VoD content assets; however, more and more service providers have started to apply real-time DRM to live broadcasts of IPTV channels. The deployment of a real-time DRM system generally involves connecting a server running the DRM software to the output of the encoder. This ensures that the content is protected prior to delivery over the distribution network.

7.4.5   Relevant DRM Industry Initiatives

The motion picture content owners, consumer electronic companies, IT equipment manufacturers, and network service providers are afraid of the "napsterization" of the video industry and have placed a huge importance on the development of standardized DRM systems. To further the work on DRM standardization a number of consortiums and industry groups have been formed, which include

Moving Pictures Experts Group (MPEG)
Secure Video Processor (SVP) Alliance
Coral Interoperability Framework (CIF)
CableLabs
ISMA
Digital Watermarking Alliance (DWA)
TV-Anytime

The following sections provide further detail on the contributions these various groups are making to the area of DRM.

7.4.5.1   MPEG   As discussed previously the MPEG group has develop a standard called MPEG-21, which adds a DRM layer to existing IPTV compression schemes, such as MPEG-2 and MPEG-4.

7.4.5.2   SVP   The SVP alliance consists of a diverse group of companies ranging from content owners and distributors to IT and telecommunication companies. The stated objective of the alliance is to promote a content protection technology called

SVP. Some of the technical and commercial features of the SVP specification include the following

- It is an open specification, which is compatible with existing copy protection standards.
- The SVP specification relies on a media processor that includes a security module, which contains the various cryptographic components and rights rules required to protect digital content. In addition to the core security functions, this processor is also involved in decoding incoming IPTV content and managing digital certificates. The content is encrypted with the 128-bit AES scrambling algorithm.
- It is capable of enforcing usage rights on a range of different types of digital content including standard definition and high definition IP video content.
- It supports the enforcement of a number of different types of content on demand business models including

   (1) Content sales: In this scenario the content is delivered with a permanent license to the SVP-enabled IPTVCD. Once delivered the purchaser owns the video asset for life.
   (2) Content rental: In this scenario the content is delivered with a lease agreement to the SVP-enabled IPTVCD. The lease time period varies between IPTV service providers. SVP also allows consumers who lease a video asset to extend that time period of the lease or else buy the content outright.
   (3) Push VoD: When this model is deployed the content along with the license criteria are delivered across the IP broadband network. The content is stored on the IPTVCD disk until a time when the IPTV subscriber decides to either purchase or lease the video asset.

A highly simplified graphical view of two SVP compliant devices communicating with each other is illustrated in Fig. 7.10.

As shown the SVP security platform is primarily based on a secure silicon chip. There is however a software driver that deals with interacting with the operating system and managing media processor functionality.

7.4.5.3 CIF   Lack of interoperability between different DRM systems is an issue that limits the portability of IPTV content. From a practical perspective if an IP-VoD title is purchased and downloaded to a DVR, this title comes with its own set of DRM mechanisms that are unique to that particular end-to-end networking environment. Without an interoperability framework it would be impossible for the end user to play this title on another destination system.

To address the issue of interoperability between different DRM systems, a group of companies have formed an organization called the Coral Consortium to progress the development of a generic DRM interoperability framework. The cross-industry group has been working over the last couple of years on developing a set of

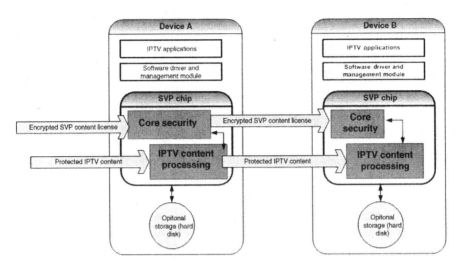

FIGURE 7.10 SVP interdevice communications

specifications that are designed to provide interoperability between disparate DRM systems. The structure at the highest level of the framework is split into of number of specifications. Table 7.7 summarizes these documents.

The combination of these specifications provides a framework for creating, distributing, and using protected content across a number of different content providers and DRM systems. CIF interfaces facilitate the exchange of data between these various systems.

In the context of an IPTV environment, this initiative will allow end users to move IP-VoD and other types of IPTV content between different protected environments. Full coverage of the Coral specifications is not possible here, but two aspects are worth highlighting in terms of importance, namely, communications between devices in the CIF framework and trust management. With regard to communication, CIF uses XML to format the interchange of messages and information. All of the messages comprise of a header and a payload. The header includes items of trust management information and the payload contains request and response messages that are encrypted using XML security algorithms. The mechanisms used by CIF to facilitate trust management are based on established security technologies such as X.509 digital certificates and public/private keys. This facility ensures that content is delivered in a format that is only viewable by IPTVCDs that are trusted by the source content owners. In addition to providing comprehensive descriptions of deploying a framework that allows interoperability between disparate DRM systems, the CIF specifications also document a number of scenarios that relate to "live" world situations that occur during the day-to-day operation of in-home digital networks.

7.4.5.4 CableLabs OpenCable Unidirectional Receiver (OCUR) is a CableLabs initiative that aims to extend CableCARD functionality to media center PCs. It facilitates the delivery of traditional broadcast and IP based digital video content services directly to a PC without the need for a digital set-top box. OCUR

TABLE 7.7   Overview of CIF Suite of Specifications

Specification Title	Brief Description of Contents
Coral Consortium Architecture Specification v3.0	This specification defines the technical architecture used by CIF to exchange content and rights between disparate DRM systems.
NEMO Architecture v2.0	This specification deals primarily with a term defined by the CIF called NEMO. Networked Environment for Media Orchestration (NEMO) is defined in the specification as a decentralized messaging ecosystem in which service providers and consumers can discover one another and interact in a trusted way.
NEMO Profiles v2.0	This specification provides further information on the use of NEMO profiles. For instance, it specifies that the main profile uses industry standard mechanisms such as HTTP, XML, and X.509 authentication within the overall NEMO architecture.
NEMO Trust Management Bindings v2.0	As the name suggests this specification describes the various trust management mechanisms used by an end-to-end NEMO networking system.
NEMO Message Bindings v2.0	This specification provides further details on the use of XML within the CIF environment.
NEMO Security Bindings v2.0	This specification deals with defining systems for performing secure message exchange between different nodes associated with the CIF. The systems make use of various techniques such as encryption, keys, and digital signatures.
NEMO Policy Bindings v2.0	As the name suggests this 27 page specification describes ways of defining requirements for protocol bindings.
NEMO Secure Messaging v2.0	This specification deals with protecting messages that are exchanged between CIF components.

compliant devices include a CableCARD slot and an IP interface, which provides connectivity to the home media server. A simplified graphical view of three OCUR enabled devices connecting via a wireless home network to a media server is provided in Fig. 7.11.

As shown each OCUR connected to the media server includes a DRM software module, which is approved by CableLabs. Some of the characteristics of the OCUR specification include the following:

- The CA and DRM subsystems are clearly separated.
- The specification supports the deployment of in-home multizone IPTV systems.

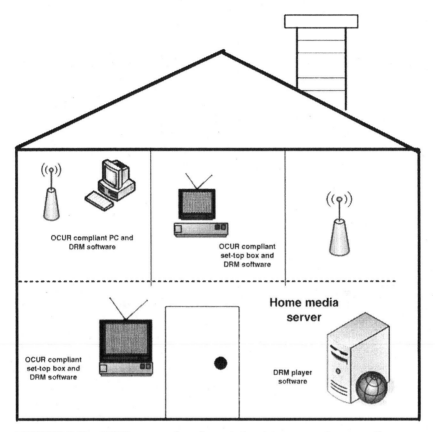

FIGURE 7.11   OCURs connecting via a wireless home network to a media server

- Output is only permitted under the protection of a CableLabs approved DRM system.

7.4.5.5 ISMA   The organization has been working in the area of IP video protection and has released a specification called ISMACryp. Also known as ISMA encryption and authentication, ISMACryp defines a DRM framework that may be applied to IPTV services.

The technical characteristics of this system include

- A REL is generally used to define the rights associated with an IPTV asset.
- ISMACryp accommodates a variety of different key management systems.
- The encryption algorithms used in an ISMACryp compliant DRM system are replaceable when a security breach occurs.
- Algorithms used by an ISMA DRM system occupy a relatively small amount of IPTVCD memory space.

- The specification states that the strength of any algorithms used in an ISMACryp compliant system must be at least 128-bits in length.
- SDP messages and a protocol called IP network multipathing (IPMP) are used for encryption signaling.
- RTP is used as the transport layer protocol and Secure Real-time Transport Protocol (SRTP) is used for authenticating real-time IPTV packets.

Note that ISMACryp provides support for a wide variety of encoding technologies, including VC-1, MPEG-2, and H.264/AVC.

7.4.5.6 DWA   As described on its Web site, the DWA is an international group of industry leading companies that was formed in 2006 to promote the value of digital watermarking to content owners, industry policy makers, and consumers. For more information visit www.digitalwatermarkingalliance.org.

7.4.5.7 Open Source DRM Systems   This relatively new and emerging category of DRM solutions uses open source software code to enforce DRM within an IPTV networking environment. One of the most high profile initiatives in this space is DReaM (DRM Everywhere Available), a project whose goal is to create an open-source, royalty-free DRM standard. Further details on this initiative are available at www.sun.com.

In addition to the various DRM industry initiatives outlined above a number of company specific DRM systems are also commercially available to IPTV providers. Further information and market analysis of the key players in the IPTV DRM space is available at www.tvmentors.com.

7.5 IPTV INTRANET PROTECTION

Most IPTV services rolled out are done so in a closed networking environment. Therefore, it is critical that these services are protected from security threats such as malicious traffic, viruses, denial of service (DoS) attacks, and spyware. Breaches on an IPTV network could result in any of the following:

- Overwhelming of multicast TV servers resulting in outages.
- Introduction of viruses.
- Loss or damage of data stored on OBSS, IP-VoD, middleware, and CA servers.
- Breaches of personal IPTV end-users privacy.
- Huge financial loss.
- Pirating and distribution of sensitive information and copy protected content to unauthorized recipients.

To mitigate against these threats IPTV service providers generally implement a tight security policy that incorporate a security tool called a firewall. An Internet firewall is a hardware or software system that implements a security policy between two networks. Firewalls can consist of a single router or may be comprised of a combination of components such as routers, computers, networks, and security software. The exact combination of components chosen to build a firewall depends on the level of security required. A firewall is normally located at the interface between the IPTV operator's private network and the wide area network connection. The company's security policy will generally stipulate that all IP traffic between these two networks is examined by the firewall. A firewall impedes the ability of outsiders to gain access to the IPTV service provider's internal broadband network. Note that in some instances a scaled down firewall or security appliance is also installed between the backend IPTV equipment and client IPTVCDs to ensure that server and encoder attacks are not launched during interactions with end users.

A common entity of most firewalls is a router, which is capable of examining each IP packet that passes through the firewall and filters them according to a set of authorization rules. The router checks the header of each IP packet and examines the payload content and addresses contained within each packet. This information is compared with a set of criteria, which have been defined by the IPTV service provider. If the packet of data is acceptable, then it is forwarded onward to its destination point; otherwise, it is discarded. It should be borne in mind that the presence of a firewall alone does not always guarantee the protection of a private IPTV network and needs to be used in conjunction with other security measures. Protecting end-users identities is another critical element of maintaining high security levels. This is typically achieved through the use of an IPTV identity management system.

SUMMARY

After seeing the affects of piracy in the music industry, TV studios are insisting that a secure CA system is in place before making the video content available to IPTV service providers. A CA system prevents unauthorized access to linear TV programming and IP-VoD services. When a CA is used the video information is encoded and digitally encrypted and is only accessible to authorized users via encrypted control words (keys) sent with the IPTV stream.

In addition to hardware centric CA systems, there are moves within the industry to deploy software centric CA systems that utilize PKI technologies to securely encrypt and decrypt video content. There are also a number of hybrid solutions available to service providers who want to combine the features of smartcard-less and hardware centric CA systems together.

Once content is securely delivered to the IPTVCD another security subsystem is required to manage how the content is used and distributed within the home. A digital rights management system is typically used for this purpose. The DRM system protects the valuable assets of content providers. A DRM system is designed to prevent unlicensed distribution of digital content once it is downloaded by an IP device. A fully implemented DRM system makes it difficult for hackers and thieves to steal digital content. Additionally, it allows content providers to securely distribute their material to IPTV telecom providers. Content providers use DRM to enforce business rules to their video assets whereas content distributors and IPTV operators use DRM systems to control access to their digital products.

DRM systems use five major tools to protect the misuse of IPTV content.

(1) *Encryption*—DRM interacts with the CA subsystem to carry out this function. The encryption function allows service providers to encrypt IPTV content at the source, preventing it from being transmitted in the clear.

(2) *Key management*—This module is responsible for generating encryption keys for the DRM system.

(3) *Watermarking*—This tool helps IPTV network operators and content owners to identify IPTV subscribers who are involved in the piracy of video content.

(4) *A DRM client module*—A DRM client is installed in the IPTVCD to decrypt the incoming IPTV stream and enforce business rules to various video asset types.

(5) *A program copy protection subsystem*—The function of this subsystem is to secure IPTV content that is outputted on a high speed digital interface. DTCP-IP and HDCP are emerging copy protection technologies used to secure content outputted from these interfaces. Security measures have also been developed to ensure the safety of IPTV content converted into analog format.

A number of major groups have been formed to work toward the release of a standardized DRM system for delivering video content over an IP based network. The delivery of an IPTV service generally takes place within a closed networking environment. If open Internet access is provided to end users than access is via the IPTV service provider's firewall and protection mechanisms.

8

MOVING IPTV AROUND THE HOUSE

A major challenge for service providers in delivering IPTV has been in distributing signals from the customer's residential gateway to other IPTVCDs located in different locations around the house. Additionally, bandwidth demands on home networking infrastructures are growing exponentially with the adoption of sophisticated Whole Home Media Networking (WHMN) applications such as HD IPTV streaming, user-generated content, and online gaming. The reliable distribution of these bandwidth intensive applications to a variety of different consumer electronic devices almost in any part of the home is only realizable through the use of next generation home networking technologies. The WHMN model consists of a number of components that may be divided into two basic categories: interconnection technologies and middleware software standards. This chapter provides an overview of the common WHMN interconnection technologies associated with the IPTV industry. These include

- GigE
- 802.11N
- HomePlugAV
- UPA-DHS
- HomePNA
- MoCA

Next Generation IPTV Services and Technologies, By Gerard O'Driscoll
Copyright © 2008 John Wiley & Sons, Inc.

In addition, the chapter describes two popular WHMN middleware software standards and finishes with a brief overview of the role that QoS plays in a WHMN environment.

8.1 ABOUT WHOLE HOME MEDIA NETWORKING (WHMN)

TVMentors classifies WHMN as a reliable method of delivering entertainment content including multiple and simultaneous IPTV streams to a variety of different consumer electronic devices in any part of the home. There are three specific requirements that a WHMN interconnection technology needs to meet to enable the delivery of WHMN applications.

(1) *High bandwidth throughput*—The WHMN interconnection technology needs to be able to support very high data rates to support the WHMN applications shown in Table 8.1. As described in Table 8.1, the WHMN interconnection technology not only needs to carry network traffic at average rates but also needs to support the peak data rates at particular times, day or night.

(2) *Reliability*—A WHMN enabling technology needs to be reliable. One of the main characteristics, often used to indicate the level of reliability of a particular type of WHMN enabling technology, is its ability to maintain packet error rates at the same level as the error rate of the incoming IPTV stream.

(3) *Ability to deal with interference*—WHMN interconnection technologies need to deal with interference that occurs within the home and in some cases originating from external sources such as neighbors. Various device types in a home have the potential to interfere with the streaming of WHMN applications, and the enabling technology needs to ensure that performance levels of the services remain high.

(4) *High QoS levels*—WHMN enabling technologies need to ensure that jitter and delays occurring over the network remain low. Additionally, the technologies need to be capable of prioritizing IPTV traffic when bandwidth capacities of the underlying physical network are exceeded. Thus, a WHMN enabling technology needs to support a QoS mechanism. Further information with regards to implementing a QoS system across a home network is described later in this chapter.

TABLE 8.1   WHMN Applications: Bandwidth Requirements

WHMN Application	Average and Peak Bandwidth Requirements
HDTV digital stream (MPEG-2 compressed)	12–18 Mbps
SDTV digital stream (MPEG-2 compressed)	3–8 Mbps
High quality digital voice	Less than a single Mbps
High Speed Internet access	2–4 Mbps
VoD IPTV digital streams	4–6 Mbps

8.2   WHMN ENABLING TECHNOLOGIES

Having established a description of WHMN services, it is first helpful to define a home network and list some of the motivating factors that are encouraging consumers to network devices in their homes before moving onto the actual WHMN enabling technologies. A home network is used to connect a number of devices or appliances within a small geographical area. A basic home network is formed when an RG and an IP set-top box are connected together. Each device is given a unique identifier, typically an IP address to facilitate interdevice communications. Some of the motivating factors that encourage home owners to network different classes of digital devices and computers together include

Leverage existing investments—First and foremost, people want to leverage investments that were made to purchase expensive devices such as computers, set-top boxes, personal video recorders, digital cameras, and cable modems. So sharing hardware resources is the number one motivating factor for consumers investing in new WHMN interconnection technologies.

Shared broadband access—The second most popular motivating factor that is fueling the deployment of home networks is shared broadband access. Home networks in conjunction with RGs allow different members of a family to simultaneously use a single broadband account.

Interconnecting subsystems—Other motivating factors include the ability of a home networking infrastructure to interconnect different types of subsystems together. For example, home security systems are also defined as a network, but instead of interconnecting devices like printers and PCs, in-home security networks connect different types of sensors with a central controller together. Integrating this type of network into an existing PC based home network helps people to expand the functionality of their security system.

Rise in demand for entertainment content—Demand for IP based entertainment services such as accessing IPTV and audio content from multiple locations in people's households is growing at a rapid pace.

The ultimate goal of a home network is to provide access to information, such as voice, audio, data, and entertainment services, from any part of the house. Achieving this goal is one of the key challenges faced by IPTV service providers. It is evident from the figures shown in Table 8.1 that enabling technologies with very high capacities are required to satisfy the requirements for WHMN applications.

There are a number of alliances and consortiums involved in the process of defining WHMN interconnection technologies, which are used by vendors to produce products for the WHMN marketplace. One of the key challenges faced by IPTV service providers is deciding on a technology that will allow them to effectively implement WHMN services across their subscribers in-home networks. A discussion of these enabling technologies is provided in the following sections.

8.3 FAST ETHERNET AND GIGABIT ETHERNET (GIGE)

Before discussing Fast Ethernet and Gigabit Ethernet WHMN interconnection technologies, it is first helpful to understand the basics of Ethernet, its technical architecture, and the various types of Ethernet specifications.

8.3.1 Introduction to Ethernet

Ethernet is a popular and internationally standardized networking technology (comprising both hardware and software) that enables various types of digital devices including IPTVCDs to communicate with each other. Ethernet was originally developed by Xerox in 1973. The Institute of Electrical and Electronics Engineers later standardized it as IEEE 802.3. As a result, people tend to use the terms Ethernet and IEEE 802.3 interchangeably. Ethernet was traditionally associated with an office environment; however, due to its increased features and capabilities Ethernet has become an increasingly important part of the digital home networking industry. The benefits of Ethernet based Home Networks are

- *Proven technology*—There are several million Ethernet users in the world today. Thus, the digital home marketplace can benefit from the global expertise of Ethernet technology built up over the past number of years.
- *Reliability*—Ethernet is very reliable and may used in a home environment to network most types of IPTVCDs including PCs, media servers, and digital set-top boxes.
- *Support for high bandwidth WHMN applications*—Adding WHMN applications to a home network dramatically increases the demand for higher transmission rates. Advanced Ethernet technologies are able to meet the growing bandwidth demands of these types of applications.
- *Wide industry support*—Ethernet is widely supported by many different vendors and is an integral part of a number of different home networking standards.

8.3.2 Ethernet Technical Architecture

The technical architecture of an Ethernet based home network may be divided into two broad categories, namely, the physical network transmission media and the Ethernet communication protocol that operates at layer two of the IPTVCM.

8.3.2.1 Physical Network Media Ethernet is very particular about proper cabling. There are three major types of cables available for Ethernet applications. We explore these network media specifications in the following subsections.

- Twisted pair
- Coaxial
- Fiber optic

Twisted pair cabling   A twisted pair cable consists of a number of wire pairs tightly twisted together to reduce the affect of electronic noise interference and to increase the bit rate throughput. The number of pairs included in a typical cable is four. There are two broad types of twisted pair cables.

UTP   Unshielded Twisted Pair (UTP) is a cable that contains four pairs of insulated copper wire, which is covered in an outer protective layer of polyvinyl chloride (PVC) material. Each of the twisted wires is marked according to a standard coloring scheme. UTP is the preferred type of cabling for home networks that use Ethernet technologies. UTP cable is graded and tested according to a set of industry standards called "categories." Factors such as resistance to electromagnetic noise, fire resistant capabilities, and bandwidth capacity all dictate the category a particular cable belongs to. Table 8.2 provides a brief overview of the various UTP categories.

TABLE 8.2   UTP Cable Categories

Category	Maximum Data Rate	Description
1	Less than 1 Mbps	Standard telephone wire, which is rarely used by structured wiring systems. The diameter of the cable is typically between 22 and 24 AWG (American wire gauge).
2	Supports data rates of 5 Mbps	This category of cable may be used by networks based on token ring technology.
3	Supports data rates of 10 Mbps	Category 3 cable supports voice and data applications.
4	Supports data rates of 20 Mbps	Primarily used by a networking technology called token ring. This category of cable is not used in the residential structured wiring marketplace.
5	Supports data rates of 100 Mbps	This category of cable has a rating of 100 MHz and supports a data throughput of 100 Mbps. It is commonly used to support Fast Ethernet based home networks.
5e	Supports data rates of 1 Gbps	Enhanced Category 5 also known as Cat 5e supports Gigabit Ethernet networking technology. Most new structured wiring installations use Cat 5e cable. Similar to Cat 5, the distance limitation of Cat 5e between a central structured wiring box and the various wall-plates is 90 m or approximately 295 conductor feet.
6	Support 10 Gigabit Ethernet over limited distances	Category 6 cabling is based on the ANSI/TIA-568-B.2-1 standard. Improved transmission performance levels combined with increased immunity from external noise over Cat 5 cables makes Cat 6 cabling an ideal media for streaming WHMN applications around a house.

Newer categories of UTP cable are also available that improve on the performance levels available across Cat 5e cables. For instance, a category 6 cable can operate at frequencies up to 250 MHz. This type of capacity supports the in-home distribution of a whole range of advanced WHMN applications.

UTP is the preferred network media type used by installers of structured wiring systems in the construction of new homes. Residential structured wiring systems generally comply with a wiring standard called EIA/TIA-568. The purpose of this standard is to specifically define structured wiring characteristics. Some of the characteristics that are directly applicable to installing a residential structured wiring system include

- The types of connectors to be used at either end of the UTP cable.
- The topology used to run the UTP cables. Under the standard, separate wires are run between the outlets and the distribution box in a star like fashion—each outlet has its own dedicated run of wire or cable that goes back to the central structured wiring hub. The structured wiring hub is where all of the outside services including broadband and IPTV enter the home. In addition, it includes a number of punch-down modules that are used to terminate all of the UTP cables that are run to each of the rooms. Using the star topology is more efficient and reliable than the traditional approach to wiring a house. It also helps troubleshooting because each cable run is independent from the others. An example of a structured wiring system designed to support the delivery of IPTV is illustrated in Fig. 8.1.

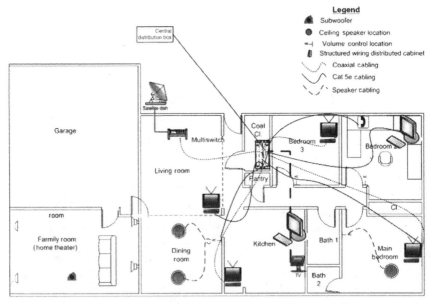

FIGURE 8.1 A structured wiring system capable of supporting WHMN applications

As shown a typical structured wiring system will also include other types of cables for routing video and audio signals throughout a home.

• The recommended distance between the UTP cable and high voltage wires. For example, a minimum of 12 in. of separation should be maintained according to the standard.

• Guidelines on how to run the UTP cable to ensure that the cable is not damaged. For example, if paths of electrical and UTP cable are required to cross, then installers are required to do so at a 90° angle. UTP cable, which is poorly installed, will affect the performance levels of an in-home network.

• Defines pin assignments for terminating UTP cables at the various wall outlets and at the punch-down modules in the structured wiring panel. UTP home networking connections are generally made through modular RJ-45 registered jacks and plugs.

There are two versions of the EIA/TIA-568 standard that describe the pin assignments and color coding for terminating UTP cables, namely, 568A and 568B. The color coding and pair assignments for both types of wiring terminations are outlined in Table 8.3.

As described in the table, the only major difference between both termination configurations is that the orange and green wire pairs are swapped around.

STP   Shielded Twisted Pair (STP) includes an additional foil shield that surrounds the four pairs of twisted cables. This helps to add additional protection against interference from external sources. STP is rarely used to support residential home networking applications because it is more expensive than UTP cable and is sometimes cumbersome to install.

TABLE 8.3   Assignment of Wire Pairs to RJ-45 Jack and Plug Pin-Outs at Wall Outlets

Pin	T568A Color Code on UTP Wire	T568B Color Code on UTP Wire	T568A Pair Number	T568B Pair Number
1	Green/white	Orange/white	3	2
2	Green	Orange	3	2
3	Orange/white	Green/white	2	3
4	Blue	Blue	1	1
5	Blue/white	Blue/white	1	1
6	Orange	Green	2	3
7	Brown/white	Brown/white	4	4
8	Brown	Brown	4	4

COAXIAL CABLE Coaxial cabling, also known as "coax" is a widely used wiring specification for distributing RF TV signals and high speed Internet content throughout homes around the globe. The cable itself consists of a single copper conductor in its center surrounded by a foam dielectric insulating layer and a protective braided copper shield. The inner parts of the cable are protected by an outer insulation sheath. Coaxial cable has a high level of immunity to electromagnetic noise and therefore supports a very high data throughput rate. Similar to UTP cabling, coaxial cable is also graded by various characteristics ranging from size and shielding levels to fire and attenuation rating. Owing to its low signal loss over relatively long distances, Radio Grade 6 (RG-6) is a common type of coaxial cable used in a variety of in-home entertainment applications, including connecting DVDs to A/V distribution systems and connecting satellite dishes to digital set-top boxes. It is important to note that the use of coaxial cable is quite pervasive in most people's homes. This is a particularly relevant point for service providers when considering the use of existing coaxial cable as a means of delivering WHMN services.

FIBER OPTIC CABLE Fiber optic cable technology converts electrical signals into optical signals, transmits them through a thin glass fiber, and then reconverts them into electrical signals. Because it carries no voltage or current, it is the least susceptible of all physical networking media to electromagnetic interference. At the time of publication the number of residential structured wiring systems based on fiber cabling was more or less nonexistent. Most IPTVCDs still use copper or wireless connections in order to interface with an in-home network. With the availability of a number of high capacity interconnection technologies capable of supporting the delivery of WHMN content, it is unlikely that the business case for installing fiber cabling in residential units will become a reality for a number of years.

8.3.2.2 The Ethernet Protocol
Ethernet adopts a layered approach to transmitting data across a home network. Ethernet defines the lower two layers of the IPTVCM, namely, the physical and data link layers. The physical layer transmits the unstructured raw bit stream over a physical medium, and describes the electrical, mechanical, and functional interface to the network. The physical layer can support a wide range of media specifications. Each of these specifications provides different data rates, media, and topology configurations. The data link layer is responsible for getting IP packets on and off the home network and can be subdivided further into the LLC (Logical Link Control) and MAC (Medium Access Control) sublayers. The LLC on the upper half of the layer is responsible for flow control and error correction while the main functionality of the MAC sublayer is to transfer IP traffic to and from the in-home network. It uses a technique called carrier sense multiple access with collision detection (CSMA/CD) to control communications with the home network. Under this mechanism an IPTVCD with an Ethernet interface who wants to transfer video content across the network first "listens" to the network to determine if there is network traffic traversing the cable.

If the home network is busy, then the IPTVCD waits until the network has cleared and then transfers its video stream onto the home network. Once streaming is in progress, a possibility exists that a collision may occur and the stream becomes corrupted. If this occurs, then the MAC layer detects the collision and stops the transmission of the stream. All other IPTVCDs connected to the home network are notified of the collision and the transmitting IPTVCD waits for a random period of time before retransmitting when network conditions have improved. So CSMA/CD is a useful mechanism for aiding the efficient transfer of WHMN data across a network.

8.3.3 Ethernet Specifications Used to Support WHMN Applications

The most popular variants of Ethernet for supporting bandwidth-intensive applications such as WHMN services are the Fast Ethernet and Gigabit Ethernet (GigE) standards.

8.3.3.1 Fast Ethernet   Some of the technical characteristics of the Fast Ethernet specification include

- It supports maximum IPTV transfer rates of 100 Mbps.
- It reliably supports the transfer of video content up to a distance of approximately 100 m or 328 ft.
- Two pairs of the Cat 5 cable are required to support its operation.

8.3.3.2 Gigabit Ethernet (GigE)   Some of the technical characteristics of the one GigE specification include

- It can transmit data at rates up to 1 Gbps. This type of capacity provides support for services such as the transmission of multiple HDTV channels throughout a house.
- It is compatible with other existing Ethernet protocol standards.
- It typically requires special high grade wiring such as Cat 5e and Cat 6.
- All four pairs of the Cat 5e and Cat 6 cables are required to support its operation.
- The distance limitation of 100 m also applies to GigE.

The speed at which Ethernet networks operate continues to increase as new specifications are introduced to the market. The introduction of the 10 Gigabit Ethernet (10 GbE) standard defined as 802.3ae in 2002 is the latest example of how Ethernet will continue to play a strong role in the digital home marketplace over the next number of years.

Although Fast Ethernet, GigE, and 10 GigE technologies offers IPTV service providers robust, reliable, and a proven solution for planning a WHMN deployment strategy, their deployment requires the installation of high grade category 5e or category 6 structured cabling systems. This is obviously problematic for many homeowners whose houses were built without a low voltage structured wiring system. Fortunately, the emergence of "no new wiring" and wireless technologies, such as IEEE 802.11n, HomePlug AV, UPA-DHS, HomePNA 3.0, and MoCA, offers prospects for solving the mass-market adoption of WHMN networking.

8.4 802.11n

A Wireless LAN (WLAN) is a data communication system that uses electromagnetic waves to transmit and receive data over the air. By minimizing the need for wires, they can be either an extension of or an alternative to an existing wired home network. Similar to other home networking technologies, devices connected to a WLAN share the media.

To date, the primary use of WLANs is to wirelessly connect notebook computers to networks at various locations ranging from peoples homes to different types of public locations and at the office. In recent years, WLAN technology has started to get more pervasive and is starting to be incorporated into different types of digital consumer devices including media servers and IP set-top boxes. Supporting applications that run on these devices such as IPTV is increasing the bandwidth and QoS requirements of existing WLAN technologies. To meet these additional demands, the IEEE has come up with a new standard called 802.11n to support emerging high capacity WHMN applications.

Before discussing 802.11n, it is first helpful to understand WLANs in the context of a home networking environment, the technical architecture, and the various types of WLAN specifications created over the past number of years.

8.4.1 IEEE 802.11 Wireless Standards

The development of any new technology is part theory and part practice. A key issue in telecommunications is the adoption of technical standards that govern the interoperability of equipment. A standard sets a norm or performance expectation on the function of the technology—not its implementation. The standard that governs the WLANs industry is the 802.11 family of standards which has been growing and evolving over the past 10 years. The original IEEE 802.3 specification was released in 1997 and operated in the 2.4 GHz frequency band for data rates of between 1 and 2 Mbps. It uses either frequency hopping spread spectrum (FHSS) or direct sequence spread spectrum (DSSS) encoding schemes:

FHSS transmissions constantly hop over entire bands of frequencies in a particular sequence. To a remote receiver not synchronized with the hopping sequence, these signals appear as random noise. A receiver can only process electromagnetic waves by tuning to the relevant transmission frequency. The FHSS receiver hops from one frequency to another in tandem with the transmitter. At any given time there may be a number of transceivers hopping along the same band of frequencies. Each transceiver uses a different hopping sequence that is carefully chosen to minimize interference on the home network.

In *DSSS*, the stream of information to be transmitted is divided into small pieces, each of which is allocated across to a frequency channel across the spectrum. A data signal at the point of transmission is combined with a higher data rate bit sequence that divides the data according to a spread-ing ratio. The redundant sequence helps the signal resist interference and also enables the original data to be recovered if data bits are damaged during transmission.

Since then IEEE has created four main specifications for WLANs: 802.11a, 802.11b, 802.11g, and 802.11n.

802.11a   802.11a operates at the 5.725 and 5.850 GHz frequency band. It uses the OFDM encoding system to achieve data rates up to 54 Mbps. This WLAN technology is ideally suited to high end applications that involve the transmission of video and large image files.

802.11b   802.11b, also known as "Wi-Fi," is a popular option for consumers seeking to install a wireless home network. It operates on the unlicensed 2.400–2.4835 GHz airwave spectrum and when installed it can support data transfer at speeds of up to 11 Mbps. In addition to DSSS, 802.11b also uses a modulation system called a Complementary Code Keying (CCK), which is less sensitive to interference and consequently supports higher data rates. Home networks that comply with the 802.11b standard are the most widely implemented wireless LAN solution on the market today.

802.11g   Like 802.11b, 802.11g operates in the unlicensed 2.4 GHz range and is thus compatible with it. It uses the more sophisticated OFDM encoding scheme to support data transmission speeds of up to 54 Mbps.

8.4.2   Introducing 802.11n Next Generation Wireless Standard

The current 802.11a, b, and g Wireless LAN standards have certain data throughput and range limitations when it comes to transporting high quality WHMN content over a home network. In response to the need for a high performance wireless technology, the IEEE approved the creation of a specialized task group called 802.11 Task Group N (802.11 TGn) in 2003 to define a technology that would provide for the transfer of high bit rate content such as standard definition and high definition IPTV channels—802.11n. Some of the technical features and charac-teristics of the 802.11n standard include the following.

8.4.2.1   Increased Data Throughput   The 802.11n standard incorporates new technologies that increase the maximum throughput rate of WLANs to approximately 600 Mbps. Rather than manufacturing all hardware devices with this data throughput capability, the 802.11n standard allows manufacturers to develop products for different target markets. For instance, many of the initial line of 802.11n networking devices support a raw data rate of 300 Mbps, which adequately supports WHMN applications. One of the features used to increase data throughput is to double the width of the channels used to carry network traffic. Although 802.11n provides support for 20 MHz, it is possible to extend the channel width up to 40 MHz as a means of improving network performance. To enhance speeds even further, 802.11n also includes a provision to bond 20 MHz channels together.

8.4.2.2 Improved Efficiencies at the MAC Layer   The transmission of IPTV packets across a wireless connection is quite challenging and requires the use of a high percentage of overhead bits to ensure the packet arrives at its destination. The overhead bits are added at different stages as packets move down through the IPTVCM. For instance, the MAC layer adds overhead by requiring that each packet is acknowledged by the receiving device. 802.11n includes a technique called aggregation to help reduce the amount of overhead required to transmit packets. By using aggregation 802.11n interfaces are able to combine the payload of multiple small packets into a single larger aggregated packet with a single overhead packet header (see Fig. 8.2). In this example, three packets are aggregated into a single packet for physical transmission across the wireless home network. This increases the proportion or percentage of the packet used for data transfer versus the number of overhead bits. In addition, it reduces the need to acknowledge every small packet that is sent across the network. The combining of packets together reduces the number of acknowledgements occurring during a particular communication session and helps to increase payload-to-overhead ratio during the session.

8.4.2.3 Improved Physical Range   The range of transmission is increased to more than double the range of the current 802.11 suite of Wi-Fi standards.

8.4.2.4 Backward Compatible with Other WLAN Technologies   Support for communications over the 2.4 and 5 GHz frequency bands ensures that 802.11n

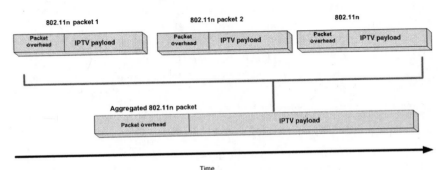

FIGURE 8.2   Using the aggregation technique to combine multiple packets into a single 802.11n packet

devices are backward compatible with legacy 802.11 wireless devices. It is important to note that network efficiency does decrease when 802.11n enabled IPTVCDs are communicating with other devices in the home network that support older 802.11 specifications. A mixed mode 802.11 environment can affect the efficiency of delivering WHMN applications.

8.4.2.5 Uses Multiple Antenna Technology   A technology called multiple-input multiple-output (MIMO) is used to extend the physical range and data throughput of 802.11n. MIMO is a technology that uses multiple antennas to maximize the data throughput over a wireless network.

The approach used by MIMO takes advantage of one of the main negative characteristics of Wi-Fi networks—*multipath signaling.* The multipath signaling process occurs during the transmission of wireless data and is caused by obstacles in the home that reflect and bounce the signal off its original intended path. Thus, the radio signals may arrive at the destination IPTVCD from multiple directions at different times, which can cancel or severely degrade IPTV stream quality. The harnessing of this behavior is called *space divison multiplexing* and is a technique that involves splitting or parsing the signal into different parts called *spatial streams.* Each stream is then transmitted from separate antennas incorporated into the 802.11n source device. On the receiving end the corresponding antennas are used to receive the various spatial streams. Once the multiple streams are received they are subsequently reassembled at the IPTVCD. Thus, each "antenna pair" is able to simultaneously receive and transmit IPTV content, which helps to improve the overall end-to-end throughput of the in-home networking connection.

Thus, the number of antennas built into the 802.11n device will dictate the bandwidth capacity of that particular device. For instance, an IP set-top box that includes two antennas capable of carrying two spatial streams will only have half the data throughput of a notebook computer that includes support for four antennas. Although the notebook computer may be capable of supporting almost double the data rate of the IP set-top box, the notebook computer has a much higher level of power consumption. In addition to using space divison multiplexing, 802.11n also extends the core MIMO technology functionality by designing support for a system that focuses the signals at the target antennas.

With MIMO, the 802.11n receiving device also uses its in-built antennas to identify the clearest path for reception and resolve the data available on multiple signal paths to improve the reception capabilities of high speed WHMN services. This is a major benefit of 802.11n because products based on previous specifications may have supported multiple antennas, however, communication only took place between two single antennas because of limitations in the silicon.

8.4.2.6 Enhanced Powersaving Abilities   802.11n supports advanced powersaving features that are used to conserve battery power. In traditional 802.11 specifications, powersaving involves shutting down the RF interface components when no traffic was on the network. In 802.11n, the MIMO technology increases the sophistication of this interface. Thus, a new approach was adopted that shuts down antennas and RF modules (antennas and corresponding RF components) that are either not receiving or transmitting data and allowing at least one RF module to stay powered up in order to monitor the wireless network. Using this power sharing mechanism significantly decreases the power requirements of 802.11n enabled IPTVCDs.

8.4.2.7   Ease of Installation   In the context of an IPTV installation, all that is needed is an 802.11n wireless access point located at the point where the broadband signals enter the household and an IP set-top box that incorporates an 802.11n chipset located next to where the TV is located. This eliminates the need to run cables, which is a major benefit for IPTV service providers.

802.11n is an important technology for IPTV service providers. Not only does it provide a credible alternative to wired home networking technologies but it also supports advanced WHMN applications, which is key to the success of the IPTV industry sector. The technical characteristics of the various IEEE 802.11 specifications are summarized in Table 8.4.

It is also important to note that 802.11n is designed to be interoperable with the upcoming IEEE 802.11s extended service set mesh networking standard. This proposed 802.11s standard provides a scalable and reliable wireless solution for peer-to-peer Wi-Fi environments such as in-home networks.

8.5   HOMEPLUG AV

Before describing HomePlug AV, it is first helpful to understand how powerline-based networking operates.

Powerline is an emerging home networking technology that allows consumers to use their already existing electrical wiring system to connect home appliances and

TABLE 8.4   Technical Summary of IEEE 802.11 Specifications

	802.11	802.11a	802.11b	802.11g	802.11n
Approval of standard	1997	1999	1999	2003	2007
Modulation scheme	FHSS	OFDM	DSSS and CCK	OFDM, DSSS, and CCK	OFDM, DSSS, and CCK
Width of channels (MHz)	20	20	20	20	40
Maximum bandwidth capacities (Mbps)	1	54	11	54	600
Frequency range (GHz)	2.4	5	2.4	2.4	2.4 or 5
Spatial streams supported	Single	Single	Single	Single	Multiple

IPTVCDs to each other and to the Internet. Home networks that utilize high speed power line technology can control anything that plugs into an outlet. This includes lights, televisions, thermostats, PCs, IPTVCDs, and alarms. Powerline communications falls into two distinct categories: access and in-home. Access powerline technologies send data over the high voltage electric networks that connect a consumer's home to the electric utility provider. The powerline access technologies enable a "last mile" local loop solution that provides individual homes with broadband connectivity to the Internet. To deliver broadband Internet to consumer's homes, a number of utility companies around the globe are using access powerline technologies to convert their infrastructures into a communication network.

In-home powerline technology communicates data exclusively within the consumer's premises and extends to all the electrical outlets within the home. The same electric outlets that provide power will also serve as access points for the network devices. Although the access and in-home solutions both send data signals over the powerlines, they are fundamentally different technologies.

Access technologies focus on delivering a long-distance solution, competing with ADSL, WiMAX, and broadband cable technologies. In-home powerline technologies however focus on delivering a short-distance high bandwidth solution that would compete against other in-home interconnection technologies such as phoneline and wireless. The advantages of in-home powerline technologies are as follows:

- It uses the existing electrical wiring in the home.
- The availability of multiple power outlets in the home eliminates the need for rewiring.
- It is easy to install and cost competitive.
- Data throughput is high and can support the delivery of advanced WHMN applications.

However, there are some technical obstacles to home powerline networking including

Not designed to support IP network traffic—Typical data and communication networks (like corporate LANs) use dedicated wiring to interconnect devices. But powerline networks, from their inception, were never intended for transmitting data. Instead, the networks were optimized to efficiently distribute power to all electrical outlets throughout a building at frequencies typically between 50 and 60 Hz. Thus, the original designs of electrical networks never considered using the powerline medium for communicating data signals at other frequencies.

Signal attenuation—Attenuation describes how the signal strength decreases and loses energy as it transmits across a medium. In the residential powerline environment, the amount of attenuation that a signal experiences is primarily a function of the signal frequency and the distance it must travel on the wire.

Overcoming signal attenuation is important for developing marketable products for the WHMN market since a consumer may want to network two IPTVCDs located at opposite sides of their home. If the signal attenuates excessively, these devices may not be able to communicate at all, rendering the technology ineffective.

Network loading—The number and type of devices that are connected to the electrical network in relation to the home networks overall size determine the network load. Even devices that are not operating and consuming electricity still inhibit the networks performance because they load the line both resistively and reactively, generating impedance mismatches that cause signal attenuation as data travels across the network. New impedance mismatches are created every time a device is plugged in or out of the network. All of these factors combine to create greater network load, which causes increased signal attenuation across the overall topology of the home network.

Home size—Just as homes vary greatly in their size and physical layout, the electrical networks that serve them vary also. The average and maximum distance between wall sockets (and subsequently node devices), the place-ment of the breaker panel and its effect on phase jumping, and the era of the home's electrical system vary greatly across the demographics of the international marketplace. Consider, for example, two homes of equal size (3000 sq. ft) — a single story and a two story. The average physical distance between outlets, and subsequently the network devices, would be much greater in the single story home, making it more susceptible to the effects of attenuation.

Signal frequency—Another variable that affects how signals attenuate in real-world environments is the frequency of signals. Generally speaking, a high frequency signal will attenuate more rapidly than a low frequency signal that is transmitted over the same length of wire.

Shared networking media—Powerline is a shared network media. Thus, on some occasions multiple IPTVCDs may need to simultaneously access the network at the same time, which can lead to an increase in latency and a degradation of signal quality.

Interference and noise sources—In power-line communications the term "noise" refers to any undesired signals on the wire that interferes with the data transmission of the original communication signal. The two common types of noise that affect the powerline environment are

- *Injected noise* (also called impulse noise) results from the switching of inductive loads, which produce strong impulses. The worst offenders for producing injected noise are devices containing alternating current (AC) motors, such as hair dryers, vacuum cleaners, and electric drills. When these appliances are turned on, they saturate the powerline with violent spikes of noise that defeat many communication methods.
- *Background noise* (also called ambient or white noise) presents an additional impediment. This quasi-steady state of noise is caused by radio

frequency interference (RFI) from noise emitters such as florescent lights that may be operating in the vicinity of the home network.

The multireflective effect—When a beam of sunlight shines at the surface of water some of the light will go through the water, and some of the light will reflect off the surface, bouncing in different directions to create the shimmering effect that people see. The fact that the air and water are physically different media causes part of the light beam to reflect back toward the source of the signal. In powerline home networks a similar phenomenon, known as the multireflective effect, happens to the data signal. Just like when some of the light reflects as it passes from the air to the water, some of the data signal reflects as it experiences impedance discontinuities on the wire. The multireflective effect comes into play anytime the signal changes impedance as it crosses a medium. Unterminated network points, changes in the physical wiring structure, and jumping phases at the circuit represent common impedance discontinuities in electrical networks. All of the electric devices in a consumer's home, like televisions, lamps, washing machines, and other appliances, combine to change the impedance of the network at various points. The multireflective effect presents an obstacle for communications because the portion of the signal that is reflected back along the origination path can interfere with the source signal. In these cases the reflected signal creates a standing wave that may result in a null, or cancellation, of the source data signal. Wall sockets create a problem both as unterminated network points (when there is no device connected) and as appliance loading points (when there is a device connected).

One of the most popular industry consortiums formed in recent years to overcome these technical obstacles and make in-home powerline networking a reality is a group called the HomePlug Powerline Alliance Inc, also known as HomePlug, established in 2000. Since its inception, it has created specifications that allow communication between electronic devices using high voltage wiring infrastructure. The specification relevant to delivering WHMN applications is HomePlug AV. At the time of writing this book, the consortium had a membership of over 70 industry-leading companies. Additional information about the alliance, including a complete listing of certified products, is available at http://www. homeplug.org. Further details on the organizations latest specification designed to support WHMN applications is described in the following sections.

8.5.1   HomePlug AV Overview

Ratified by the HomePlug Powerline Alliance in 2005, the HomePlug AV (Audio Video) is a global standard that enables the delivery of next generation WHMN based content such as HD and SD digital video channels, voice over IP, and whole house audio content over standard home power cables. It is the successor to the HomePlug 1.0 international standard and has been designed from the ground up to support the high bandwidth and low latency needs of next generation WHMN applications. Some of the characteristics of the HomePlug AV specification include

No additional wiring required—As the technology will run over the standard power-line network, there is no requirement for additional wiring. Home networks that utilize HomePlug AV technology can control anything that plugs into an outlet. This includes lights, televisions, thermostats, and alarms.

Advanced QoS support—HomePlug AV supports advanced QoS features that help to improve experience of accessing WHMN content. 802.1q VLAN tags are used to prioritize traffic types in order to guarantee different levels of QoS.

Includes DRM technologies—HomePlug AV provides support for DRM to ensure that WHMN content is not pirated while traversing the home network.

Backward compatible with older HomePlug products—The HomePlug AV technology is backward compatible with HomePlug 1.0 products. Thus, HomePlug AV and HomePlug 1.0 products are interoperable and coexist on the same home network.

Enhanced security capabilities—HomePlug AV uses a powerful security mechanism to ensure privacy of WHMN data exchanged over the in-home network. For instance, a security policy is supported that only allows IPTVCDs and other digital devices with a Network Membership Key (NEK) to join the home network. An NEK is a HomePlug AV specific term and is used by the 128-bit AES encryption technique to secure network traffic, while traversing the powerline network. It is important to note that HomePlug AV also provides support for the industry standard 56-bit DES encryption mechanism. The provision and issuing of keys varies between networks.

High data throughput—The raw channel bit rate for carrying WHMN services is about 200 Mbps. This is a significant increase when compared to previous versions of the specification. Table 8.5 summarizes the data rates of the different generations of HomePlug technologies.

TABLE 8.5   Maximum Data Rates for Different Generations of HomePlug Products

	HomePlug 1.0	HomePlug 1.0 with Turbo	HomePlug AV
Maximum Data Rate	14 Mbps	85 Mbps although realistic rates in a live IPTV environment are significantly less.	200 Mbps

8.5.2 HomePlug AV Technical Architecture

HomePlug AV is built on a layered approach that corresponds to the IPTVCM. A simplified block diagram depicting the HomePlug AV protocol stack is depicted in Fig. 8.3 and explained in the following paragraphs.

8.5.2.1 PHY Layer   HomePlug AV networking is built on an underlying PHY protocol and uses a tailored version of the OFDM coding mechanism to transmit data over the powerline circuits. The HomePlug AV PHY layer operates in the 2–28 MHz frequency range. Enhancements to OFDM including long symbols in conjunction with a high number of tones and a flexible guard interval help to provide the necessary bandwidth requirements of WHMN applications. The PHY layer can receive up to three different types of input from the upper MAC layer, namely, raw data, HomePlug AV control information, and HomePlug 1.0 control information. Each input type is processed differently by the PHY layer.

8.5.2.2 MAC Layer   The MAC layer seamlessly integrates with the PHY layer and supports two types of communication services, namely, connectionless and connection-orientated. Similar to HomePlug 1.0, connectionless based network sessions use the CSMA/CD access mechanism. As discussed previously this mechanism requires HomePlug AV enabled devices to sense the network media for

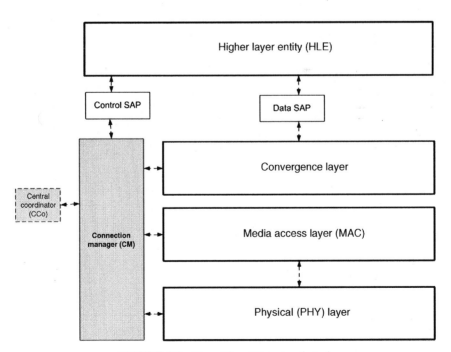

FIGURE 8.3   HomePlug AV protocol stack

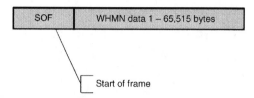

FIGURE 8.4   Format of a HomePlug AV MAC layer frame

network traffic. If the home network is busy then the device waits a finite time before transmitting. Otherwise, the data is placed onto the home network. The connection orientated mechanism is particularly relevant to WHMN services because it ensures that sufficient bandwidth is available to reliably transport content across the home network. It is based on the well known TDMA mechanism. The TDMA technique assigns traffic to a certain portion of time on a designated frequency. The MAC layer receives input from the convergence layer in the form of IEEE 802.3 Ethernet frames. Additional information is then appended to the frame. The structure of a MAC layer packet is shown in Fig. 8.4.

As depicted the packet consists of a Start of Frame (SoF) and WHMN content payload. The encapsulated frames are subsequently inserted into a serialized stream and placed onto the power-line network.

8.5.2.3   CCo   As the name suggests, the central coordinator (CCo) is responsible for a number of control functions:

- Management of both connection orientated and connectionless network sessions.
- Managing the various IPTVCDs and other device types that make up networking units defined by HomePlug AV as AV Logical Networks (AVLNs).
- Implementing security policies for AVLNs.
- Providing bandwidth management services including allocating time slots on the network to different connections.
- Building and maintaining topology maps of the AVLN network.

8.5.2.4   Connection Manager   The connection manager interfaces to all other modules in the HomePlug AV protocol stack. Connection Specification (CSPEC) messages are the main inputs to this module. CSPECs are generated at the higher layers of the IPTVCM and include specific QoS requirements:

(1) Guaranteed bandwidth requirements
(2) Latency limitations
(3) Jitter parameters

Once the CSPEC is received the connection manager analyses the request included in the message. If the CSPEC is valid the connection manager requests a contention free TDMA slot from the CCo. Once the time slot is granted the network session commences. While the communication session is operational between two HomePlug AV network interfaces it is the responsibility of the connection manager to ensure that enough bandwidth is available to transport the WHMN content.

8.5.2.5 Convergence Layer As illustrated the convergence layer provides an interface between the upper layers of the HomePlug AV protocol stack and the lower PHY and MAC layers. Communicating to the upper layers is done through service access points (SAPs). The type of traffic handled at this layer is all Ethernet. The convergence layer provides the following functions:

- It associates time stamps with IP packets.
- Provides traffic smoothing services such as minimizing network jitter.
- Provides the connection manager with QoS information.
- Associates IP packets with a specific connection.

8.5.2.6 SAPs The SAPs are used by the upper and lower layers of the HomePlug AV protocol architecture to request certain services off each other. For instance, the Control SAP identifies communications of control data between the connection manager and the upper layers of the protocol stack.

8.5.2.7 Higher Layer Entity (HLE) This is a term that is specific to the HomePlug AV specification and equates to the application and presentation layers of the OSI networking model. Interactive IPTV applications such as games and Web browsing are considered to be part of the HLE layer.

In 2004, an international organization called the Universal Powerline Alliance (UPA) established itself to develop a rival specification for developing powerline based WHMN products. An overview of the specification called Digital Home Standard (DHS) is described below.

8.6 UPA-DHS

The UPA released the first version of its DHS high speed powerline networking specification in Q1 of 2006 and is primarily based on a technology created by an EU sponsored Powerline project (EC/IST FP6 Project No 026920) called OPERA—Open PLC European Research Alliance. At the time of this writing, the UPA were close to filing the specification with the IEEE P1901 task group on broadband power-line standards. Technical and operational characteristics of DHS-UPA include the following.

8.6.1   Data Rates That Facilitate The Delivery Of WHMN Applications

DHS supports room-to-room transfer rates of up to 200 Mbps across an in home powerline wiring infrastructure. These data rates support the distribution of triple-play and WHMN applications.

8.6.2   Supports Delivery of WHMN Services Over Powerline and Coaxial Cable

In addition to operating across powerline networks, the PHY layer included in the DHS specification can also support the transfer of data across coaxial cables.

8.6.3   Uses a Layered Approach

Similar to other protocols, the DHS specification bases its technology on a series of layers that align quite closely to the IPTVCM. The layers defined by DHS are illustrated in Fig. 8.5 and summarized in the following section.

- The *PHY layer* deals with transmitting IPTV content across the network media. Reliability is a key characteristic of WHMN applications. As such the DHS specifications employ a specialized system called High Ultra Reliable Transmission for OFDM (HURTO) for reliably carrying content across the in-home network.
- The *MAC layer* defines mechanisms for allocating time slots on the home network for content transmission. The mechanism used by DHS for optimizing the distribution of WHMN content is called Advanced Dynamic

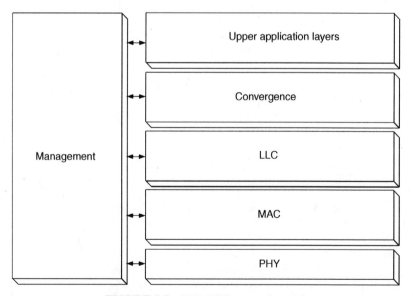

FIGURE 8.5   UPA-DHS protocol model

Time Division MAC (ADTDM). Under this mechanism the network provides IPTV applications with collision free access to channels. The process of controlling and allocating channels is determined by an in-home networking device called an "access point." In addition to access points the DHS specification also specifies the roles of two other device types, namely, repeaters and end points. As the name suggests the repeater receives and retransmits WHMN packets, while the end points are standard devices that do not provide access point or repeater functionality.

- The *LLC layer* provides error correction services that are used during communication between IPTVCDs. An acknowledgement scheme is used by DHS to achieve this type of functionality.

- A *convergence layer* is also defined. Its main function is to map Ethernet frames to the lower layers of the DHS protocol stack. This is achieved by encapsulating the frames inside power-line packets. In addition to encapsulating packets, the convergence layer also supports the implementation of VLANs across the in-home digital network. This approach allows IPTV service providers to separate the traffic into different logical subnetworks.

- The specification also employs a *management layer* to control the overall operation of the DHS protocol.

Ensuring high QoS levels is another important section of the DHS specification. Traffic classification and centralized bandwidth management techniques are used by DHS to ensure the smooth delivery of WHMN services under difficult networking conditions. The final section of the DHS specification deals with security. Under the security architecture supported by DHS, network identification systems and encryption are used to maintain the integrity of WHMN content as it traverses the in-home network. The identification system only allows communication between digital devices that have the same identifier. With regard to encryption, the UPA has decided to use a 168-bit triple play DES algorithm for securing communications between various types of IPTVCDs.

Other groups promoting and developing separate standards for powerline networking include the Consumer Electronics Powerline Communications Alliance (CEPCA) and the PowerLine Communications Forum.

8.7 HOMEPNA™ 3.1

Originally, home networks were based on Ethernet and depended on special Cat-5 cables to link PCs, A/V equipment and peripheral devices together. These cables often required professional installation and could be expensive and problematic if the hardware components were in different rooms of the house. A number of technologies including HomePNA (also known as HPNA) that work over existing wires were developed to address this issue however, with the development of 802.11 Wireless LANs, customers could connect their PCs to each other, peripherals and the Internet, the problem of installing new wires was largely eliminated. Now, thanks to recent market developments, strong new interest in home networking over existing wires and especially over coaxial cables and phone wires has emerged. This interest is coming from service providers such as telephone companies that are using the technology as a key component of their new "triple play" voice–video–Internet access service offerings.

A major requirement of this new generation home network solution is that it allows homeowners to share computing resources and provide connectivity between the service provider's network and consumer equipment such as television set-top boxes and DVRs without installing additional wiring. HomePNA networking meets this requirement by using the preexisting home telephone and coaxial wiring to distribute computer and entertainment data throughout a household. A basic reference topology used by HomePNA 3.1 in a three bedroom house is presented in Fig. 8.6.

As shown the installation of a HomePNA network is straightforward and involves using HPNA to Ethernet adaptors to connect networked equipment to a telephone (RJ-11) or coaxial (Type F) wall connector in the home. Creating a technology that can deliver high speed multimedia services over a standard coaxial and phone wire is a challenging task.

An industry consortium called the HomePNA Alliance was created back in the late 1990s to develop specifications that supported the simultaneous operation of normal telephone service, Internet connectivity, and home networking functions over in-home phone wiring systems. The technology was significantly enhanced in 2004 with HomePNA version 3.0, which added much higher data rate capacities

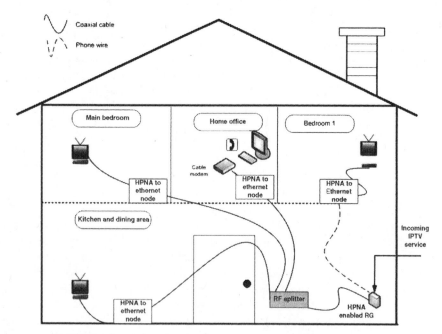

FIGURE 8.6   Basic HomePNA 3.1 enabled home networking topology

TABLE 8.6   Data Rates for Different Generations of HomePNA Products

	Home PNA 1.0	Home PNA 2.0	HomePNA 3.0	HomePNA 3.1
Data rate	1 Mbps	16 Mbps	128 Mbps with optional extensions reaching up to 240 Mbps	320 Mbps
Networking medium	Phone wires	Phone wires	Phone wires	Phone wires and coaxial cable
QoS	Priority based	Priority based	Guaranteed	Guaranteed

compared to its predecessors and the guaranteed QoS capability required by commercial VoIP and IPTV services. The most recent HomePNA 3.1 version, HPNA 3.1, again increases the data rate while also standardizing operation over coaxial cable as well as phone wires. Table 8.6 summarizes the data rates of the different generations of HomePNA technologies.

For the purpose of this chapter, the focus will be on the latest HomePNA specification—Version 3.1. HPNA 3.1 technology was standardized by the ITU in 2007. Some of the technical features of the HPNA 3.1 specification include

Operates over both phone wires and coaxial cable—This technology supports the simultaneous operation of normal telephone service, Internet connectivity, and home networking functionality, over the same phone wiring system. In addition, the technology coexists with TV signals on the same coaxial cable in the home because each service uses a different set of fre-quencies. The ability of the latest version of HomePNA to transport signals across coaxial cables expands appeal of the technology worldwide such as in parts of Europe that may have phone wires or coaxial cables but not both.

Provides support for random configuration of topologies—Rather than the hub structure of business networks, the topology of a HomePNA network is a random "tree" like configuration. Something as simple as plugging in a telephone or disconnecting a fax machine changes the topology of the network. For this reason, the protocol is designed to adapt very quickly (measured in milliseconds) to changes in network topology, an ability that is critical in IPTV and VoIP applications where long delays can affect customer satisfaction.

Compensates for signal attenuation—The random tree network topology of phone-line wiring system can cause signal attenuation. The attenuation on coaxial cable is normally caused by splitters. HomePNA 3.1 is designed and tested to operate over a wide range of existing wire configurations and has margin to compensate for this attenuation in typical installations.

Signal noise—Entertainment appliances, heaters, air conditioners, and telephones can all introduce signal noise onto the phone wires. The HPNA 3.1 protocol is designed to compensate for this with features such as automatic retrans-mission. Operation over coaxial cables is less problematic.

QoS guarantees—The home network must be able to function reliably and deliver consistent levels of service despite changes that result from someone picking up the phone, accessing a Web site, or an answering machine recording a message. The fast adaptation and other features described above compensate for these disturbances; however, IPTV and VoIP services impose new and much stricter requirements on the home network. For these services, "guaranteed" QoS is required that guarantees that the amount of delay, jitter, and errors of the data streams will not exceed a specified amount. HomePNA 3.1 meets these requirements.

FDM support—It utilizes a technology known as Frequency Division Multiplexing (FDM) that essentially divides the data traveling over the phone and coaxial wires into different frequency bands separating voice, high bandwidth Internet access and networked data. These frequencies can coexist on the same telephone wire without impacting on one another.

Complies with Ethernet technologies—HPNA 3.1 is based on standard Ethernet technology, adapting it where necessary to overcome the challenges presented by the home phone-line environment. The fact that Ethernet has been chosen as a networking technology allows HPNA 3.1 to leverage the tremendous amount of Ethernet compatible software, applications, and existing hardware in the market today.

Compliance with the IPTVCM—As shown in Fig. 8.7, the HPNA 3.1 aligns with the two lower layers of the IPTVCM.

As mentioned above HomePNA will also operate over 75-ohm coaxial cable. Coaxial and phone wire implementations use different impedance matching devices to connect to the wiring. The impedance matching device, often

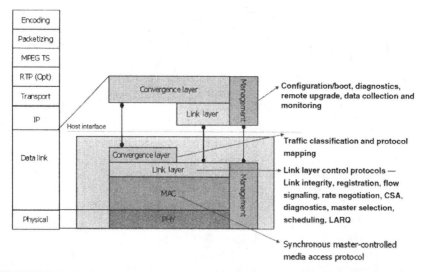

FIGURE 8.7 HPNA 3.1 protocol stack mapped to the IPTVCM data link and physical layers (Courtesy of HomePNA)

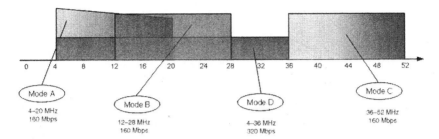

FIGURE 8.8   HPNA 3.1 spectral allocations

called a "hybrid" is composed of a small number of passive devices. It is low cost and some manufacturers include both phone wire and coax hybrids on their products allowing the same device to connect to either coax or phone wire. Other manufacturers specify use of a low cost external phone wire to coax impedance matching device. The physical makeup typically consists of an RJ-11 connector, an F-Connector, and a few passive components. The internal electronics of these devices are made up of filters that route to the appropriate interfaces. With regard to operating frequencies there are four bands specified by the HPNA 3.1 standard. As shown in Fig. 8.8, HPNA can use the 4–20 MHz, 12–28 MHz, as well as the 36–52 MHz and 4–36 MHz bands, which will allow multiple HomePNA bands to coexist on the same cable.

8.8   MULTIMEDIA OVER COAX ALLIANCE (MOCA™)

MoCA is an open, industry driven initiative promoting the distribution of digital video and entertainment through existing coaxial cable in the home. MoCA technology provides the backbone for whole home entertainment networks of multiple wired and wireless products.

The goal of MoCA is to create specifications and certify products that tap into the vast amounts of unused bandwidth available on in-home coaxial cables without the need for new connections, wiring, and truck rolls. In addition to developing an open WHMN specification, the alliance has instituted an ongoing certification process that allows companies to validate their products as interoperable with other MoCA enabled products. Since the certification process commenced, a number of products have successfully passed and have been awarded MoCA certification. Some of the technical features and characteristics of the MoCA specification include

Operates across multiple splitters—The MoCA network transmits high speed multimedia data over the in-home coaxial cable infrastructure at speeds of up to 270 Mbps by "jumping" backwards through existing splitters in the home. The in-home coaxial network generally comprises of a number of splitters and

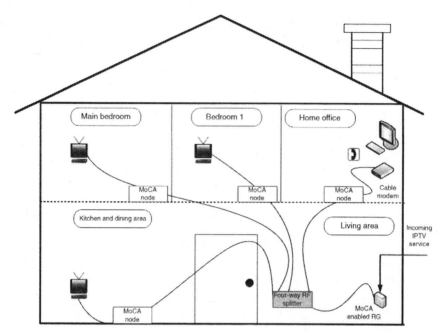

FIGURE 8.9   Basic MoCA enabled home networking topology

several runs of RG-59 or RG-6 cable. A basic reference topology used by
MoCA in a three bedroom house is presented in Fig. 8.9.

This reference network is using coaxial cable as the physical network media to
distribute IP video, data, and voice traffic throughout the building. A splitter is
also used in the configuration to illustrate MoCAs ability to transfer IPTV
traffic in both directions through splitters. This capability allows all coaxial
outlets in the home to access high quality IP based video content.

Transmits data over long distances—The maximum distance for transporting
WHMN content via MoCA networking technology is 300 ft. Achieving this
maximum distance depends on attenuation of the cable and the number of
splitters connected to the coaxial network.

Coexists with cable, terrestrial, and satellite RF TV signals—A single MoCA
channel occupies 50 MHz of channel bandwidth and resides flexibly between
850 and 1.5 GHz. As shown in Fig. 8.10, the MoCA speci-cation divides this
spectrum into defined bands and describes how they may be used.

The exact location of the channel bandwidth differs between geographic locations.
Owing to the fact that MoCA operates in the upper end of the frequency
spectrum, it coexists with standard cable, terrestrial and satellite TV signals.

Uses advanced encryption technologies—It uses the DES security mechanism
for privacy.

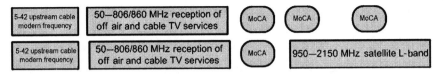

FIGURE 8.10   MoCA spectrum allocation

High data rate—The throughput and PHY level bit rates at the maximum transmission power as measured in the MoCA field tests are shown in the following curves of Fig. 8.11.

Provides remote management capability—It supports SNMP agents, which allows service providers to remotely manage a coax centric home network. MoCA is also specified as part of the DSL Forum's TR-069 remote management requirements that allow telecommunication providers to manage devices and run diagnostics on their subscriber's home networks.

Does not interfere with other services—MoCA supports the delivery of WHMN services without disrupting core home networking functionalities.

8.8.1   Examining the MOCA PHY and MAC Layers

High data rates are a key requirement for WHMN applications. This requirement, given the topology of the in-home coax infrastructure and its associated channel characteristics, has influenced the development of the MoCA protocol. In particular, special attention has been given to maximizing network robustness while ensuring that the packet error rate is kept to a minimum. This is achieved primarily through the use of a full-mesh preequalization layer using a form of OFDM modulation at the PHY referred to as Adaptive Constellation Multitone (ACMT). The signal is preequalized using a technique called bit loading, or water

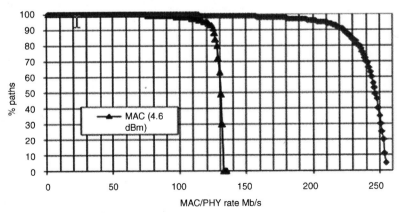

FIGURE 8.11   MoCA MAC and PHY performance percent of paths versus bitrate (Courtesy of MoCA)

filling, which assigns an independent QAM constellation to each subcarrier based on the SNR of that subcarrier. Forward error correction is used at the PHY layer to maintain the required bit error rate.

With regard to accessing a MoCA based network the MAC layer uses synchronous beacons to allow new nodes to gain admission to the network. Beacons are scheduled every 10 ms to allow for quick admission. Access packets are transmitted approximately every millisecond by a network coordinator (NC) to schedule transmissions on the medium. The NC schedules beacon, control, data and probe transmissions. Data transmissions are scheduled based on reservation requests from each node. Since the access cycle is short, latency in the network is kept to a minimum. Special probe packets are used to estimate various channel parameters in order to maximize throughput by preequalization techniques, as well as optimize throughput by reducing preamble sizes. Preambles are optimized for different packet types. The various types of MAC layer packets are shown in Fig. 8.12.

There are some more options on the market that IPTV professionals need to be aware of when creating a WHMN strategy, namely, 1394 over coaxial cable, ultrawideband (UWB) and of course wireless USB.

8.9 WHMN MIDDLEWARE SOFTWARE STANDARDS

A WHMN middleware plug-and-play technology is often used to simplify the implementation of in-home networks. Two WHMN middleware industry standards have emerged in recent years to address this growing market—Digital Living Network Alliance (DLNA) and High Definition Audio–Video Network Alliance (HANA). The following sections provide an overview of these two standards.

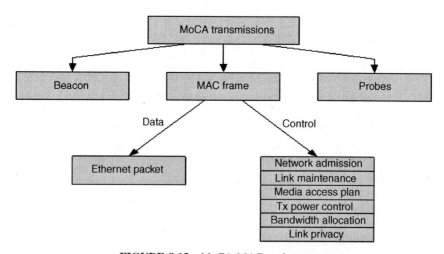

FIGURE 8.12   MoCA MAC packet types

8.9.1 Digital Living Network Alliance (DLNA)

DLNA is a cross-industry organization of consumer electronics, computing industry, and mobile device companies. The organization has published a number of interoperability guidelines that define the building blocks of a digital home ecosystem. Some of the main characteristics of the DLNA guidelines include

Strong interconnectivity support—Transparent connectivity between various types of IPTVCDs and other devices in the digital home is implemented at the data link layer of the IPTVCM.

Advanced discovery mechanisms—DLNA allows all devices connected to an in-home network to initialize, discover, and understand each others capabilities. This helps to simplify networking of a diverse range of digital media devices and IPTVCDs in the home.

A wide range of popular media formats are supported—These formats are classified into two broad categories, those that are generally used by portable mobile devices and those used by WHMN applications. Table 8.7 summarizes mandatory and optional media format types used by DLNA based WHMN applications.

It is important to note that the media formats listed in Table 8.7 were defined in the DLNA guidelines at the time of this writing. It is expected that this list of optional media formats will grow in the coming years.

Promotes interoperability between different DRM systems—DLNA addresses the issue of protecting usage rights of commercial content when being transported or stored on a home network. This is a complex issue for DLNA to address because a typical home network may have multiple DRM technologies in use to protect different types of content including IPTV assets. In an effort to resolve the issue DLNA is proposing to define a DRM interoperability framework that allows content protected by one type of DRM system at the source to be viewed by another device, which uses a different DRM environment. This needs to be achieved while adhering to the copyright limitations associated with the IPTV asset. As an initial step into the area of content protection the DLNA extended the existing guidelines in 2006 to include a section on link protection technologies. One of the main security technologies mandated by this extension is DTCP-IP.

TABLE 8.7   DLNA Media Formats for WHMN Applications

Media Format	Mandatory Support	Optional Support
Imaging	JPEG	GIF, TIFF, and PNG
Video	MPEG-2	MPEG-1, H.264, and VC-1
Audio	LPCM	MP3, WMA9, AC-3, and AAC

8.9.1.1 DLNA Technical Architecture   The DLNA technology consists of a number of interoperable building blocks, which are based around established standards. These building blocks are used by vendors to build end products for the WHMN marketplace. The technical architecture of a DLNA certified IPTVCD is shown in Fig. 8.13.

As depicted the DLNA architecture is focused on referencing existing standards and specifications and defining specific implementation details so that products of different vendors will interoperate. Standards such as 802.11a, b, g, and n plus wired Ethernet and more recently BlueTooth are also used at the network connectivity layer of the DLNA protocol stack. The DLNA communication model uses protocols, such as IPv4, HTTP, and RTP to transport content, while the various supported media formats are assigned to the top of the protocol.

Universal Plug and Play (UPnP) version 1.0 has been chosen to provide an interoperability infrastructure for plug-and-play functionality, whereas the UPnP AV (audio and video) layer deals with distributing WHMN applications. The role of UPnP 1.0 and UPnP AV technologies in DLNA is described in the following sections.

UPnP 1.0   UPnP is an extension to the plug and play initiative that was introduced by Intel, Compaq and Microsoft back in 1992. It uses open Internet

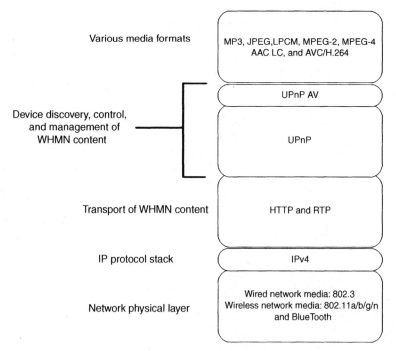

FIGURE 8.13   DLNA certified IPTVCD technical architecture

communication standards to transparently connect various types of consumer electronic devices and PCs together. So rather than concentrating on one particular device type, UPnP interconnects all types of devices used in digital homes, including PCs, PC peripherals, new smart home appliances, gateway devices, home control systems, and Web connectable devices. The result of this pragmatic and relatively simple approach is that implementing UPnP on an in-house network requires very little work and human intervention. UPnP defines a set of common interfaces that allows consumers to plug an IPTVCD directly into their in-house network without the need to configure various technical settings or install specialized driver software modules. Additionally, UPnP makes it possible to initiate and control the transfer of IPTV streams from any IPTVCD on an in-home network. Some of the features and characteristics of UPnP, which make it particularly suitable for transporting WHMN content across an in-home network, include:

USES OPEN INTERNET STANDARDS UPnP was developed within the context of existing industry standards, including IP, TCP, UDP, HTTP, and XML. Because it is based on standard Internet protocols, UPnP can work with a broad range of devices, from large PCs to small consumer electronics devices.

SUPPORTS PLUG AND PLAY FUNCTIONALITY UPnP is based on straightforward, innovative mechanisms for discovery and connectivity that provide a basis for the concept of zero configuration for end users. For instance, by using plug-and-play functionality, it is possible to connect an IPTVCD to a home network, automatically receive an IP address, find out about the other IP devices on the network, and immediately commence network communications without intervention from the home owner.

LOW HARDWARE FOOTPRINT Unlike traditional PC based solutions, consumer electronic appliances and some types of IPTVCDs have radically less system resources at hand. Typically, they are based on low cost microprocessors and small quantities of memory resources. Adding UPnP functionality to these device types requires a very small amount of system resources.

SMOOTH INTEGRATION WITH LEGACY SYSTEMS AND NON-IP DEVICES Although IP internetworking is a strong choice for UPnP, it also accommodates home networks that run non-IP protocols such as IEEE 1394 based entertainment networks.

NON-PC-CENTRIC ARCHITECTURE The configuration of an UPnP based network can be based on a peer-to-peer network architecture, which basically means that a home network can function without a PC.

AGNOSTIC TO UNDERLYING NETWORK MEDIA Another advantage of UPnP is its independence of the physical network media. It is compatible with existing networks, such as standard Ethernet and new networking technologies that do not require installation of new wiring systems in existing homes — HomePNA, 802.11n, UPA-DHS, MoCA, and HomePlug AV.

Networking UPnP enabled IPTVCDs   UPnP classifies home networking nodes such as IPTVCDs into two broad categories, namely, control points and devices. The *control point*, also known as a controller, describes an embedded piece of software that facilitates communication between itself and a number of devices on a home network. Examples of IPTVCDs that could function as a control point include a standard PC or a digital set-top box.

A *device* in a UPnP based home networked environment represents a container for home networking services. A device generally models a physical entity but can also represent a logical entity. A digital set-top box emulating the traditional functions of a PC would be an example of a logical device. Examples of home network nodes that could function as controlled devices include VCRs, DVD players, security systems, and automated light controllers. Devices can contain other devices. The primary difference between a control point and a device is that the control point is always the initiator of the communications session. UPnP technology defines a networking framework for implementing protocols in a number of different layers.

It is possible to use this framework to simplify the installation and management of WHMN services and IPTVCDs. The deployment of the UPnP Networking Framework involves five specific processes. The following sections describe these steps in the context of an example that involves the addition of an IPTVCD to a live home network, which uses a media server as its control point.

STEP 1: IP ADDRESSING AND DISCOVERING OTHER IPTVCDs. Getting an IP address is the very first step that occurs when an IPTVCD is connected to a home network. The allocation of IP addresses on a home network is the responsibility of the Dynamic Host Configuration Protocol. In addition to assigning IP addresses, the DHCP also automates the assignment of other IP configuration parameters such as subnet masks and the default gateway address.

The use of this system simplifies administration of the in-home network because the software keeps track of IP addresses rather than contacting the IPTV service provider to manage the task. This means that new IPTV consumer and network devices can be added to a home network without the hassle and time of manually assigning new addresses. In a home networking environment, the DHCP server software required to dynamically lease IP addresses generally runs on the RG whereas the client DHCP software is embedded into the various digital consumer devices connected to the network. Once a UPnP enabled IPTVCD boots up and connects into the home network a DHCP request is sent to the RG. The DHCP server running on the RG replies to the client with various IP parameters required to communicate over the home network. The IP address assigned to the UPnP enabled IPTVCD expires after a predetermined period of time. When this time period, also known as a lease, expires the DHCP client and server renegotiate renewal of the lease for another fixed time period. The DHCP server ensures that the leased address does not get allocated to any other IPTVCD during the lease period. Note DHCP runs over UDP, utilizing ports 67 and 68.

TABLE 8.8   DHCP Message Structure as Defined in RFC 2131

Field Name	Size (bits)	Description
Op	8	Short for operation code, this field defines the message type. So, for instance, a value of one indicates that the message is a request whereas a value of two indicates that the message is a response.
HType	8	This defines the type of physical hardware technology used on the network. The value of one is generally used by home networks to describe an Ethernet topology.
HLen	8	HLen is an abbreviation for hardware length and identifies the length of the hardware addresses used by the IPTVCDs connected to the network. For Ethernet based home networks this value is typically one.
Hops	8	This is normally set to a low value because the number of network hops between a source and a destination device on a home network is minimal.
XID	32	Short for transaction ID, this field is used by the IPTVCD to match requests with responses.
Secs	16	The "seconds" field measures the time required for a digital consumer device to acquire or renew an IP address lease.
Flags	16	The majority of this field is reserved with one bit set as a broadcast in the event that an IPTVCD does not know its own address.
Ciaddr	32	This field contains the client IP address.
Yiaddr	32	Short for "Your IP address," this field holds the address that the RG DHCP server is assigning to a IPTVCD connected to the home network.
Siaddr	32	Short for "Server IP address," this field holds the address of the DHCP server.
Giaddr	32	Short for "Gateway IP address," this field is used to route BOOTP messages within the home network.
Chaddr	128	Short for "Client Hardware Address" this field stores the MAC address of the in-home IPTVCDs.
Sname	512	Short for "Server Name" this field stores the name of the DHCP server. This is optional as it adds significant overhead to the message.
File	1024	This field stores directory path details for a BOOT file.
Options	Variable	As the name suggests this field stores parameters that are used to support the operation of DHCP across a live home network.

Technical details of the DHCP protocol are published across a number of RFCs including RFC 2131 and RFC 2132. As described in these RFCs, the DHCP client software supports the communication of a number of message types that comply with a specific fixed structure. Table 8.8 explains the structure of DHCP messages.

There are a number of different types of messages, which may be used by DHCP, some of the popular ones used by WHMN systems include

- *DHCP Discovery* messages are requests for a DHCP IP address. The information stored in these message types is limited to the MAC address of the IPTVCD and the name if available.
- *DHCP Offer* is sent by the server when an IP lease request is received from an IPTVCD. The message itself contains information such as the IPTVCDs MAC hardware address, leased IP address, subnet, and duration of the IP address lease.
- *DHCP Lease* messages are generated when the lease at the IPTVCD is over half finished. The contents of this message identify the requesting IPTVCD and the lease time period.
- DHCP acknowledgement messages are used to complete the IPTVCD IP address leasing process. Message contents include lease details and associated parameters.

It is important to note that separate DHCP address pools need to be established for each of the triple play services—one for each Internet access, VoIP, and IPTV services. Intercommunication between these subnets is facilitated via an RG that supports routing functionality at layer 3 of the IPTVCM.

If a DHCP server is unavailable on the UPnP home network, the IPTVCD must use a feature called "Auto IP" to get an address. Auto IP allows the IPTVCD to obtain an address in a network without a DHCP server. It should be noted that Auto IP is only applicable to devices that are running the DHCP client service. Consider the simple example, where an IP set-top box (a popular type of IPTVCD) is connected to a home network that does not have a DHCP server. The following steps follow the sequence of events for autoconfiguring an IP address on the IP set-top box:

(1) The IP set-top box uses the DHCP protocol to request an IP address.
(2) A DHCPDISCOVER message is sent out.
(3) If after a specified period of time, no valid IP addresses are received, the IP set-top box is free to autoconfigure an IP address.
(4) Once the IP set-top box determines that it must autoconfigure an IP address, it uses an algorithm for choosing this address. This value must use the 169.254/16 IP address range, which is registered with IANA.

(5) When an address is chosen, the IP set-top box tests to see if the address is already in use. If the network address appears to be in use, the set-top box must choose another address, and try again.

It is important to note that even though the IP set-top box has been assigned a new address, it keeps on checking the home network for an active DHCP server.

At this stage the IPTVCD has been physically connected to the network and assigned an IP address. The next step in the UPnP process is to allow the IPTVCD to advertise the types of services that it is offering to other network nodes including a control point. In this instance, a media server is used to provide control point functionality.

The advertising of services is done through the use of discovery messages that are sent to a standard multicast address using extensions to the multicast variant of HTTP — Simple Service Discovery Protocol (SSDP) and General Event Notification Architecture (GENA). The information contained in the discovery message is generally quite limited and contains a URL to find out more detailed information on the IP set-top box and an expiry time for the discovery message. These messages are examined by the media server. Once the device has established itself on the home network, it is then capable of discovering new services.

STEP 2: DESCRIBING THE UPnP ENABLED IPTVCD As mentioned previously, the information contained in the discovery message sent from the IPTVCD to the media server is basic. Thus, the next step of the UPnP process involves retrieving further details. This is achieved by issuing a HTTP GET command on the URL contained in the discovery message to locate and retrieve a more detailed description of the IPTVCD. XML is used to format the detailed description. The type of information contained in this file includes vendor specifics and further detail on available IPTVCD services.

STEP 3: CONTROLLING THE UPnP ENABLED IPTVCD The next step in the UPnP networking cycle is control. At this stage the media server has retrieved detailed information about the IPTVCD and is in a position to invoke commands to the IPTVCD. These commands are formatted as XML messages and exchanged across the home network using the Simple Object Access Protocol (SOAP).

STEP 4: RESPONDING TO EVENTS DURING LIVE OPERATION From time to time certain status changes may occur at the IPTVCD that warrant the need to generate an events message. These types of messages are expressed in XML and structured according to the HTTP notification architecture — GENA. Once formatted the event messages are sent onward via HTTP over TCP/IP to a software module called an event subscription server that runs on the media server.

STEP 5: PRESENTATION OF A USER INTERFACE (UI) If the IPTVCD supports a presentation Web server, then a browser is used by the media server to directly access the IPTVCD across the home network. The graphical Web interface may be used to access IPTV content stored on the device, view device status, or control the IPTVCD.

About UPnP AV   UPnP AV sits in between the upper media format and UPnP version 1.0 layers of the DLNA protocol stack. As the name implies, UPnP AV is particularly suited to the in-home distribution of both unicast and multicast IPTV content. In addition to providing the core interoperability functionality of UPnP, the UPnP AV specification extends on the earlier version of the technology through the following enhancements:

- Allows the use of multiple DRM systems across a DLNA based home network.
- Similar to the original UPnP specification the AV version is independent of the underlying transport protocols, IPTV compression formats, and hardware platforms used by the in-home network.
- A wide range of different IPTVCD types are supported by UPnP AV, including digital televisions, IP set-top boxes, portable media players, RGs, and media servers.
- A number of media formats are supported ranging from MPEG-2 and H.264 to JPEG and VC-1.

UPnP AV is used by DLNA to provide media management and control services. The following sections define the technical framework required to support advanced communication between IPTVCDs that include support for UPnP AV.

Networking IPTVCDs Using UPnP AV   The specification defines an overall technical framework that supports the delivery of WHMN services between UPnP AV enabled IPTVCDs. The underlying UPnP AV framework has three separate components that work together in order to deliver WHMN applications to end users in the home (Fig. 8.14).

The role of each component is discussed in the following sections.

A UPNP CONTROL POINT   The main role of the control point is to allow WHMN end users to locate desired IPTV content and select a DLNA certified device to view the material. The interaction between end users and the control point is normally facilitated by a browser. The browser allows end users to view detailed meta-data about each WHMN content asset, including title, description of the asset, length of the asset, and other types of pertinent information.

A DIGITAL MEDIA SERVER (DMS)   This component is responsible for streaming WHMN content. The content is either stored locally or on some type of network addressable storage device that is accessible to the DMS. The type of content processed by DMSs varies from IP-VoD content to digital pictures and audio files.

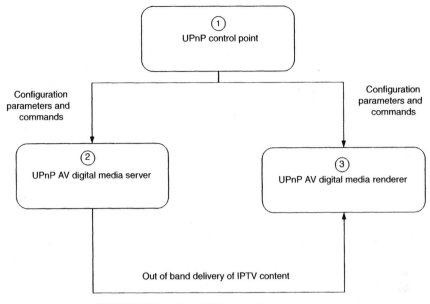

FIGURE 8.14 UPnP AV networking components

In addition to allowing end-users browse different types of WHMN content, some of the more advanced models may also include additional features:

- The ability to deliver multiple IPTV streams across the network at the same time.
- The ability to transcode or change the format of certain media types into a format that is more suitable for the target media rendering device.
- Recording capabilities are also supported by the recent version of UPnP AV.
- The ability to implement a service that allows the control point execute such actions as fast forwarding and rewinding particular streams.

A DIGITAL MEDIA RENDERER (DMR)   This UPnP AV component renders or plays the IPTV content, which is received from the media server. Digital picture frames and MP3 players are typical examples of UPnP AV based DMRs. The content types that are supported by DMRs vary between products. For instance, a digital picture frame may only be able to display JPEG pictures; whereas, a digital TV with an in-built decoder may be able to render IPTV content coded using various types of next generation compression algorithms. Similar to the DMS, some rendering products may also include services that allow end users to avail of extra functionality, such as the ability to define how incoming WHMN content is rendered in terms of viewing and audible characteristics.

As illustrated the interaction with the control point is different compared to the traditional approach used by non-AV based UPnP devices. In the standard UPnP networking architecture, the interactions occur between IPTVCDs and the control point directly. In the UPnP AV architecture, the control point is responsible for configuring the DMS and various DMRs. Additionally, the control point issues commands to commence a streaming session between both device types. It does not however get directly involved in the actual transfer of IPTV content because this takes place directly between the DMS and DMR over a separate out-of-band network channel. Thus, the majority of interactions that take place are initiated by the control point. It is however possible for either the DMS or the DMR to initiate a communication session with the control point if a change of state occurs and an event message needs to be sent.

It is important to note that the UPnP AV specification allows all three components to be incorporated into a single physical IPTVCD. A digital TV is an example of an IPTVCD that supports all three components – the display renders the content, the tuner or IP interface acts as a source for the video content, and the EPG acts as a control point. An industry group called the UPNP Forum is responsible for developing and promoting UPnP technologies. Further details are available at www.upnp.org.

8.9.2 HANA

High-Definition Audio-Video Alliance is an organization primarily made up of consumer electronic, cable, semiconductor, and IT companies. It develops solutions that focus on the interconnection of high definition, home entertainment products and services. Similar to DLNA, the HANA specification is based on existing standards. Although HANA concentrates its efforts on the upper layers of the IPTVCM, it has decided to initially use the IEEE 1394b protocol as a basis to support its networking capabilities. The following sections give a brief overview of the IEEE 1394 protocol when used in a HANA based networking infrastructure.

8.9.2.1 Introduction to IEEE 1394
IEEE 1394 widely referred to as FireWire is another high speed interface, which is vying for a share of the WHMN marketplace. The FireWire Bus standard was originally created by Apple Computer to provide a low cost, consumer-oriented connection between digital-video recorders and personal computers. It grew into a standard called the IEEE 1394. In 1994, an organization called 1394 Trade Association (TA) was formed to support and promote the adoption of the 1394 standard and in 1995 they formally released the 1394 specification. Since 1995, a couple of new versions have been developed. For example, 1394a was introduced in 1998 and 1394b was introduced in the following year. 1394b is fully backward compatible with previous specifications. Each revision of 1394 has some added features and capabilities.

IEEE 1394 has become a mainstream consumer electronics interface and is found in a wide variety of home entertainment devices including MP3 players, digital video cameras, digital televisions, set-top boxes, DVRs, and DVD players/recorders. Some of the characteristics of the IEEE 1394 networking standard include the following.

- An excellent technology for transporting high definition video streams.
- It can operate over multiple network media types.
- The IEEE 1394b standard is designed to support the transportation of HD based WHMN content across a home network at very high speeds. The latest version of 1394 supports a maximum data rate of 800 Mbps at cable distances up to 100 m.
- 1394 supports plug and play. As a result, consumers can add or remove devices without resetting the home network. Upon altering the bus configuration, topology changes are automatically recognized.

8.9.2.2 1394 Technical Architecture   To understand how 1394 operates, it is important to briefly examine the elements of a 1394-based home network. The components that form a IEEE 1394 based home network may be classified as follows:

- The IEEE 1394 protocol stack
- The cabling system
- The architectural design of the network

1394 Protocol Stack   Similar to other high speed networking systems, IEEE 1394 is a layered transport system. The three layers used by the IEEE 1394 are illustrated in Fig. 8.15.

The physical layer provides the electrical and mechanical connection between the 1394 device and the cable itself. Besides the actual data transmission and reception tasks, the physical layer also provides arbitration to insure all devices have fair access to the bus.

The link layer takes the raw data from the physical layer and formats it into two types of recognizable 1394 packets:

Isochronous—Isochronous data transfer concentrates on the guaranteed timing of the data, rather than its delivery. This is especially important for time-critical IPTV streams where just-in-time delivery eliminates the need for buffering. Isochronous transfers are always broadcast in a one-to-one or one-to-many fashion. No error correction or retransmission capabilities are available for isochronous transfers. Multiple channels of isochronous data can be transferred simultaneously on the 1394 bus.

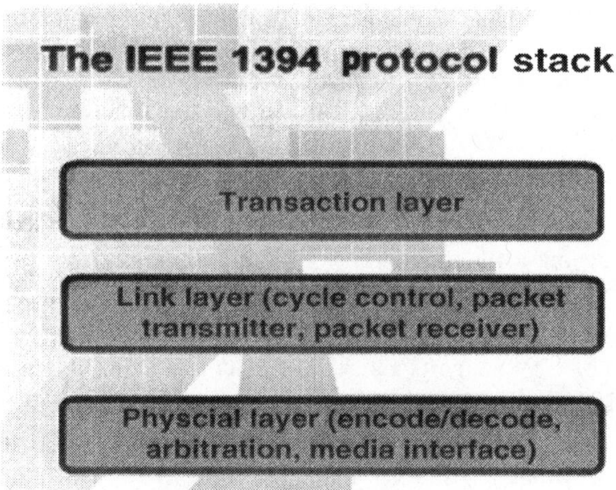

The IEEE 1394 protocol stack

Transaction layer

Link layer (cycle control, packet transmitter, packet receiver)

Physcial layer (encode/decode, arbitration, media interface)

FIGURE 8.15 1394 protocol stack layers

Asynchronous—In contrast, asynchronous data transfer is the traditional method of transmitting data between computers and peripherals, data being sent in one direction followed by acknowledgement to the requester. Asynchronous data transfers place emphasis on delivery rather than timing. The data transmission is guaranteed and retries are supported.

Providing both asynchronous and isochronous formats on the same interface allows both nonreal-time critical applications, such as printers and scanners, and real-time critical applications, such as WHMN applications, to operate on the same bus.

The third layer used by the IEEE 1394 protocol is called the transaction layer and is responsible for managing the commands that are executed across the home network. The transaction layer receives the packets from the link layer and presents them to the home networking application.

The Cabling System   IEEE 1394 requires the installation of new wires. The IEEE 1394 based home network typically uses Cat 5e, Cat 6, or fiber optic cabling, similar to that of Ethernet to interconnect devices.

Proprietory 1394 cables are also available with six-pin connectors that are used to physically attach an IPTVCD to the home network. IEEE 1394 cable connectors are constructed with the electrical contacts or pins inside the structure of the connector thus preventing any shock to the user or contamination to the contacts by the user's hands.

The six-pin connectors have two data wires and two power wires for devices that derive their power from the 1394 bus. Each signal pair is shielded and the entire cable is shielded.

Architectural Network Design   The topology of a 1394 based home network can be a star, daisy chain, or tree-like structure. This type of topology means that there is no need for a PC dedicated to the role of administering a home network. It is important to note that IEEE 1394 will also operate over coaxial cable, which is an important feature of HANA.

8.9.2.3 1394 over Coax   At the time of writing, a task group within the 1394 TA was working on a specification that enables 1394 to deliver audio and video over coaxial cable at very high bandwidth capacities. This feature is enabled through the use of a technology called ultra-wideband (UWB), which is used to modulate the signal before passing HDTV streaming content over coaxial cable. UWB is a short-range low power radio technology that operates over both wireless and coaxial cable connections. Characteristics of UWB include the following:

- It is capable of delivering data rates in excess of 1 Gbps, which allow the technology to support WHMN applications—the transmission of high definition video and multichannel audio streams in real-time across an in-home network.
- It occupies a wide frequency band when compared to other wireless technologies. The IEEE 1394 over coax standard for instance uses UWB to modulate signals in the 3.3–4.7 GHz frequency band. Thus, it is capable of coexisting with existing cable TV channels operating across the same coaxial networking medium.
- It produces a minimum amount of interference and coexists quite well with other WHMN technologies.
- It is designed to traverse coaxial cable splitters and may also be used as a backbone technology for home networks.

There are a couple of different variants of the UWB radio platform:

MB-OFDM™ (multiband orthogonal frequency division multiplexing) defined and supported by the WiMedia™ Alliance. This technology has been selected by the USB Implementors Forum (USB-IF) as a basis for building wireless USB products. The performance levels for a USB interface that includes a WiMedia UWB radio ranges from 480 Mbps at a distance of 3 m to 110 Mbps at 10 m.

The direct-sequence UWB(DS-UWB) is defined and supported by the UWB forum.

8.9.2.4 Upper Layer HANA Protocols   At the upper layers of the HANA protocol stack, open Internet standards such as IP, HTTP, and xHTML are used for addressing and executing various control instructions. Additionally, HANA uses three home networking protocols defined by the Consumers Electronics Associations Home Networking Committee (R7):

CEA 2027-A— This CEA home networking standard allows IPTV end users to use a TV or a PC Web interface to control other devices connected to an in-home network. The commands issued from the user interface are transported across the network using the IP suite of protocols. The user interface design varies between manufacturers. From a practical perspective, the implementation of this particular standard allows the viewing, pausing, and recording of HD digital content in any part of the house with a single remote control.

CEA 931-B— This CEA home networking standard, published in 2003, is used to deliver remote control commands over an IP based home network. This helps to improve the interoperability between devices connected to a WHMN. Thus, a single remote control may be used to control all IPTVCDs on the network.

CEA 851-A— This home networking standard specifies an advanced networking architecture and suite of communication protocols.

These protocols work in synchronization to allow consumers seamlessly manage a number of different IPTVCDs using a simple TV user interface. It is also important to note that HANA has defined proxies that will accommodate the use of non-HANA compliant and legacy IPTVCDs.

8.10 QOS AND WHMN APPLICATIONS

Ensuring that video signal quality is not degraded while traversing a home network is a major priority for WHMN interconnection technologies. Sufficient bandwidth capabilities are required to provide this level of service. While increasing the amount of available bandwidth on the home network to support WHMN content might decrease the need for a QoS, it may not always be a viable option.

As a result, all WHMN interconnection technologies have built in functionality that deals with QoS. The term QoS in the context of a WHMN environment refers to a system, which is capable of differentiating traffic types according to different parameters such as packet delays, packet losses, and bandwidth requirements. Once differentiated, the traffic is delivered in a manner that does not degrade the quality of the signal received from the incoming broadband access network.

With regard to the technical requirements of a home networking QoS, the first prerequisite is the ability to support whatever QoS mechanism is used on the IPTV distribution network. Thus, the RG will need to be able to map the QoS features contained in the traffic arriving at and leaving the in-home network. The level of QoS functionality varies between WHMN technologies; however, most are based on the IEEE 802.1Q standard. As discussed previously in this book, the 802.1Q marks IP packets with different DiffServ code points. Once marked using the 802.1Q coding system IP packets are classified. This classification determines how

TABLE 8.9   Next Generation Home Networking QoS Classifications

QoS Classification	Action Required by WHMN Technology	Applicable Types of Network Services
Best effort	No need to apply QoS to packets assigned to this category.	E-mail and Internet browsing
Prioritized	A certain level of priority treatment is applied to packets assigned to this category.	VoIP, IP-VoD and video conferencing
Parameterized	Packets that are assigned to this category require a level of guarantee for a particular parameter. For instance typical guarantees required by IPTV multicast streams are low latency and jitter levels	Multicast IPTV

the IP packets are handled and processed by the various home networking devices and IPTVCDs. Table 8.9 explains some of the most popular QoS classifications used by next generation home networks.

SUMMARY

Using technologies such as GigE, IEEE 802.1n, HomePlug AV, UPA-DHS, HomePNA 3.0, and MoCA, IPTV end users can create high speed multimedia home networks that are able to carry various types of WHMN services. Table 8.10 shows a comparative study between the various types of networking technologies used to enable WHMN services.

As consumer demand for distributing, sharing, and engaging with digital content grows, so too does the need for a home networking middleware platform that provides these features. The DLNA and HANA organizations have both developed solutions to address these requirements.

In an effort to grow the WHMN marketplace, the DLNA has released a set of industry design guidelines that focus on ensuring interoperability between various types of fixed and mobile home networking products. UPnP and UPnP AV protocols are used to support the seamless delivery of WHMN services between DLNA compliant devices.

The HANA organization has also developed a standards based framework that provides interoperability between IPTVCDs and supports the carriage of WHMN content over home networks. In addition to IP, HANA also provides support for CEA 2027-A, CEA 931-B, and CEA 851-A upper layer protocols. The technical characteristics of DLNA and HANA are summarized in Table 8.11.

The main benefit of both DLNA and HANA is their ability to run WHMN applications without worrying about the underlying hardware configuration of the network.

TABLE 8.10 Comparative Study Between GigE, IEEE 802.1n, MoCA, HomePNA 3.0, HomePlug AV and UPA-DHS

Parameters	GigE	IEEE 802.1n	HomePlug AV	UPA-DHS	HomePNA 3.1	MoCA
Network media	Cat 5e or 6 structured cabling system	Air	Powerline cable	Powerline cable	Standard phoneline and coaxial cables	Coaxial cable
Maximum data rates (actual throughputs levels are lower)	1,000 Mbps	Approximately 600 Mbps	Approximately 200 Mbps	Approximately 200 Mbps	Approximately 300 Mbps	Approximately 270 Mbps

TABLE 8.11   Characteristics of DLNA and HANA WHMN Middleware Software Standards

Characteristics	DLNA	HANA
Recommended PHY and MAC layer protocols	Ethernet and Wi-Fi	IEEE 1394
IPTVCM upper layers	UPnP 1.0 and UPnP AV	CEA 2027-A, CEA 931-B, and CEA 851-A
Mandated security technologies	DTCP-IP	HDCP

WHMN interconnection technologies often include QoS mechanisms that allow service providers to prioritize traffic types that flow through their subscriber's home networks.

9

VIDEO-ON-DEMAND (VoD) OVER IP DELIVERY NETWORKS

Extensive coverage on next generation video technologies such as IPTV and HDTV has already been presented in earlier chapters. This chapter examines another technology that is changing the way people watch and interact with television— *video on demand(VoD)*. VoD is particularly suited to two-way RF and IP based broadband networks. It allows consumers to browse a library of digitally stored movies and shows, select, and instantly view the title. Additionally, VoD allows consumers to view TV on their schedule rather than depending on the schedule defined by the network service provider. It has been widely deployed by the cable TV industry sector and is used to compete with offerings from satellite and telecommunication providers.

Spurred by the emergence of next generation broadband networks combined with improvements in the price and performance ratios of IPTVCDs, the worldwide deployments of VoD systems based on IP networking technologies has grown significantly in recent times. This chapter examines IP-VoD systems in terms of the types of services and its associated networking technologies. The final section of the chapter covers integrating IP-VoD into an existing networking infrastructure.

9.1 HISTORY OF PAY-PER-VIEW

The whole business of paying for TV content has come a long way over the past three decades. As shown in Fig. 9.1, the story of paying for video content begins in the early seventies. The pay-per-view, often abbreviated as PPV, model of supplying

Next Generation IPTV Services and Technologies, By Gerard O'Driscoll
Copyright © 2008 John Wiley & Sons, Inc.

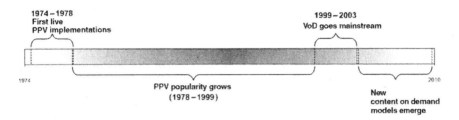

FIGURE 9.1   From PPV to everything on demand

movies and special one-time only events in return for fees was inaugurated by the U.S. cable industry. The first account of a PPV live implementation was in a city called Columbus, in the state of Ohio in the United States. The service was operational in 1974 and over the next two decades network operators all over the world launched PPV services.

In the 1980s many cable operators in particular upgraded their networks to support reverse path signaling. This helped to simplify the process of impulsively paying for PPV content.

9.2 UNDERSTANDING PPV

A PPV system only allows authorized users to watch an event such as a movie or a football game. Ordering a PPV event can only take place within a designated time window. The window deployed by most operators' opens a couple of weeks before the event start time and 10 min after the event starts. Placing an order outside this window will result in either no event or the wrong event been ordered.

A dedicated PPV interactive TV application is typically used to order PPV events. It is either stored locally by the set-top box or else downloaded from the network into memory. The PPV application typically contains different menus that allow the subscriber to search for events on the current day, the week ahead, or coming soon. The look and feel of the menu screen layout is customizable and is unique to each service provider. To order, the subscriber selects the event using a remote control and enters a secret code. Once an accurate code has been entered the order is placed and the rights to watch the event are returned to the viewer. The PPV channel is descrambled and the viewer proceeds to watch the event. The use of an interactive TV application automates the process of ordering PPV events. The PPV application allows subscribers to pay for the event either using prearranged credit that is stored on a smart card or else using a standard bankcard—located in the second smart card reader. The amount can also be included in the subscriber's monthly bill. PPV applications can operate in two different modes.

(1) *Impulse PPV (IPPV)*—IPPV allows subscribers to store electronic tokens, which have been purchased from the pay TV network provider locally on a

set-top box. A section of the smart card called the electronic wallet is used to store and manage the purchased tokens. The subscriber can then use these electronic tokens to impulsively purchase PPV events without using the return channel. Once an event is purchased, the quantity of tokens contained within the wallet will be decremented by the amount of tokens that are related to the purchased event. Some providers can also use this system to implement a type of credit system, whereby a customer would be able to purchase PPV events, even if the number of tokens stored in the wallet were not sufficient. Once the customer exceeds a predefined credit limit, the provider accordingly bills the customer for the amount that is due. A simple telephone call to the pay TV service provider is normally required to refresh the tokens stored on the smart card. Most IPPV systems also support a report back or call back feature. This feature ensures that providers automatically receive accurate information about the subscribers IPPV purchases. In a traditional cable or satellite TV networking configuration the smart card uses a modem built into the set-top box to send information back to the provider at predefined times (i.e., at the end of every week).

(2) *Pre-booked PPV*—This mode of PPV operation allows TV subscribers to book upcoming events in advance. A prebooked event can be placed not only for a single broadcast event but also for multiple PPV events.

A relatively low grade analog or digital set-top box is needed to access and run PPV applications.

9.3   THE EMERGENCE OF RF AND IP BASED VOD

Viewing PPV video content is based on the computing model, where the content is broadcasted at scheduled intervals. This model for delivering content became known as the push delivery mechanism, where the provider decides when the content is available to subscribers. This is one of the main limitations of PPV and is frustrating for subscribers who prefer to base their TV viewing around their own personal schedule. This service limitation combined by a need by cable operators to differentiate their video product offerings from satellite TV providers helped to drive deployments of first generation VoD systems in the late nineties. Having started small during this period the VoD industry sector has grown rapidly over the last couple of years. The two main factors, which have helped to propel this sector are

(1) *Improved VoD server price and performance metrics*—For the last couple of years, cheaper, faster, and smaller VoD servers have been coming out every 12–18 months, just as Moore's law predicted; the law states that the number of transistors on a chip doubles about every two years.

(2) *Installation of next generation network infrastructures*—Continued invest-ment into upgrading networks to support two-way interactivity over the past

half a decade has made VoD a valuable revenue generating service for the cable TV industry.

A VoD system is the "holy grail" of viewing TV and enables an individual subscriber located in geographically dispersed locations to demand a program or movie when and where they want it. This contrasts to the traditional approach to watching television, in which programs and shows are transmitted according to a predefined time schedule.

Therefore, VoD is categorized as a pull-mode service. The library of video content is stored on a server and accessed using a VoD capable IPTVCD. The selection of a VoD title by an IPTV end user is relatively easy and typically comprises the following five steps.

(1) The subscriber selects a VoD title from the interactive TV application.
(2) The IPTVCD accepts the command and sends this instruction to the headend or data center.
(3) The conditional access system is checked to verify that the subscriber is authorized to view the particular VoD title.
(4) Once authorization is complete, a unicast video stream is forwarded to the regional office and onward to the IPTVCD.
(5) The IP stream is controlled by the subscriber.

The key benefits for an end user of a VoD system include the following abilities:

• Watch any program at a time convenient to them and obviates the need to make a trip to a video rental store.
• Control to watch, what they want, when they want. Additionally, VoD allows viewers to stop, start, fast forward, or rewind the video content.
• Provide instantaneous access to a wide variety of compelling and high quality content.
• View content on a variety of networked consumer devices throughout the end-users house.

Although VoD has its origins in the cable TV industry sector, it is an ideal application for deployment across two-way IP based broadband networks. IP-VoD services are enabled though the deployment of a range of technologies, which are combined together to form an end-to-end delivery system.

9.4   TYPES OF IP-VOD SERVICES

VoD provides IPTV end-users with a broad spectrum of on-demand type applications.

9.4.1   Push VoD

Push VoD is based on the concept that video content gets pushed from the VoD server to an IPTVCD hard disk. This transfer of content normally takes place during off peak times, such as late at night and during weekends. The thinking behind this model is to reduce bandwidth bottlenecks that may occur when a number of VoD IPTVCD clients are concurrently viewing VoD content streams. The amount of VoD content stored on the set-top box is dependent on the compression format used and disk size. VoD content type and refresh rate is based on the personal TV preferences defined by the subscriber during registration for the service.

9.4.2   Movie-on-Demand (MoD)

MoD is clearly the most common type of VoD application and is defined as the on-demand delivery of DVD-quality video over a digital network with support for VCR type controls. The subscription paid or the service provider's business model will typically determine the downloadable list of movies and the availability time frame.

9.4.3   Subscription VOD (SVoD)

SVoD uses the same distribution networking infrastructure as MoD. As the name suggests this flavor of VoD is a subscription based service, where subscribers are charged a fixed monthly fee. This charge gives subscribers the ability to view a package of on-demand movies on their own schedule. For example, the entire programming content for the Irish national channel, Radio Telefís Éireann, could be recorded and stored on a VoD server for a particular period of time. The subscriber to the SVoD service would then be allowed to watch the content of the channel at their own convenience.

9.4.4   Television-on-Demand (ToD)

ToD is simply a new method for watching TV. ToD is implemented by recording real-time broadcast TV programs. These recorded programs are encoded and stored on a cluster of video servers. Once stored on the server, the broadcast TV channels can be viewed at any stage. From a service providers perspective the implementation of this on-demand application requires additional disk storage space and hardware to encode the digital signals from various sources in real time.

9.4.5   High Definition VoD (HDVoD)

As the name implies HDVoD allows subscribers to download a range of HD video assets for viewing on HD flat screen display panels or televisions.

9.4.6   Subscription Music on Demand (SMoD)

A monthly subscription to a music streaming service is offered by some telecom operators as part of an IPTV offering. In addition to creating a significant library of music titles, the main challenge for IPTV service providers is to differentiate SMoD from the music stores available on the Internet.

9.4.7   Network-Based Digital Video Recording (NDVR)

nDVR is a technology that allows consumers to record video programming and play back at their convenience. The difference between nDVR and VoD is that the storage of video content and manipulation takes place at the IPTV data center and not the IPTVCD. In effect, it centralizes the functions of a DVR. Thus, consumers can use standard DVR type controls such as fast-forwarding, rewinding, pausing, and recording.

When an nDVR platform is deployed the subscriber's desired content is stored on a central server, which negates the costly requirement of an IPTVCD hard disk. There are a couple of different approaches to deploying nDVR:

(1) Hosted DVR space—Provide subscribers with hard disk space on the VoD server. The hard disk space provided is typically in the tens of gigabytes and varies between operators. Once a request is received by the IPTVCD interface software to record a particular movie, the content is recorded and stored in the subscriber's share of the server's hard disk.

(2) Record the linear channels—The second approach involves recording the entire program lineup and making it available to subscribers for a fixed period of time, typically a week.

The nDVR delivery mechanism eliminates the costs associated with supplying subscribers with hard-disk enabled IPTVCDs. nDVR does have a drawback, namely, the large amount of storage space required at the service provider's IPTV data center.

9.4.8   Free on Demand (FoD)

FoD provides consumers with free access to a library of content that can include programs, localized content, movies, or music videos. The use of FoD is seen by many service providers as a competitive edge and a means of reducing churn of their customer bases.

9.4.9   Bandwidth on Demand (BoD)

This is a relatively new on-demand service that allows consumers to increase network capacity "on the fly" to support specific applications such as downloading high definition programming content.

9.4.10   Internet VoD (IVoD)

With the increased adoption of broadband in recent years the use of the public Internet as a means to access on-demand video content is growing. The library of content used by IVoD systems is developed through partnerships with producers and distributors of movies. The content is typically provided to consumers on a pay-per-view basis and is accessed via a standard PC or a dedicated digital IP set-top box.

9.4.11   Advertising on Demand (AoD)

AoD is a whole new approach to TV advertising and involves the splicing of advertisements directly into VoD content. The length of advertisements range from ten seconds to a minute and are positioned before and after viewing of the VoD asset. In addition to inserting adverts, many of the modern AoD systems can report on advertisement impressions and other marketing data that is relevant to the advertisers.

9.4.12   Extended Video on Demand (EVoD)

Under this delivery model, the on-demand content is initially sent to an IPTVCD, which in turn diverts the content onward to another device at a separate location. In Fig. 9.2 a simplified delivery model used to support EVoD applications is illustrated.

As shown the IP-set-top box is connected to a separate hardware device, which in turn is connected to the RG or internal home network. This EVoD device takes

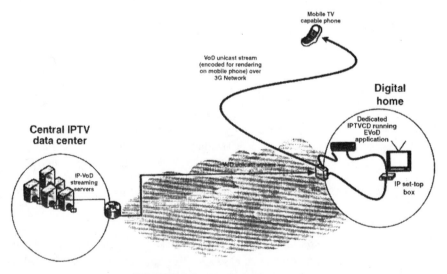

FIGURE 9.2   EVoD delivery model

the output from the IP set-top box and encodes into a format that is suitable for delivery over a number of different types of broadband access technologies. In this example the VoD content is encoded and streamed across a 3G network to a phone that supports the viewing of mobile TV services. This type of VoD application allows consumers to remotely view on-demand content while on the move. Additionally, some of the more sophisticated EVoD systems allow mobile IPTV end users to remotely control A/V equipment such as DVRs.

9.4.13 Everything on Demand (XoD)

Ultimately all subscribers will be able to take full control of their television viewing experience. The move to XoD will enable subscribers to use their IPTVCDs to purchase an asset of digital content, receive it over the network, and view it anywhere. In the not so distant future, it is expected that most of if not all TV programming and movie content will be available on-demand to consumers around the world.

9.4.14 Near Video on Demand (NVoD) Services

NVoD refers to a system that starts the same program on a number of different channels with a time interval between start times (e.g., every 10 min). The concept of NVoD is graphically illustrated in Fig. 9.3.

As shown in Fig. 9.3, the starting times of the same video asset, in this case the all Ireland Gaelic Football Final, is staggered across four channels. This allows subscribers to experience the concept of fast forwarding and rewinding. In this case a subscriber is able to jump channels to rewind and fast forward by 10 min. Additionally, a subscriber can use a simple on screen menu or the EPG to select a convenient viewing time to watch the event. At the specified time, the set-top box automatically tunes to the appropriate channel and begins decrypting the TV event.

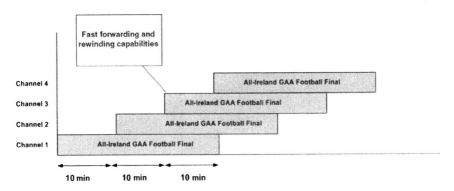

FIGURE 9.3 NVOD system model

With NVoD, subscribers have the convenience of renting a video asset without leaving their home.

An end-to-end NVoD system comprises of a number of hardware and software subsystems that are used to store, schedule, and deliver full motion video content. NVoD does not however require as many hardware and software resources as a complete end-to-end VoD system. It is seen by many service providers as an alternative to true VoD. One of the main drawbacks of an NVoD system is the fact that end users need to wait a finite amount of time before the video title becomes available for viewing. This decreases the potential for impulse purchases.

9.5   UNDERLYING BUILDING BLOCKS OF AN END-TO-END IP-VOD INFRASTRUCTURE

In addition to a high capacity two-way broadband network, the deployment of IP-VoD services also require a number of other logical and physical technology blocks:

- IP-VoD streaming server(s)
- IP-VoD transport protocols
- An Interactive IP-VoD client application

We explore these technology blocks in the following subsections.

9.5.1   IP-VoD Streaming Servers

In addition to the various processes that are executed in IPTV data centers, namely, encoding, multiplexing, and modulation, a number of high capacity computer servers are also installed to allow the delivery of IP-VoD services to various types of IPTVCDs. The main function of these VoD streaming servers is to retrieve and deliver on demand video content to a distribution network.

IP VoD servers are built using standard off-the-shelf computer and electronic components. Lots of processing power, fast input/output (I/O) systems, and large amounts of storage space are the main hardware characteristics of VoD streaming servers. Fortunately, for IPTV service providers the prices for these components continue to fall, while their capacities are continuing to rise. A video server typically includes redundant components such as power supplies, hot swappable hard drives and multiple CPUs. It acquires VoD assets from a number of content sources and is typically able to handle a range of different types of compression formats including MPEG-2, H.264/AVC, and VC-1. A Gigabit Ethernet output provides connectivity to the access network.

A large database of video movies and program assets resides on the server. The server receives requests for a particular content asset from VoD capable IPTVCDs, identifies the item, and streams the requested item to the IPTVCD. Table 9.1 lists the types of on-demand content that could be stored on a VoD server.

TABLE 9.1   Types of Content Stored on VoD Servers

Asset Category	Asset Description
Entertainment	Movies, dramas, music videos, sitcom episodes, and comedy programs
Children	Cartoons and kid shows
News	Local and international news channels
Weather	Local and international weather channels
Advertisements	General and interactive
One off events	Political broadcasts and yearly ceremonies
Sports	Various sporting events

The size and capabilities of the video server varies from supplier to supplier; however, popular servers currently on the marketplace will support most of the following characteristics:

Advanced streaming capabilities—Streaming involves retrieving the content, transferring across the internal server bus, and forwarding via a network interface port to the IP distribution network. Streaming VoD server manufactures have optimized their platforms to maximize the efficiency of this process.

Resilience—Hardware and software resilience is a core feature of streaming VoD servers. Service providers have very high reliability standards because the VoD service must be robust and available 7 days a week, 24 hours a day. Therefore, fault tolerant hard disk arrays, redundant internal bus architectures, redundant fiber channel controllers, and multiple dedicated processors are all used to improve the resilience of VoD servers. Therefore, single failures of server components can be handled without affecting the performance of the VoD service delivered to subscribers.

High capacity storage capabilities—Advanced VoD servers can store hundreds of terabytes of digital content and are typically able to scale to support a growing base of video content.

Monitoring—Server software is generally included as part of the configuration to measure performance of the end-to-end VoD system and issue alerts once particular performance levels are not achieved.

Scalable—VoD servers are able to provide simultaneous access to several hundreds or in some cases thousands of subscribers in real-time, each seeking to view, rewind, pause, and fast forward their video content. High capacity networking interfaces are incorporated into VoD server hardware designs to handle thousands and ultimately millions of compressed video streams.

Multiple formats—VoD servers integrated into IPTV networks are able to stream various types of content such as MPEG-2, VC1, and H.264/AVC across a broadband distribution network.

Interoperability—The video server is normally integrated with a number of other key elements on the network including:

- The billing system to allow charging for movie purchases.
- The conditional access and digital rights management systems to protect the stored content.
- The encoders to compress incoming digital TV signals.

Real-time ingesting of content—VoD servers allow the addition of on-demand programming to their servers by ingesting and recording in real-time the video assets from various content providers and storing onto a storage subsystem. The latest servers to come on the market are able to capture, record, and stream live television to VoD enabled clients within seconds.

The hardware architecture of a streaming VoD server consists of four major technologies, namely, storage, processing and memory, network connectivity, and software. Interaction between these components involves the following: The on-demand content gets archived by the storage subsystem and is retrieved by the processing and memory subsystem once a request for an IP video asset is received; the networking subsystem then packetizes the video data into either ATM cells or IP packets and streams over the network. The software subsystem is responsible for managing the overall process. We explore these subsystems in the following subsections.

9.5.1.1 Advanced Storage

9.5.1.1 Advanced Storage   Video servers are used as centralized storage platforms for VoD services. As the uptake of on-demand services rises, so too will the need for improved and enhanced storage systems to handle this demand for increased volumes of video and music files. VoD storage systems are typically quite scalable and store vast video libraries. There are three primary storage systems used by VoD servers to aggregate and stream on-demand content—mechanical hard drives, solid-state memory disks (SSMDs), and hybrid storage solutions.

Mechanical Hard Drives   VoD servers will generally include support for a number of multigigabyte hard disk drives capable of storing thousands of hours of on demand content. Most VoD servers support a certain level of scalability that allows service providers to add more disk arrays to increase storage capacities. These large disk based storage systems are required to simultaneously deliver hundreds and possibly thousands of video streams to a large group of IPTVCDs.

Hard disks can fail, which can have serious repercussions for live on-demand systems. Therefore, a technique called Redundant Array of Independent Disks (RAID) is used to improve resilience in the event of hardware fault. RAID technology has been used for years by the IT industry to improve the I/O performance and protect mission critical systems against the loss of a single drive. This technology has now been adopted by the VoD industry to reduce the possibility of bottlenecks occurring when accessing video content. RAID is a type of file management system that allows the sharing of digital assets across multiple hard disks and even multiple servers within a cluster. Thus, a portion of each video asset is stored across the disk array. From an end-user's perspective RAID combines multiple drives into an array and this array appears as a single file system on a single hard disk. There are a number of different levels of RAID, but only RAID level 5 (RAID 5) is heavily deployed by the VoD server industry. The RAID 5 technique is widely used by VoD servers because it makes efficient use of disk space, provides a high level of performance when accessing video files, and delivers good fault protection. The storage capacity of video servers that use hard disks can be easily increased by adding new disks to the chassis of the VoD server.

SSMDs   The primary functions of a hard disk are to store and retrieve electronic files. In the context of a VoD system, the frequent reading and writing of content can prove to be extremely demanding on hard disks used in video servers. Therefore, some VoD server platforms use SSMDs based storage technologies. Unlike mechanical hard disks SSMDs based devices contain no moving parts and use flash memory chips for storage instead of rotational magnetic disks to store the on-demand content. So when a request for a stream is received by the server the SSMD fetches the video title and commences the streaming process.

The ability to access content at much higher rates when compared to a hard disk access system is the main technical advantage of these types of storage systems. This translates into an increase in the number of titles that can be delivered over a network for a particular video stream.

Hybrid Storage Solutions   The hybrid approach to storage is based on using a mix of SSMD and mechanical components. On servers that are configured to support hybrid storage solutions the SSMD subsystem is typically responsible for streaming the most frequently accessed video titles over the network, while the hard disks are reserved for storing less popular titles. The decision of what titles are stored on the SSMD comes down to a number of factors, including the video title age, historical popularity, and pricing.

In addition to the above, technologies such as optical drives, network attached storage units, and storage area networks are also used for archiving purposes.

9.5.1.2 *Processing and Memory Technologies*   High powered processors are used in video servers to cope with concurrent access and delivery of numerous video streams. The number of processors required in the server grows exponentially with stream capacity. In addition to a number of processors supporting large computational power, a video server also includes a memory subsystem, namely RAM, to provide buffering and caching space for the VoD content. Advanced mathematical algorithms are normally used by the VoD server to determine what titles are stored in memory.

9.5.1.3 *Network Connectivity*   Up until recent times VoD servers used traditional and legacy ports such as DVB enabled Asynchronous Serial Interface

(ASI) and ATM ports for transporting compressed MPEG-2 video streams across a network. These interface types have worked well for deployments of VoD systems in the early days; however, they had some limitations including scalability and a limited capacity to handle large amounts of video streams. Therefore, VoD server platforms have more or less transitioned away from using these types of ports and replaced with GigE and 10 GigE enabled interfaces. An entry level GigE based port for instance can support data transfer rates of 1000 megabits per second and is capable of concurrently carrying up to 240 video streams with an average bit rate of 3.8 Mbps. This compares quite well to an ASI port, which is only capable of concurrently handling 40 streams of MPEG-2 video. The additional bandwidth supported by a GigE port allows operators to offer bigger libraries of on demand content to their customers. The output from this port is then fed directly into the distribution network.

Although the popularity of integrating DVB-ASI interfaces with IP-VoD servers has decreased in recent years, a certain level of ASI functionality needs to be retained in some deployments to ensure interoperability with existing DVB-ASI enabled networking devices. A typical cable TV network architecture using a mix of GigE and ASI technologies is presented in Fig. 9.4.

As shown a specialized hardware module is used to translate GigE based VoD signals into ASI format prior to modulation and transmission over the cable TV network.

9.5.1.4 IP-VoD Back-office Server Software
A cluster of IP-VoD servers incorporate operating systems and application software that optimize the process of streaming video assets and simplify the day-to-day operations of the overall system. The operating systems used by VoD servers vary between vendors. The application software typically performs a number of critical functions.

Management of Digital Streams The setup, teardown, and control of VoD streaming sessions is a core function of the VoD server software. A graphical illustration showing the steps associated with establishing an IP-VoD stream

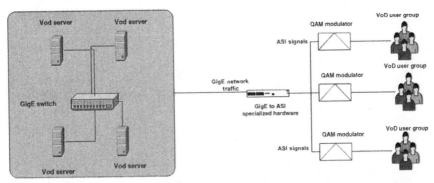

FIGURE 9.4 Cable TV VoD network with support for GigE and ASI technologies

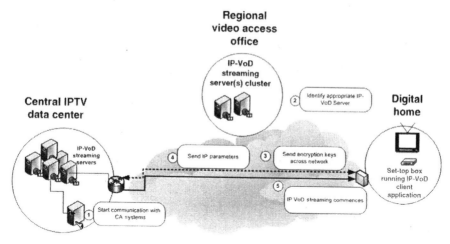

FIGURE 9.5   Setting up a VoD stream

between a client IP set-top box and a backend VoD server is shown in Figure 9.5 and explained in the following sections.

(1) *Start communication with the CA System*—On receipt of a request from the IP set-top box, the server software communicates with the CA system to determine if the IPTV subscriber has been authorized to view the requested IP VoD asset.

(2) *Identify appropriate IP-VoD Server*—Once authorization has been verified, the software identifies a suitable VoD server cluster to fulfill the request. The location of this server or cluster of servers will depend on the IP subnet of the requesting set-top box.

(3) *Send encryption keys across network*—Once the server cluster has been allocated, the backend software or the CA system sends a decryption key to the IP set-top box to facilitate decryption of the IP VoD content.

(4) *Send IP parameters*—IP transport protocol parameters and the IP address of the VoD server is also sent to the IP set-top box.

(5) *IP VoD streaming commences*—Bandwidth is allocated and streaming of the IP VoD asset commences. The IP set-top box uses a protocol called Real-Time Streaming Protocol (RTSP) to manage the flow of the stream.

While the stream is live on the network, it is the responsibility of the server software to ensure that if a fault occurs that the fail over system is activated and the stream continues without interruption.

Updating Digital Content   A VoD software infrastructure has to have the capability to automatically manage the updating of video content. Relying on a manual system to keep VoD content libraries up to date is problematic. The backoffice software ensures that all the digital assets included in the VoD library are current. This is achieved by automatically loading files from the content reception system into the VoD cluster of servers.

Replication Management   The backoffice software manages the replication of digital assets across a distributed networking infrastructure. In the event that new content is made available, it is the responsibility of this software to ensure that the asset is propagated to edge and cache servers around the network.

Management of Metadata   A centralized database is used to store the various attributes or metadata that are associated with each of the digital assets. Metadata is generally formatted as an XML file and provides descriptive data about each video asset. Metadata is typically used to search and browse VoD content. The types of metadata associated with a standard movie include

- Movie producers name
- Description
- Date when the movie was created
- Movie summary
- Parental rating
- Run time of the movie
- Actor and director details
- Genre
- Licensing details
- Royalty details

The backoffice software is responsible for ensuring the integrity of this metadata. Industry groups such as CableLabs have defined a standard for VoD metadata.

Search Capabilities   The indexing capabilities of video server software allow IPTV end users to carry out searches for digital assets on the backend IP-VoD servers.

Managing Access of IP-VoD Digital Assets   The server software provides an external interface to the service provider's back office components — the conditional access, digital rights management, and billing systems. Inputs from this interface are used to control and manage access to the digital VoD assets.

Interfacing with IPTVCDs   The back-office software is also responsible for interfacing and communicating information about VoD assets that are available for purchase to the VoD client software application installed on various types of IPTVCDs.

9.5.1.5   Video Server Architectures   As the demand for VoD content increases over time a single VoD server is often not capable of supporting the complete demands of an end-to-end VoD infrastructure. Therefore, it becomes necessary to add additional servers to the network. Spreading the processing of requests from the

IPTVCDs among several servers ensures that all requests from end users will be performed in the most efficient manner possible. The location of these additional servers can be centralized or installed in remote regional sites that are physically closer to the IPTVCDs. Both of these approaches are discussed below.

Centralized—One possible solution to distributing IP video content over a network is to locate all the video delivery servers in a central location, which is typically the IPTV data center. From a technical standpoint a centralized video server architecture combines a number of individual video servers into a cluster. Each of these servers is interconnected via a bidirectional high speed networking technology, such as Fast Ethernet. Figure 9.6 shows an example of a centralized server architecture used to deliver VoD services.

The diagram shows four servers networked via a multi-port Gigabit IP network switch. This configuration enables high speed server-to-server transfers of digital data. The cluster configuration also includes an interface to a high capacity data tape archive. The servers within the cluster act as a single computing platform to deliver VoD streams. So when a request to play a title arrives at the cluster, any one of the servers connected to the cluster will be able to service that request and commence streaming the content over the network.

The use of clustering technology in a VoD infrastructure allows servers to be added or removed from the network without affecting the operation of the VoD delivery system. The number of servers connected in a cluster will depend on a number of factors:

(1) The size of the content library;
(2) The rate of subscriber growth;
(3) The frequency and patterns of peak loads;
(4) The average number of concurrent users;

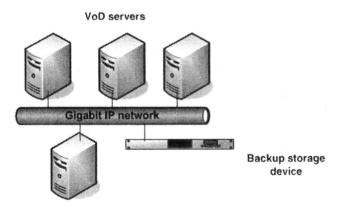

FIGURE 9.6   Centralized VoD server architecture

(5) The policy for service availability;

(6) Load balancing between servers.

Centralizing the servers can ease video content management, improve overall resilience, and availability of the VoD system. Additionally, it can in some circumstances save money for network operators by reducing the number of servers required to deliver the service. Companies that choose this type of architecture do however require a high bandwidth capacity network to effectively deliver the video content to their subscribers IPTVCDs.

Distributed—As the popularity of high definition on-demand content services grows and sizes of content libraries expand, some service providers are starting to engineer distributed video architectures, which maintain a cluster of IP-VoD servers at the IPTV data center while installing additional servers at a number of remote sites. These remote sites house various types of last mile equipment such as DSLAMs, CMTSs, and aggregation routers.

Distributed IP-VoD servers are typically used to provide caching services for popular on-demand content. Replication and propagation to these servers takes place once new VoD titles are added at the IPTV data center. The availability of locally cached video content is a useful mechanism of decreasing the IPTV distribution network's bandwidth requirements.

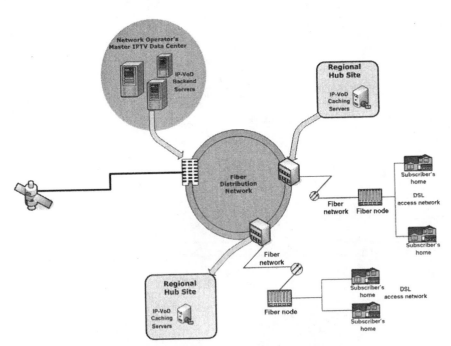

FIGURE 9.7   Distributed VoD server architecture

A sample distributed server architecture operating over, an IP network is shown in Fig. 9.7 and includes backend VoD servers that feed a number of local caching servers at different geographic locations. The IPTVCDs establish dedicated links to these servers and stream the on-demand content directly from them. If the requested content is not available at the local IP-VoD server, then the IPTVCD request is redirected to another regional cluster of IP-VoD servers or else sent back to the central IPTV data center.

Locating VoD servers close to the end-user reduces the amount of network traffic traversing the core IP backbone. The goal of these servers is to service the majority of service requests for VoD titles. This type of server architecture requires more storage when compared to a purely centralized approach because the video content is replicated and cached at the remote servers. Replication of content between the IPTV data center servers and the clusters of IP-VoD servers at the regional offices can occur either when the content is being streamed to the client IPTVCD or at another time of the day when traffic levels are low across the network. It is up to each IPTV provider to decide on which is the most appropriate architecture for delivering IP VoD services to their subscribers.

9.5.2   IP VoD Transport Protocols

IP VoD servers typically use the Real-time Transport Protocol (RTP) and Real-time Control Protocol (RTCP) to stream video data to an IPTVCD. The Real-Time Streaming Protocol is subsequently used to control these streams. The following sections give a brief overview of these three protocols when used in an IPTV infrastructure.

9.5.2.1   Overview of RTP and RTCP   As the name suggests RTP is a native part of the IP suite of communication protocols. It was published as a standard by the ITU-T and has been specifically designed to carry signals for a wide range of real-time applications.

In addition to IPTV there are several other applications that use the RTP protocol ranging from video conferencing to VoIP. Applications that run RTP usually do so on top of the UDP and IP protocol layers. This is because RTP provides appropriate QoS mechanisms and is able to recover from problems that go undetected by UDP. The next section reviews some of the basic technical characteristics and functions of RTP.

Technical Architecture   RTP includes two closely linked parts, a data part and a control part. The data portion is a thin protocol that maintains real-time properties like timing, reconstruction, time stamping, delivery monitoring, security, content identification, and loss detection.

The protocol works well for continuous media streaming and supports a wide range of video and audio formats. When audio and video are used in an IPTV stream, RTP transmits this information as two separate streams. When this occurs a time stamp is normally applied to each RTP stream to ensure that the IPTVCD is able to synchronize multiple streams from the same IPTV source server. In other words, the clocks from the various streams are aligned to ensure that the IPTV subscriber receives a good picture and that the audio is in synchronicity with the image.

RTCP or the control part of RTP monitors the quality of real-time IPTV services. Its main function is to work with protocols such as UDP to provide feedback information to the IPTV data center systems about the quality of data delivery and reception. The types of feedback information ranges from how many IPTV packets were lost while traversing the network to delays in the delivery of IPTV packets. From a technical perspective this information is contained inside different types of RTCP packets.

If the feedback from the RTCP protocol indicates that there are viewing problems, then the interactive IPTV application will make appropriate adjustments to improve signal quality. So unlike TCP, RTP does not automatically reduce transmission rate across the network once an issue is reported. Instead, it passes issues to the higher layers in the protocol stack and allows the Interactive IPTV application to decide on the best approach to dealing with the problem. Solutions to viewing problems can range from slowing down the transmission of video frames from the IPTV server until the IPTVCD buffer has freed up to examining the possibility of using a higher compression rate when preparing the content for transmission.

RTP and RTCP are agnostic to the technologies running on the underlying network. However, the majority of IPTV deployments use the UDP/IP protocol combination to carry video content across the network (see Fig. 9.8).

With regard to port numbers, RTP data is channeled through the IPTVCM using even UDP port numbers, while RTCP packets are carried on the next highest odd port number. These port numbers are not standardized; however, as a rule of thumb RTP dynamically selects a port within the range 16384–32767. Although RTP has huge benefits when used in an IPTV deployment, it does have a couple of drawbacks. For a start, RTP does not have the necessary capabilities such as flow and congestion control to guarantee time of delivery. Nor does it have the necessary means to make sure that other standards of service are maintained. These tasks are performed by the lower layer protocols. Nondelivery of packets, out-of-order delivery, and the sequence of delivery lie outside the purview of RTP. Also RTP does not support multiplexing and therefore makes use of multiplexing and checksum services available at the UDP and IP networking layers. Figure 9.9 shows how RTP packets are typically encapsulated inside IP packets.

9.5.2.2 RTSP Overview   Real-Time Streaming Protocol is an application level protocol belonging to the IP communication model that enables IPTVCDs to establish and control the flow of IPTV streams. The specification for RTSP was

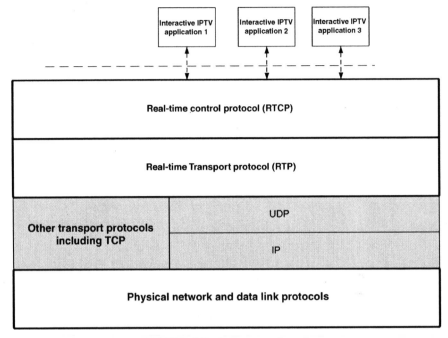

FIGURE 9.8   RTP protocol stack

published in 1998 and can be found in RFC 2326. It allows IPTVCDs to issue VCR style commands to an IPTV streaming server. Details of the primary commands used in a typical RTSP session are listed in Table 9.2 and graphically depicted in Fig. 9.10.

In addition to VCR type functionality, RTSP also allows an IPTVCD to request and retrieve a particular item of IPTV content. The fulfillment of this request involves the inclusion of the appropriate VoD servers IP address inside the request command issued by the IPTVCD. If however the request is for a live broadcast TV channel, then the IP address of the multicast group is included in the requesting message. The key characteristics of RTSP include the following.

Client-Server Computing Model   RTSP operates using a client-server computing model. Under this model three separate connections are established to provide

IP header	UDP header	RTP header	IPTV content

FIGURE 9.9   RTP packet encapsulation

TABLE 9.2   RTSP Commands

Command	Functionality
SETUP	This command contains some important items of information such as the URL of the IPTV asset and an identifier to specify the port to be used for content transportation. The sever replies to this command and allocates appropriate resources to stream out the requested content
PLAY and RECORD	Once the SETUP command has been processed the PLAY command is used to start transmission of the specified IP-VoD asset. In normal circumstances the video content is played across the network from beginning to end
PAUSE	As the name suggests the server temporarily pauses transmission of the IP-VoD stream. An acknowledgement is typically sent to the IPTVCD to confirm the status of the Pause instruction
RECORD	This command is used to record IPTV content on to a particular type of storage media
TEARDOWN	Under this command the IP-VoD streaming session is terminated and any resources assigned to maintaining the stream are freed up. A message confirming teardown is also sent to the IPTVCD
ANNOUNCE	This command is generally used to convey channel and IP service information. Note that the data included in the ANNOUNCE message defined in DVB-IPI is presented to the IP-VoD client in XML format
DESCRIBE	This is another RTSP command that is specified by DVB-IPI and is used to carry XML formatted channel and IP-VoD descriptors

communications between the RTSP client running on an IPTVCD and the IP-VoD server. The three connection types are shown in Fig. 9.11 and explained in the following section.

(1) An out-of-band connection is established to carry RTSP control information. The transport layer communication protocol used by this connection type is based on either UDP or TCP. In the case of DVB networks, a persistent TCP connection is generally used. In addition to passing RTSP control type

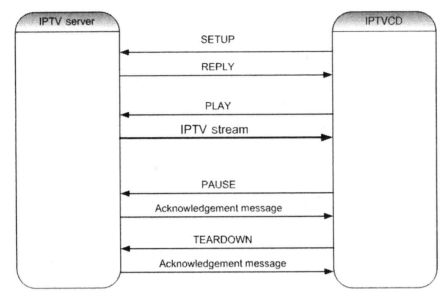

FIGURE 9.10   Use of commands in an RTSP session

information this connection type may also be used to carry and interleave IPTV content.

(2) A separate RTP over UDP connection is established to carry encoded IPTV content.

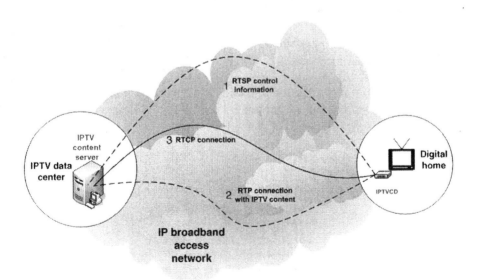

FIGURE 9.11   Client-server RTSP communication model

(3) The third connection carries RTCP over UDP synchronization information. This stream provides feedback to the server on the quality of the stream been delivered to the IPTVCD.

Similar Operational Functionality to HTTP   The operation of RTSP and HTTP are similar. Both have comparable addressing formats and operate in request/ response mode when communications are occurring between devices. They do however differ in three key areas:

- The protocol identifier is different. So rather than using the "http://" identifier, RTSP uses "rtsp" at the start of a URL to locate a particular IPTV channel or IP-VoD asset.
- Another distinction between both protocols is the state of operation. RTSP servers generally operate in a "state" mode, while HTTP servers operate in a "stateless mode."
- HTTP is primarily an asymmetric protocol; in other words, the bulk of the communication takes place between the IPTV server and the IPTVCD. Whereas RTSP allows both device types to issue commands.

Support for Both Unicast and Multicast Traffic   RTSP is able to control both live multicast TV and on-demand unicast video streams. Although trick mode operations such as rewind and fast forward are not applicable to multicast channels.

Independent of Underlying Transport Protocols   RTSP leverages existing transport protocols and operates over connection or connectionless orientated systems such as TCP and UDP.

Works in Conjunction with RTP   RTSP and RTP work closely together to deliver IPTV content across a network.

RTSP Message Formats   The messages used by RTSP may be broadly classified into two categories: requests and responses.
    The general syntax for an RTSP request message is

{method name} {URL} {Protocol Version} CRLF {Parameters}

The general syntax for an RTSP response message is

{Protocol Version} {status code} {reason phrase} CRLF {Parameters}

The best way to understand about RTSP message formats is to look at a practical example. In this case, the client IP set-top box is sending a request to view a movie about Irish history called "The Wind That Shakes the Barley," which is stored on an IP-VoD server at the service provider's IPTV data center. A simplified graphic showing communication messages associated with viewing this on-demand movie is depicted in Fig. 9.12 and described in the following steps.

(1) Setup a connection between the client IP set-top box and the server

First and foremost a TCP connection between the IP set-top box and the server are initially established. In this example the default RTSP listening port 554 is used by the server.

(2) Issue a DESCRIBE request message

The first message to be sent contains the DESCRIBE method:

Example Request

```
DESCRIBE rtsp://192.168.1.25:554/mpeg4/movies/windthatshakesbarley.mpg
RTSP/1.0
```

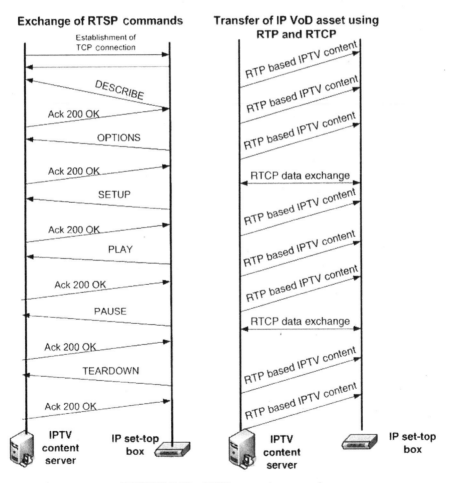

FIGURE 9.12   RTSP messaging example

```
CSec: 101
Accept: application/sdp
```

This is a request to the IPTV content server to send a description of the IP VoD asset windthatshakesbarley.mpg, using SDP.

The following is an example of the server response to the DESCRIBE command back to the IP set-top box.

Example Response

```
RTSP/1.0 200 OK
CSec: 101
Content-Base:  rtsp://192.168.1.25:554/mpeg4/movies/windthatshakesbarley.
mpg
Content-Type: application/sdp
Content-Length: 320
<SDP Data..........>
```

The response message includes a full description of the movie "The Wind that Shakes the Barley," which has been formatted by SDP.

(3) Issue an OPTIONS request message

The next message to be sent by the IP set-top box is the OPTIONS Command

Example Request

```
OPTIONS rtsp://192.168.1.25:554/mpeg4/movies/windthatshakesbarley.mpg
CSec: 102
```

This message queries the server on the types of commands that are supported.

Example Response

```
RTSP/1.0 200 OK
CSec: 102
Public: DESCRIBE, OPTIONS, PAUSE, PLAY, SETUP, TEARDOWN, ANNOUNCE
```

The IPTV streaming server responds with a list of RTSP commands that are supported

(4) Issue a SETUP request message

The next command instructs the IPTV server to allocate some resources.

Request

```
SETUP rtsp://192.168.1.25:554/mpeg4/movies/windthatshakesbarley.mpg
CSeq: 103
Transport: RTP/UDP;unicast;client_port=4042-4043
```

In this command transport details are specified. In this case the IP set-top box wants to use the unicast mechanism in combination with the RTP over UDP protocol suite.

Response

```
RTSP/1.0 200 OK
CSeq: 103
```

```
Session: 1234567891;timeout=10
Transport: RTP/UDP;unicast;mode=play; client_port=4042-4043;
server_port=5072-5073
```

The response from the IPTV server contains a session ID or identifier, which is used to tag or recognize a particular session. The requested mechanism for streaming the content has also been confirmed in this message. Additionally, the message includes server port information.

(5) Issue a PLAY request message

PLAY is the next command issued by the IP Set-top box.

Request

```
PLAY rtsp://192.168.1.25:554/mpeg4/movies/windthatshakesbarley.mpg
RTSP/1.0
CSeq: 104
Session: 1234567891
```

This play command is used to start the streaming of the IP-VoD asset. It can also be used to restart a paused IPTV stream. The message contains the URL of the movie and session ID. As shown on the diagram, the RTP protocol is used for the actual delivery of the movie over the network. The delivery of information occurs in one direction — from the IP-VoD server to the IP set-top box. The RTCP protocol is also activated while the movie is transferring across the network. This protocol communicates information in both directions and is used to monitor stream quality.

Example Response

```
RTSP/1.0 200 OK
CSeq: 104
Session: 1234567891
```

Another 200 OK message is received to confirm that the streaming of the IPTV content has commenced.

(6) Issue a PAUSE request message

A pause command is then issued by the IP set-top box.

Example Request

```
PAUSE rtsp://192.168.1.25:554/mpeg4/movies/windthatshakesbarley.mpg
CSeq: 105
Session: 1234567891
```

This command puts in a request to the server to pause the delivery of the IPTV content.

Example Response

```
RTSP/1.0 200 OK
CSeq: 105
Session: 1234567891
```

The server responds with another 200 OK acknowledgment indicating that the requested action has been completed.

(7) Issue a TEARDOWN request message
The final command issued by the IP set-top box in this example is the teardown command.

Request

```
TEARDOWN rtsp://192.168.1.25:554/mpeg4/movies/windthatshakesbarley.mpg
RTSP/1.0
CSeq: 106
Session: 1234567891
```

This request asks that the IPTV server terminates the delivery of the IP-VoD asset to the IP set-top box.

Response

```
RTSP/1.0 200 OK
CSeq: 106
Session: 1234567891
```

The server responds with another 200 OK acknowledgment indicating that the requested action has been completed and transmission of the IPTV stream is ceased.

9.5.3 Interactive IP-VoD Client Application

A deployment of a generic IP-VoD system architecture is based on a client/server-computing model. This means that a dedicated point-to-point connection is setup between the IPTVCD and the server at the data center. To interoperate with the data center server over this connection the IPTVCD requires a VoD client software application that is capable of receiving IP based VoD streams. The client application presents the IPTV end user with a menu of video assets and associated descriptions on their television screen. The contents of the menu are based on the programming package that the consumer has subscribed to. The technical gathering of this information is called service discovery and varies between networks. For instance, on a DVB network the system used to discover information about available on-demand services is a standard XML record, which has been defined in section 5 of the ETSI TS 102 034 standard. Details of this record are available in Table 9.3.

In addition to allowing subscribers to select and download specific movies a typical IP-VoD client application also supports content searching and stream control functionality. An implementation of the client IP-VoD application is typically implemented through an embedded HTML browser.

9.6 INTEGRATING IP-VOD APPLICATIONS WITH OTHER IP BASED SERVICES

Another important aspect to deploying IP VoD applications is how to integrate into a network that is already carrying VoIP and high speed Internet services. The

TABLE 9.3   Video and Content on Demand Discovery Record Structure as Defined in ETSI TS 102 034

XML Attribute	Description	Comments
Catalogue@ID	This identifies a VoD server	Mandatory attribute
Name	This gives the name of the overall VoD offerings and may be presented in a number of different languages	
Description	This field details the catalogue which may be presented in a number of different languages	
Locator	This provides a URL that points to a location where the VoD content descriptions are held	

approach taken needs to be guided by the transport mechanisms used by traditional IP voice and data services. One differentiator is the mechanism used by various service types to authenticate and identify end users. For example, the Point-to-Point Protocol over Ethernet (PPPoE) defined in RFC 2516 is generally used to authenticate and authorize Internet access and VoIP sessions but is not used by IPTV services.

As the name suggests PPPoE takes a PPP packet, which includes support for a number of network features and encapsulates it inside an Ethernet frame for delivery across an IP broadband network. PPPoE enforces traditional authentication techniques such as logins and passwords, which are a critical part of authenticating Internet access services. In the context of the IP protocol stack, PPPoE sits just above the Ethernet or physical layer and it is here that the authentication takes place.

PPPoE is not used to authenticate IPTV services such as IP-VoD because the authentication of subscribers typically occurs at the application layer of the IPTVCM. Additionally, PPPoE does not support multicast traffic.

Differing approaches to authentication, encapsulation, bandwidth usage, and in some cases packet routing protocols will often require the creation of separate logical topologies for each of the IP services that run across a broadband network. These logical topologies are known as virtual LANs (VLANs) and ATM based Private Virtual Circuits PVCs.

9.6.1   Introduction to VLANs

VLANs are the most popular method used by service providers to integrate IPTV services such as VoD onto their existing network infrastructure. A VLAN is defined as a mechanism for creating a series of independent logical networks. Each logical network consists of a number of IPTVCDs that are connected at physically disparate sections of the IP network infrastructure. As such all of the IPTVCDs associated with a particular VLAN can behave as if they were directly connected together.

Under this approach a number of different VLANs can coexist together within the same networking infrastructure. The creation of VLANs to support an IPTV implementation is typically done with management software that interfaces directly with the various network components. The main benefits of using VLANs to deploy both IPTV broadacast and on-demand services are reducing interference with existing services, scalability, use of distributed management techniques, and flexibility.

Minimise interference with existing IP services—Separating IPTV traffic into a logical network removes the chances of interference with existing IP based products such as high speed Internet access and VoIP. This is due to the fact that creating multiple VLANs reduces the broadcast domain size and lessens the possibility of packet collisions. This in turn helps to meet the minimal packet loss requirements of a QoS architecture

Scalable to support large numbers of VLAN users—VLANs can be created to span across a network that supports thousands and even hundreds of thousands of IPTVCDs

Distributed Network Management—The creation of VLANs allows network operators to separately manage IP address ranges, routing protocols, and topologies for each service available on the network. For instance, by using VLAN technology, network administrators can configure a separate DHCP server at the IPTV data center that will only deal with assigning IP addresses to IP-VoD client devices.

Flexibilty—With the increasing mobility of people nowadays it has become important that different types of IPTVCDs can easily connect with their IPTV services. VLANs provide support for mobility strategies by minimizing the need to make changes to hardware configurations when accessing IPTV services on the move.

9.6.1.1 Types of VLANs
VLAN membership can be configured in four different ways:

(1) *Port-based*—With a port-based VLAN, a port on a network switch, router, CMTS or DSLAM is configured as part of the VLAN. Once a port is configured in VLAN mode, all IPTVCDs connected to that interface port belong to the same VLAN.

(2) *MAC-based*—Under this configuration, membership to a particular VLAN is based on the MAC or hardware address of the IPTVCD. In order to

accomplish this, a database is used by the VLAN networking equipment to map the various hardware addresses to the associated VLANs.

(3) *Protocol based*—As the name suggests membership to a particular VLAN is based on the protocol identifier within the Ethernet packet. This mode of VLAN operation is not used by triple-play service providers because most if not all services nowadays use the IP protocol. Additionally, the inspection of packets to identify protocol related information can be time consuming.

(4) *Authentication based*—This membership mode requires IPTVCDs to be authenticated prior to gaining connectivity to a particular VLAN.

9.6.1.2 Communicating VLAN Information   The 802.1Q specification defines the protocol used to communicate information across Ethernet based VLANs. Unlike many other protocols 802.1Q does not apply encapsulation to data packets but instead inserts identifiers or tags directly into an Ethernet packet to identify which VLAN it belongs to. In addition to inserting new information into an Ethernet packet, one of the values in the header called the Ethertype is also changed to identify the packet as belonging to the 802.1Q communication process. Details of how a standard Ethernet frame carrying IPTV content is modified by 802.1Q is depicted in Fig. 9.13.

As depicted the Ethertype field has been changed to 0x8100. In addition, a VLAN tag has been inserted into the frame and consists of three new fields:

- *User priority field*—This 3-bit field stores the priority level for the IPTV Ethernet frame.

- *Canonical format indicator* (CFI)—This is a 1-bit indicator that is always set to a value of one for Ethernet based networks.

- *VLAN identifier*—This 12-bit field identifies which VLAN the packet belongs to. In this case the packet is belonged to the IPTV VLAN because it is carrying video content in its payload. The fact that this field is 12 bits long means that a network can be configured to support 4095 different VLANs. If this field is configured with a value of zero, then the packet does not belong to any VLAN.

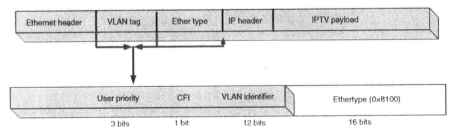

FIGURE 9.13   802.1Q tagging of Ethernet frame

9.6.1.3 VLAN Network Design Options The design of the VLAN infrastructure deployed on a network is dependant on the type of traffic being carried across the connection. There are three main options.

Subscriber Centric VLANs Under this type of network design, a dedicated VLAN is established between the RG at the subscriber's house and the distribution router back at the IPTV data center. Intermediate devices such as DSLAMs and network switches are also part of this type of end-to-end VLAN solution. Figure 9.14 shows a simplified example of a network designed to support dedicated VLANs to each subscriber.

The maximum number of VLANs supported by the above design between a DSLAM and an edge router is the maximum number contained in the VLAN identifier field, which is 4095; this equates to 4095 subscribers. It is possible to use a process called VLAN stacking to increase subscriber numbers above 4095. VLAN stacking involves including multiple tags inside an Ethernet frame. This facility allows the edge router to insert one VLAN tag identifying the port of the router and a second tag to identify each of the DSLAMs. In this example VLAN stacking is used as a means of increasing capacity at the regional office. Although subscriber centric VLANs have a number of advantages they do not however support multicast replication. Thus, these VLAN types work satisfactorily for delivering unicast applications such as VoD but are problematic when it comes to the delivery of broadcast television to large numbers of IPTV subscribers.

Service Centric VLANs In this network design, each VLAN corresponds to an IP service that runs across the network. So to deliver a multiple-play product offering

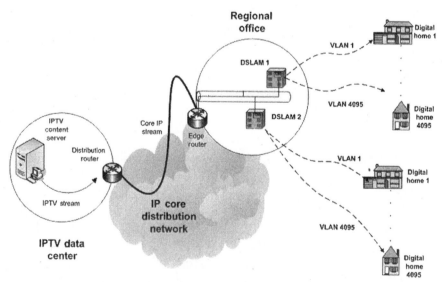

FIGURE 9.14 Simplified example of a subscriber centric VLAN design

to consumers, a separate VLAN is created for each of the IP services, namely, voice, IPTV, and broadband. This allows subscribers to choose and mix the types of IP services they want delivered to their homes. Additionally, it allows existing telecommunication companies to incrementally add new services such as IPTV to their networks without having a negative impact on their existing telephony and broadband services. Figure 9.15 depicts a service-based VLAN model.

As shown, the backend servers do not connect with multiple LANs. However, IPTVCDs can be configured to connect with multiple VLANs once authorization has been granted by the service provider. So in a typical IPTV deployment all subscribers would have access to the broadband VLAN, while a smaller percentage will have access to VoIP and broadband VLANs, while yet another portion of subscribers will have subscribed to all three services and are hence connected to all three VLANs.

A Hybrid Design   The hybrid VLAN network design combines the benefits of both architectures. In this hybrid model, subscriber centric VLANs are built to carry high speed Internet access, VoIP, standard definition and high definition VoD content, and other types of unicast data. In addition to subscriber VLANs, a service centric VLAN is also deployed to carry any multicast linear TV programming content.

9.6.1.4 VLAN Configuration   Once the decision is made to use VLANs the next stage is to actually configure each network component to support VLAN functionality. A typical triple-play forwarding architecture with a DSL access network is presented in Fig. 9.16.

As depicted the interfaces connecting the four DSLAMs with the three downstream routers are configured to support separate 801.1Q VLANs. A VLAN is assigned to each one of the triple-play services—VoIP, Internet access, and IPTV. Note the VLAN assigned to delivering video is capable of carrying both unicast VoD streams and IP multicast traffic.

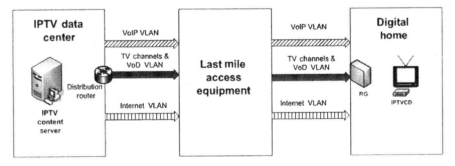

FIGURE 9.15   Service based VLAN network design

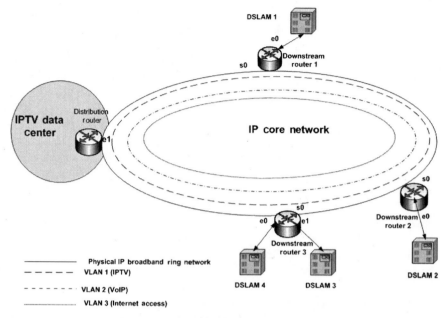

FIGURE 9.16   Triple-play VLAN configuration

9.6.2   ATM PVC

In an ATM networking architecture, different data types are transported using PVCs. ATM PVCs can support demanding applications that require high bandwidth and low transmission delays. Using an ATM infrastructure allows network operators to assign each triple-play service to a separate PVC. Although a number of networks use ATM as a mechanism for transporting data, there is a move toward the use of IP and Ethernet networks for delivering services.

9.7   PROTECTING IP-VOD CONTENT

Copyright protection of video content is a major issue for the delivery of IP-VoD services. Since the IP protocol has no inherent copy protection built into the technology, conditional access and DRM systems are used to protect the rights of the original content producer. The objectives of these security systems is to ensure that only authorized paying subscribers can access IP-VoD assets. These systems interface quite closely with the OBSS system to determine which IP VoD titles are associated with which product packages. This then allows these systems to determine the authorization rights associated with the various requests. Chapter 7 provides further details about CA and DRM systems.

SUMMARY

The traditional method of ordering pay-per-view video content is changing and evolving to meet consumer demands. This evolution is resulting in the widespread deployments of a technology called video on demand. VoD is an interactive TV application that enables consumers to view video content when they want. VoD provides IPTV end users with a broad spectrum of on-demand type applications ranging from HDVoD and MoD to AoD and EVoD. The primary components of an end-to-end IP-VoD infrastructure include IP video servers, transport protocols and client software.

The video server(s) store and implement real-time streaming of IP-VoD content. The assets are physically stored across the servers in the form of computer files. There has been a tremendous growth recently in the demand for advanced storage technologies to support the growing popularity of on-demand products. The storage technologies used by VoD servers fall into two broad categories: hard disks and SSMDs.

Hard disks—Many VOD systems in deployment around the world today rely on hard disks for both streaming and storage of VoD assets to IPTVCDs. RAID technology protects data in the event of drive failure and ensures that VoD servers remain up and running at all times.

SSMDs—High performance SSMDs help to improve performance levels of IP-VoD applications and accelerate storage access rates for IP-VoD titles.

Some VoD servers use a mix of these technologies where the SSMD storage space serves requests for popular VoD titles, while the hard disk arrays are used for storage purposes only.

The move to GigE networking interfaces and architectures that support multiple processors allow VoD servers to support a higher density of video streams. IP-VoD servers also include a management software platform capable of managing stored content assets. Specialized software stored on the VoD set-top box helps to enhance the experience of downloading various types of VoD content. This application, which is generally integrated into the set-top box communicates directly with the VoD server.

There are two approaches to designing a network to support the delivery of IP video content, namely, centralized and distributed. In the centralized approach the servers are kept in a central location whereas a design based on a distributed architecture locates the servers in different locations throughout a particular geographical region.

The protocols used to facilitate communication between IP-VoD servers and client IPTVCDs vary between network architectures but typically include the use of RTP, RTCP and RTSP.

- RTP is a part of the Internet suite of communication protocols that carries IPTV streams over broadband networks.

- RTCP is used by IPTV networks to monitor the quality of the communication session between client and server devices.

- RTSP is an applications level protocol that allows for control of real-time and on-demand video content. RTSP has been widely adopted by the IPTV industry and is used by many service providers to deploy IP-VoD services.

A number of challenges exist when attempting to integrate IPTV services into a network that is already carrying VoIP and high speed Internet services. Segmenting IPTV traffic into separate VLANs or ATM PVCs is the most popular approach to deploying multiple IP services across the same networking infrastructure.

10

IP BASED HIGH DEFINITION TV

Next generation high definition TV (HDTV) channels are revolutionizing the transmission and reception of video services. HDTV has finally arrived and is here to stay. Factors such as consumer demand and improved adoption of digital networking technologies are enticing multiservice network operators to start adding HDTV channels to their program lineup. The main focus of this chapter is to define HDTV and explain how the technology works.

10.1 OVERVIEW OF SDTV AND HDTV TECHNOLOGIES

SDTV is a digital TV broadcast system that provides consumers with better pictures and richer colors than analog TV systems. Digital technology is also capable of producing HDTV pictures with more image information than conventional analog signals, producing an effect that rivals the movie theater for clarity and color purity.

Although both systems are based on digital technologies, HDTV generally offers viewers nearly six times the sharpness of SDTV based broadcasts. In addition to classifying digital TV signals into HDTV and SDTV, each of these technologies has the following subtype classifications.

10.1.1 Resolution

Resolution is defined as the amount of detail or pixels contained in an image displayed on a television set. This characteristic is one of the main reasons why people

Next Generation IPTV Services and Technologies, By Gerard O'Driscoll
Copyright © 2008 John Wiley & Sons, Inc.

experience an improved viewing experience while watching a HDTV in comparison to standard analog based TVs. A television display has display capabilities of 644 × 483 pixels, although most people refer to it as 640 × 480. This again is not a true description of the display capabilities of a standard TV. For technical reasons, most viewers are unable to see the margins that surround the perimeter of their television screens. Consequently, most industry experts calculate the actual viewing area by subtracting the marginal areas. Such calculations result in an actual viewing area on a standard TV of 620 × 440 pixels. It is also possible to measure the resolution of a television display by the number of video lines available on the screen. The video lines are developed with a type of electronic gun that operates inside the television tube. This gun emits a beam of light that moves back and forth across the surface of the screen. As the beam moves across the inner surface of the screen, it creates an image, line by line. As a general rule, the more lines created by the gun, the higher the resolution. The resolution of an IP HDTV program is dependent on the size of the TV display and the number of video lines present on the screen. For example, clearer pictures are available to owners of new HD televisions because they are able to display 1080 lines of resolution versus 525 lines displayed on an ordinary television sets.

10.1.2   Interlaced and Progressive Scanning

The process of scanning defines how the lines are displayed across a TV screen. There are two different scanning methods, namely, interlaced and progressive. Interlacing scanning is a system that was designed by TV engineers to overcome the inability of early television tubes to draw a video display on the screen before the top of the display began to fade. With interlacing the image is created by instantaneously applying light to every second line on the screen. Once this is completed the remaining lines are illuminated almost immediately. This happens so fast that the human eye is unable to detect. Interlace scanning is popular in analog televisions and low resolution computer monitors.

Progressive scanning, also known as noninterlaced video, refers to a process of a television screen refreshing itself line by line. This is done at the same time therefore the image appears smooth to the human eye and stays sharp during motion in the video pictures. Progressive scanning is popular in HDTVs and computer monitors.

10.1.3   Aspect Ratio

Aspect ratio defines the relationship between the horizontal width and vertical height of a TV screen. The aspect ratio of a traditional TV screen is 4 : 3. In other words, for every four units of measurement across the length of the TV screen the height will consist of three units. HDTV uses an aspect ratio of 16 : 9.

These classifications are used by broadcasters and TV manufacturers to define a notation for identifying different types of HDTV products. The notation used in the United States is based on lines of resolution followed by a letter that represents the type of scanning used. An "i" represents interlaced and a "p" represents progressive. The notation also includes a figure for the number of frames or fields

TABLE 10.1   Popular HDTV Formats

HDTV Format	Pixel Resolution	Description
720i	1280 × 720	This format provides IP HDTV subscribers with a line resolution of 720 and uses the interlaced scan type. The aspect ratio for this format is 16 : 9.
720p	1280 × 720	This format provides IP HDTV subscribers with a line resolution of 720 and uses the progressive scan type. The aspect ratio for this format is 16 : 9.
1080i	1920 × 1080	This format provides IP HDTV subscribers with a line resolution of 1080 and uses the interlaced scan type. This format also uses a wide-screen format of 16 : 9.
1080p	1920 × 1080	This format provides IP HDTV subscribers with a line resolution of 1080 and uses the progressive scan type. This format also uses a wide screen format of 16 : 9. 1080p is currently the highest possible HD resolution and its viewing quality is unmatched.

per second. Examples of some of the most popular HDTV standards are shown in Table 10.1.

The differences between two popular HDTV formats (720p and 1080p) are illustrated in Fig. 10.1.

10.2   HDTV OVER IP DEFINED

The acronym IP HDTV is typically used to describe the delivery of high definition video content over a broadband connection. It is a relatively new technology and is steadily taking off as a value added service for telecommunication companies who want to differentiate their IPTV services from existing cable and satellite offerings. There are a number of reasons why service providers are upgrading their network architectures to support the transmission of HDTV content. The following is a short list of some of the most compelling ones.

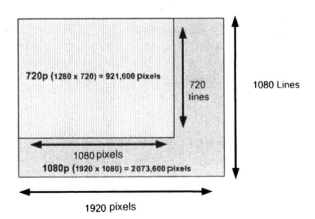

FIGURE 10.1   Comparison of HDTV formats

10.2.1   Improvement in Video Quality

The evolution from traditional analog TV to digital TV has already provided consumers with sharper, richer, and more engaging video content. Continuing this evolution from standard digital TV content to high definition broadcasting brings the TV viewing experience closer to resembling the sharper and crisp video quality provided by cinemas.

10.2.2   Better Color Resolution

The color resolution supported by HDTV is far superior to the resolution of regular SDTV video content.

10.2.3   High Quality Levels

By using advanced compression technologies in conjunction with high speed networking technologies, it is now possible to provide consumers with IP HDTV services at reasonably high quality levels. The delivery of HD content over an IP broadband platform comes with a number of challenges including

- HDTV has greater bandwidth requirements compared to SDTV video streams. As such the networking infrastructure needs to be upgraded to provide sufficient bandwidth to deliver multiple HDTV streams into people's homes. The amount of bandwidth required depends on the compression algorithm applied to the content at the IPTV data center. Higher compression rates require less bandwidth; however, final picture quality is degraded.
- The reception system at the IPTV data center may need to be upgraded to support and manage larger volumes of video data. Additionally, extra storage resources are also required to accommodate increased file sizes, which is a "feature" of HD video content.
- The internal networking infrastructure of the IPTV data center needs to support very high transfer rates. This may require an upgrade to the existing cabling infrastructure.
- IPTV end users need to either lease or buy HD capable TVs and set-top boxes to enjoy the full benefits of IP HDTV.

10.3   AN END-TO-END IP HDTV SYSTEM

In general the protocols used to transport SDTV signals across an IP network are the same used in the delivery of HD video content. The specific hardware requirements of producing and viewing HD video content are shown in Fig. 10.2.

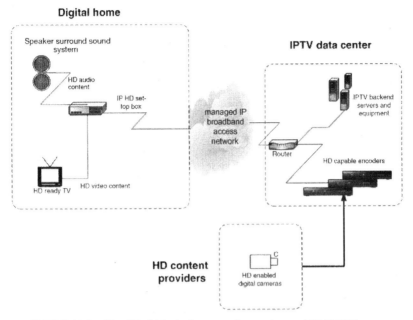

FIGURE 10.2 Simplified block diagram of an end-to-end IP HDTV system

10.3.1 HD Enabled Digital Cameras

HD cameras are been used by a growing number of film production companies. These types of cameras shoot video content up to five times the resolution of SDTV. The content from a HD camera typically gets digitally transferred to a server using fiber optic cable.

10.3.2 HDTV Servers

The HDTV server is connected via a high speed broadband network to an HD capable IP set-top box or a standard PC. Its main function is to digitally store HDTV content on a high capacity storage device.

10.3.3 Encoders

MPEG-2 encoders are the most common format for compressing SDTV. The processing of a high definition MPEG-2 video channel can use up to six times the amount of bandwidth of a standard channel. A number of telecommunication providers have limited bandwidth capacity on their networks to support MPEG-2 compressed high definition content. Therefore HD capable encoders play an important role in the HDTV environment and are responsible for achieving the best picture quality at the lowest bit rate. Most encoders used to deliver IP HDTV content are based on advanced encoding algorithms such as MPEG-4 and VC-1. These encoding technologies promise to deliver better video quality at lower bandwidth capacities.

10.3.4   A High Capacity IP Network

A HD video file can include up to five times the information of an SDTV file. Therefore, a high speed digital cable, satellite, DSL broadband, or terrestrial network with sufficient bandwidth capacity is required to transport HD content from the IPTV service provider's data center to the HDTV home viewing system.

10.3.5   Home Viewing System

A HDTV viewing system consists of three main components—a HD capable IP set-top box, a display unit (typically an LCD or Plasma screen), and a surround sound system.

10.3.5.1   Overview of HD Capable IP Set-top Boxes   HD capable IP set-top boxes are a fast growing sector in the consumer electronics marketplace. HD capable set-tops typically deliver enhanced HD video and audio to consumers. These types of IPTVCDs not only allow viewers to watch high quality HD transmissions but also standard definition programming content as well. HD capable IP set-top boxes are typically classified as follows.

HD Multicast Only Boxes   These set-tops are capable of receiving HD channels via a number of different types of IP broadband access networks.

HD Set-tops with Hard Drives   This type of IP set-top box is able to replicate PC-like functionality and allow consumers to personally record HDTV video content and cache HD multicast programming content.

HD Set-tops with Integrated Recording Unit   HDTV consumes huge amounts of network capacity. A typical HDTV MPEG-2 stream is transmitted at over 20 Mbps. To record 1 h of this stream, over 6 GB of storage would be required. Despite the high capacity storage required, the demand from consumers for solutions, which allow recording of HD content, is rising. Traditional storage media devices such as CD-ROMs are not capable of archiving these volumes of data. Therefore, manufacturers have started to build HD capable set-top boxes that include an optical recording unit that allows consumers to securely record content onto a hard disk or writeable DVDs. These disks can then be played on other devices such as PCs, notebooks, or portable DVD players. In addition to recording material these units are also able to play off-the-shelf DVDs and CDs. These devices support next generation optical disk formats, namely, Blu-ray and HD-DVD.

BLU-RAY   Blu-ray, also known as Blu-ray Disc (BD), is a next generation storage format that is used by optical disks to store up to 50 GB of data. The standard is developed and promoted by a group of consumer electronics, personal computer, and media manufacturers called the Blu-ray Disc Association (BDA). HD capable IP set-top boxes that are equipped with Blu-ray compliant units are able to directly record, rewrite, and store large amounts of high quality HD video content.

HD-DVD Some HD capable IP set-top boxes come with a DVD recording unit that supports a format called HD-DVD. Similar to Blu-ray the HD-DVD storage medium also allows the recording, delivery, and storing of HD digital video content onto optical disks.

All of the above IP HD set-top boxes receive the HDTV signal from a IP broadband network. The set-top box then decodes the signal and streams the output to a display. The format of the signal outputted depends on the capabilities of the display device used in the consumer's household. For instance, advanced flat panel displays such as plasma screens can support HD broadcasts in their original studio format, whereas a standard TV will only be able to view a scaled down version. As HDTV services become more advanced, the set-top industry needs high performance technologies to support the decoding of high definition video streams.

An IP HD capable digital set-top box hardware platform has similar features to cable, terrestrial, IP, and satellite set-top boxes. HD capable IP set-top boxes do however have some additional and unique characteristics that are used to deliver services such as high definition streaming video, improved digital audio, and connectivity to other in-home digital devices. These characteristics include advanced connectivity ports and a sophisticated security system for protecting content outputs.

Digital Video Connectivity Interfaces   HD capable set-top boxes support a range of interfaces that are used to connect with TVs that support high definition viewing. In addition to the IEEE 1394 interface covered in Chapter 8, a number of other interfaces are available to manufacturers for outputting IP HDTV content. The technical details of these audio and video interfaces follow.

DIGITAL VISUAL INTERFACE (DVI)   In the late 1990s, most video displays sold to consumers were connected to a TV service via an analog interface on the set-top box. The presence of analog interfaces acted as an impediment to the adoption of flat panel display technologies, so set-top manufacturers started to support advanced digital interfaces such as the DVI. The specification for the DVI was created by a consortium of companies called the Digital Display Working Group (DDWG). From a technical perspective, DVI is defined as a high speed serial interface that provides digital connectivity between devices such as set-top boxes and digital displays. It is designed to transfer uncompressed real-time digital video and supports the resolution settings of HDTV. DVI uses a protocol called transition minimized differential signaling (TMDS). Its principle of operation is based on the conversion of the eight bits of a 10-bit serial state during transmission between the HD capable IP set-top box and the digital display. The DVI receiver chip in the flat panel display then deserializes the video data, converts back to a standard 8-bit format and displays the crisp images on the screen. The use of DVI to connect a HD capable set-top box to a flat screen display is shown in Fig. 10.3.

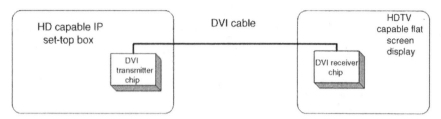

FIGURE 10.3   Using DVI to connect a HD set-top box to a flat screen display

   There are a couple of variants of DVI connectors that may be incorporated into HD capable IP set-top boxes, namely, DVI-Digital (DVI-D)—supporting digital displays only and DVI-Integrated (DVI-I)—supporting digital displays ensuring backward compatibility with analog displays. The incorporation of DVI into HD capable set-tops provides manufactures and IPTV service providers with the following benefits.

- DVI allows increased control over image quality.
- The interface supports copy protection and can be used by providers to secure video content.
- It enables subscribers to connect their set-top box to a variety of displays including traditional TVs and digital flat-panels.

For more information visit www.ddwg.org.

HIGH DEFINITION MULTIMEDIA INTERFACE (HDMI)   IPTVCD manufacturers have started to add HDMI digital outputs to their product ranges in recent years. The technology behind HDMI is quite similar to DVI. It is a standard, which was developed by the HDMI working group whose founders included Hitachi, Matsushita Electric (Panasonic), Philips, Silicon Image, Sony, Thomson, and Toshiba. The historical evolution of the HDMI technology over the past 4 years is illustrated in Fig. 10.4.
   In addition to interconnecting audio/visual devices such as DVD players and digital set-top boxes with HD ready television sets, HDMI also have some additional key features and functionalities that are not available on DVI interfaces.

- Improved levels of copy protection for video content outputted from the set-top box.
- The connector is smaller than a DVI connector, which improves aesthetics. It is also well suited to consumer electronic devices with small form factors.
- Digital audio and video are transported in the one cable. Therefore, a separate audio cable between the HD capable set-top box and television set is not required.
- HDMI is backward compatible with HDMI enabled TVs and flat panel displays.

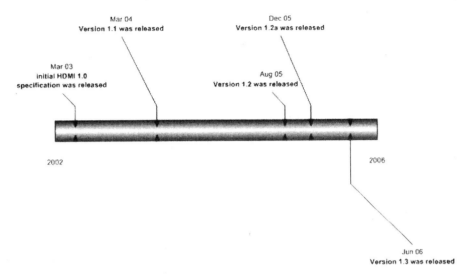

FIGURE 10.4 HDMI specification evolution

- It supports all of the international digital TV standards.
- The latest version of the technology is able to carry up to eight channels of uncompressed digital audio.
- The Consumer Electronics Control (CEC) channel feature allows consumers to use a single remote control for managing multiple HD enabled IPTVCDs.
- Data rates of 10.2 Gbps are supported by HDMI interfaces. As such HDMI is able to transmit video content in an uncompressed format.

From a technical perspective HDMI uses the same standards included in the DVI specification and also supports the TMDS protocol for serializing the data while transferring digital data. HDMI is an emerging connection technology and will become popular in HD compliant set-tops over the coming years. For more information visit www.hdmi.org.

DISPLAYPORT The DisplayPort is an open standard, which has been developed by the Video Electronics Standards Association (VESA). The key operational characteristics of DisplayPort 1.0 include the following:

- It allows the transfer of HD quality video and audio signals over a single cable.
- Similar to HDMI, DisplayPort supports plug-and-play interoperability.
- The High Bandwidth Digital Content Protection security mechanism, which is described in Chapter 7 of this book, is supported.
- It also features a small, user-friendly connector optimized for use in PCs and IPTVCDs.

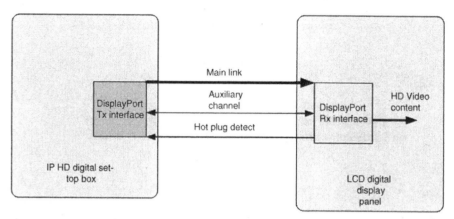

FIGURE 10.5   Using DisplayPort to communicate between an IP HD set-top box and a LCD screen

The technical architecture of an end-to-end DisplayPort link consists of three separate channel types that operate at the PHY Layer. Fig. 10.5 shows these channels.

The role of each DisplayPort channel follows:

Main link—This channel carries isochronous streams of audio and video content. According to VESA this channel supports a data transfer rate up to 10.8 Gbits/s.

Auxiliary channel—This two-way channel is used for management and control purposes. It provides for 1 Mbps of data rate over supported cable lengths of up to 15 m.

Hot plug detect—Used to provide a communication channel to support plug-and-play functionality.

At the link layer of the IPTVCM the following functionalities are provided by DisplayPort:

Isochronous transport of digital content—The link layer maps the audio and video streams from the upper layers of the protocol stack onto the main link channel. Correct mapping is essential to ensure that the streams are properly reconstructed at their destination point.

Providing link and device management services—In addition to mapping digital content onto the PHY layer, the link layer also takes care of manag-ing the connection between source and destinations. Device discovery and config-uration are other functions provided by this layer. This layer also has the ability to deal with implementing a content protection technology if required.

As described above the DisplayPort interface is designed to scale and accom-modate the requirements of transporting HDTV signals between an IP HD set-top box or some other type of IPTVCD and a flat screen display panel. For more information visit www.vesa.org.

UNIFIED DISPLAY INTERFACE (UDI)   UDI is another interface, which may be used to carry IP HD content between various types of IPTVCDs. An end-to-end UDI connection consists of a source, which is responsible for transmitting UDI signals and a sink, which receives UDI signals. As illustrated in Fig. 10.6 a, UDI connection consists of two separate links, which provide transport for control information and actual data.

The UDI data link carries the video data in an isochronous manner from the source to its destination point, whereas the control link carries management instructions between both devices. As depicted the interface also supports two other signal connections that indicate power availability and event details. The main features of UDI are

- interoperable with HDMI technology architecture
- provides support for repeaters
- support for multiple video formats
- capable of carrying HDCP security information
- may be implemented in a wide range of PC and consumer electronic devices

For more information visit www.udiwg.org.

10.3.5.2 HDTV Display Technologies   Consumers who want to watch an IPTV channel in HDTV format will need to purchase a new digital television screen capable of processing HD content. The main differences between the new HD capable digital televisions and the current analog TVs include

- wider screens
- higher resolution

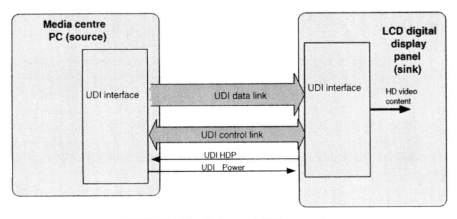

FIGURE 10.6   End-to-end UDI connection

Commonly called wide-screen, HD capable digital televisions are nearly twice as wide as they are high. The screen itself is of a slim depth and looks more like a movie screen. Standard analog televisions use square looking screens that are four units high and three units wide. For example, if the width of the screen is 20 in., its height is about 15 in. HDTV screens are about one-third wider than a standard television with screens that are nine units high by 16 wide. The screens of current analog TV sets contain approximately 300,000 pixels, while the HDTV televisions have up to 2 million pixels. Wide screen based digital TVs allow for an expanded view of events without sacrificing quality. In fact, pictures are displayed in a format that is closer to the way the director of the movie originally intended.

To be classified as a fully enabled HDTV household a HD capable display is also required. A whole host of new display technologies has emerged to support HD programming over the past number of years. The two most popular types of displays best suited to connecting with HD capable IP set-top boxes are plasma display panels (PDPs) and flat panel liquid crystal displays (LCDs).

Plasma Display Panels (PDPs)   PDPs are the largest variations, in terms of size, of flat panel display units currently available on the marketplace. The principle of operation of a PDP is based on passing a precise electrical voltage through a gas, which consists of a number of positively charged particles. Once the voltage is applied the gas becomes charged and produces a light. The types of gas used in the panel determine the color of light emitted. A substance or a powder called phosphor is used to display multiple colors across the screen. The physical architecture of the screen is made up of a large matrix of miniature cells or pixels, which are "sandwiched" between two sheets of glass. Each of these pixels includes electrodes and a coating of phosphor applied to its face. When a pixel becomes energized with a voltage signal the gas stored inside the pixel ionizes and converts to a plasma substance. This substance is made up of electrons and ions that serve to enhance the conductivity of the material. Once in this state the highly conductive plasma material emits ultraviolet (UV) light, which strikes the red, blue, and green phosphor deposited on the face of the pixel. This in turn generates a glow of the appropriate color at each pixel, which is viewed by the IPTV end user. It is important to note that the architecture of modern displays actually subdivide the pixel into three separate subpixels—one for each of the primary red, green, and blue phosphors. In addition to generating pure colors, plasmas need to control brightness and color shading. This is achieved by applying a technique called pulse code modulation (PCM) to the pixels. With PCM, a number of pulses are applied at a rate, which is unrecognizable by the human eye. The main advantages of PDPs connected to an HD IPTV service are:

- A PDP can achieve perfect focus producing a good picture for the viewer.
- A PDP is typically only 3 or 4 in. in thickness and is easily hung on a wall. The plasma has become synonymous with how TV will evolve in the future.

- The image quality of PDPs continues to improve. In fact, some of the high end plasma screens can display 16.77 million colors.
- The cost of PDPs has also come down nicely over the past few years, and this will continue.

PDPs do however have their downsides. For instance, some of the older models experience a process called gas pixels burning. This has resulted in the creation of black spots on the screen.

Liquid Crystal Displays (LCDs)   Since its advent in the early 1970s, LCD technologies have become an integral part of products ranging from video cameras to computer monitors. LCD screens have a number of advantages over traditional TV sets including

- No bulky and heavy weight picture tube is required.
- LCD screens are esthetically more pleasing when compared to their CRT counterparts.
- They use a lot less power, which is important for energy conscious consumers.

The construction of an LCD screen involves placing liquid crystal material between two pieces of glass (or quartz). Once the glass sheets are in place, a number of electrodes are attached to one glass in a horizontal configuration (rows on the display panel) and to the second sheet of glass in a vertical configuration (columns on the display panel). This creates a series of intersection points, which are filled by a liquid crystal substance. These tiny intersection points are called pixels. The final step in the construction of the LCD screen involves the addition of a special plate called a polarizing filter on the outer surfaces of the two ultra-flat glass substrates. The display works by allowing varying amounts of light pass through the glass substrates, liquid crystals, and polarizers. The creation of an image commences when a voltage is applied to the crystals. Once this occurs the liquid crystal molecules change their alignment in a particular direction, which polarizes the light. Twisting of the molecules allows light to pass through the display, producing a single dot onscreen. Thus, the electrodes allow the display's electronics to independently turn any pixel fully on or fully off. The coloring affect on an LCD screen is achieved by adding a three-color filter to the panel. Each pixel in the display corresponds to a red, blue, or green dot on the filter. In addition to turning pixels on and off, the electrodes are able to create different levels of color brightness. This is achieved by changing the voltage levels applied to the pixels. The liquid crystals change alignment according to the voltage variation. Thus, an increase in the strength of the voltage signal increases the amount of light passing through the pixels. The source of light for the display typically comes from some type of fluorescent or halogen bulb(s) placed at the back of the display. The light sources required for LCDs feature a high wattage in order to ensure that the light is clear and is evenly spread across the display panel. The combination of all these factors make LCDs a popular choice for consumers who want to enjoy the full benefits of viewing IP HDTV content.

The advantages of flat panel technologies combined with price decreases and increased demand for IPTV HD content will ensure that technical innovations in this sector will remain strong in the coming years.

10.3.5.3 Multichannel Surround Sound Capabilities HDTV signals also contain multi-channel surround sound to complete the home theater experience. A HDTV will include up to six different channels:

Two front channels—The speakers driven by these channels are optimally located on the left and right hand sides of the display unit. They generally carry soundtrack music.

The center channel—The speaker connected to the center channel resides at slightly above or below the display unit. It carries the dialogue from the actors.

Two rear channels—The speakers for these channels are normally mounted at the back of the room. Locating the speakers behind the viewer increases the impact of the sound affects.

A Low-Frequency Effects (LFE)—Low frequency sounds such as base signals are carried on this channel. The LFE channel is typically terminated in a device called a subwoofer.

Note a digital output connection is used to carry these channels to an amplifier, which provides subscribers with surround sound home theater like experience.

SUMMARY

IPTV is steadily taking off as an alternative, or addition to existing digital terrestrial, cable and satellite services, but the high definition versions are taking a little longer to arrive. There is however a growing number of telecommunication companies around the globe starting to launch IP based HD services in an effort to generate additional revenues from existing customers.

HDTV provides a digital picture of greater clarity and richer sound than the analog signals that have been broadcast for decades. HDTV is a next generation digital TV standard. Benefits of HDTV over SDTV range from improved picture quality to providing consumers with home theater style audio. Its characteristics range from very high picture resolutions to a wide screen 16 : 9 aspect ratio display. Both of these factors combined allow end users to experience a true movie like viewing experience. One of the key rationales for producing HD video content is that it offers general consumers a better viewing experience when compared to SDTV. In addition to upgrading the networking infrastructure to handle bandwidth demands of HD video the equipment used to produce HD content and IT infrastructure at the IPTV data center also needs change. Optical fiber or high end cabling systems for instance may need to be installed to interconnect the various types of HDTV equipment.

Although an IP set-top box capable of processing HDTV signals can be connected to a standard television set, consumers should also own a HD ready display device to fully enjoy the main advantages of HDTV programming. LCDs and plasma panels are the two most popular forms of flat panel displays used by consumers to view HD IP based video content.

11

INTERACTIVE IPTV APPLICATIONS

In addition to providing linear programming and access to on-demand video content, IP based technologies also allow telecom operators to offer end consumers a range of advanced interactive TV (iTV) applications. The deployment of iTV across an IP based networking platform is considered a natural progression for IPTV network operators. iTV changes the TV set from being the source of one-way, passive entertainment to two-way active entertainment, infotainment, and communication. Through iTV, viewers are empowered to control their viewing and to chat with friends, surf the net, and read the latest sports updates while watching TV. This ability to deliver two-way interactive TV services to consumers is a differentiator for IPTV operators. IPTV has the potential to bring interactive applications to the masses. iTV can provide such offerings as EPG, IP VoD, IPTV browsing, IPTV e-mail, DVR, walled garden portal, instant IPTV Messaging, IPTV-commerce, remote management, caller ID for TVs, IPTV advertising, localized video content, gaming on demand, parental control, and so on. The following sections in this chapter describe these iTV applications in greater detail.

11.1 THE EVOLUTION OF iTV

iTV first emerged as a way of enhancing the experience of the viewer by making it less passive and more interactive. It was also seen by broadcasters as a means of making their programming more attractive to increase audience size and thereby,

Next Generation IPTV Services and Technologies, By Gerard O'Driscoll
Copyright © 2008 John Wiley & Sons, Inc.

ratings. Although, the idea of iTV was conceived in the early 1950s, wide deployment of iTV did not take place until the late 1990s and the early 2000s. To understand iTV in context and to gauge the future developments of iTV, it is important to consider its history. The history of iTV spans over the past five decades. Its evolution is a result of many developments both outside and within the TV industry. Developments in areas related to television and external to it have been both influencers and enablers of iTV. iTV involves the convergence of ~~five~~ four industries:

- Cable and satellite service providers
- Consumer electronics
- Personal computers and software industry
- Telecommunication operators

iTV began almost 50 years ago in 1953 with the "Winky Dink and You" programme. The CBS presenter, Jack Barry, prompted young viewers to assist the animated onscreen character "Winky Dink" in his adventures by drawing pictures on acetate sheets stuck by static to their TV screens (see Fig. 11.1).

These were part of the "Winky Dink and You Magic TV Kit", which sold very well and was a successful debut for iTV. Four years later, remote controls were introduced to enable viewers to change channels without having to move from their seats. Also in the 1950s, Alliance Internationales de la Distribution par Cable was established in Europe and cable operators started to offer more programming choices by acquiring access to distant signals, and multiple system operators (MSOs) emerged in the United States.

By the 1960s, cable systems increased from 800 to 850,000 subscribers. To stem the growth of cable and to stabilize the faltering broadcasting industries, the FCC imposed restrictions on cable systems so that they no longer had unlimited access to near and far TV signals.

FIGURE 11.1 "Winky Dink and You" programme

In the 1960s, Massachusetts Institute of Technology (MIT) engendered and experimented with the concept of large networks, which were put into practice initially in universities as the basis for testing and development. The ARPANET network remained the focus for technological innovators for many years; it grew to become the network of networks and became more complex and efficient every year. It became known as the Internet and was eventually unleashed to the public in the 1990s as a revolutionary communication and information tool. The Internet is the manifestation of research and development on networking, hardware, software, communication technologies, and global advertising and marketing. Advances in technology, innovations, and discoveries made for computers and the Internet have resulted in rich pickings for iTV that used IP based networks. The Internet is the interaction leader that interactive IPTV begins to emulate. The enthusiasm, with which the Internet was created and is received, is encouraging for the future of the interactive IPTV industry sector. The 1970s saw the emergence of satellite television, the launch of Home Box Office (HBO), the first set-top box, teletext, closed captions, the first service-wide iTV experiment, and the first two-way set-top box. In 1972, the FCC introduced a policy of gradual cable deregulation. Satellite TV emerges as a reliable form of broadcasting that also facilitated point to multipoint broadcasting. Deregulation and an inexpensive alternative for distributing video, satellite benefited Cable TV greatly. Service Electric offered HBO, the first pay TV network, over its cable system in Pennsylvania. The system used satellite technology to distribute its programming content. Subscribers receive the first set-top boxes. By distributing by satellite, HBOs signal was available to cable operators throughout North America. The national satellite distribution system is created as a result. HBO is an early version of NVOD. Teletext and closed captioning are products of the VBI. Teletext was devised as a means of enhancing one-way, standard broadcast TV and Cable TV. BBC engineers invented teletext in the 1970s and it appeared as static text overlaid on broadcast television. Closed captioning, U.S. invention also emerged in the 1970s.

In 1977, iTV in the sense that it is used now, is given a trial run by Warner Cable in the form of a network called Qube, in Columbus, Ohio. The objective for Qube was to entice subscribers, and so, improve competitive advantage. Qube saw the introduction of the first two-way set-top box, thereby enabling subscribers to select pay-per-view films, participate in live polls, and to obtain additional information while watching a program, and the concept is well received. However, due to the expense of the equipment and other factors, it was discontinued. The Qube experiment provided useful insights for future iTV developments:

- Consumers responded well to interactive offerings.
- Consumer and provider costs should not outweigh iTV advantages.

Additionally, the Qube experiment helped to accelerate advances in iTV-associated technology. It also resulted in a pool of iTV developers, innovators, and enthusiasts who nurtured its development throughout the following decades. A

Japanese invention, HDTV is introduced to the United States. It requires 20 MHz of bandwidth. At the end of the 1980s, the number of homes with cable increased to over 50 million and one-way set-top boxes become prevalent. This is facilitated by the 1984 Cable Act, which enables deregulation. During the 1990s rapid advancements were seen in many areas of the TV industry:

iTV experiments—Throughout the 1990s many trials and experiments in iTV were conducted in the United States and in Europe, chief among these was Time Warner's Full Service Network (FSN). In 1994, FSN offered a full iTV service including VoD, Internet access, and interactive programming. The technology and equipment FSN used to create this service proved too costly and diminishes the attractiveness of the service. The cost of the digital set-top box was $10,000.

PCTV, EPGs, DVRs, and DVDs emerge—Through browser plug-ins TV content was transmitted to PCs via MPEG-compressed data streams creating PCTVs in 1997. Cable and Satellite service providers offered hundreds of channels and electronic programming guides evolve to improve viewer navigation. The DVD player was introduced in March of 1997 bringing more interaction and user control to video watching. DVRs emerge in 1999 to enable viewers to record programmes minutes after they are broadcast for more control. They can rewind, forward, pause, and stop programming in this way. DVRs were adopted by cable and satellite providers and offered as an additional service.

U.S. Telecommunications Act—The U.S. Telecommunications Act of 1996 opened up the telecommunication market and established deadlines for the digital transition culminating in 2006. Additionally, the FCC approved an ATSC standard for DTV in the United States. Canada, South Korea, Taiwan, and Argentina adopt the ATSC standard subsequently.

The ATSC sanctioned the conversion of digital signals to analog ones in set-top boxes using MPEG-2.

Proliferation of Digital TV (DTV)—True digital television refers to the combination of digital camera-recorded and other digitized content, a digital signal, and a DTV or HDTV set. DTV and HDTV sets did not become freely available till the late 1990s and their costs were prohibitive. Therefore, cable and satellite service providers provided their services through a compressed digital signal that was then decompressed and converted by MPEG-2 in set-top boxes and displayed on analog TVs. In this way, they continue to provide subscribers with most of the advantages of a digital signal but with an inferior picture quality. HDTV, which provides quality picture and sound, was introduced to the consumer market in the United States in 1998.

The Internet explodes—The Web is born in 1991 and an experimental system called multicast backbone (MBONE) starts broadcasting audio and video over a portion of the Internet gets underway. At this time, the Internet was composed of over 1 million hosts. In 1993, Mosaic, the first graphics-based Web browser, becomes available. In 1994, the facility to order pizza online becomes possible through Pizza Hut and online shopping malls established a presence, the first online bank opened, and banner advertisements appeared for the first time. By 1996, the Internet had increased to over 10 million hosts distributed worldwide (almost 150 countries). The Internet is considered a valuable information and communication tool.

Cable and satellite television expand—The telecommunication reform law was enacted in 1996. This opened wireline and wireless telephone and data service markets to cable and satellite companies. It also helped to promote growth and to encourage service providers to invest in new infrastructure and services. Additionally, it facilitated these providers to deliver online services, data, and Internet access. By the end of the decade, when digital satellite and cable became viable, iTV services emerged. Because of the consumer demand for high quality and diverse programming, cable and satellite networks became more involved in developing and funding programming, interactive programming, and interactive Web sites. Digital satellite TV emerged in 1996 and by the end of the decade 12 million 18-in. dishes were sold. By 1998, digital cable is delivered to over a million households and by the end of the decade that number increased to over 5 million.

DVB is adopted in Europe—The specification for the transmission of DVB signals in cable networks was developed between August 1993 and February 1994. In the spring of 1994, this was recommended by ETSI as a European transmission standard.

In addition to TV developments there were many advances in broadband technology in the 1990s that facilitated the beginnings of the IPTV industry sector. The PC revolution brought with it software that favorably impacted the production of film and television enabling digitally enhanced content and improved task management. In conjunction with PCs and the Internet, devices such as CDs and DVDs, wet consumers' appetites for higher quality sound and picture and for user control and interactivity.

In the 2000s, iTV deployment has become more widespread. By the end of 2001, over 20 million homes have boxes capable of some form of interactivity. In 2001, DSS reached 20% household penetration. HDTV sales accelerated thanks to increased HDTV programming and reduced costs. DVRs sell well. Between the late 1990s and early 2000s cable and satellite networks and service providers invested billions of dollars in equipment and in building digital networking infrastructures. Manufacturers of devices such as DTVs, HDTVs, PVRs, navigational devices, and set-top boxes subsidize some of the cost of these appliances to enhance their appeal to consumers.

These industries' main concerns in the early 2000s were to remain in competition and to attain return on investment. Thanks to the increased content capacity that a digital signal affords, to governmental and organizational promotion of digital television, and to a desire of television providers to improve their offerings and increase their revenues, iTV is a reality that will endure. The technology and infrastructure available today makes iTV a more viable option for IPTV providers.

11.2 ABOUT iTV

As the transition from all things analog to digital continues, iTV is becoming more and more prevalent. Thanks to increased bandwidth in new networking infrastructure and the great carrying capacity of the compressed digital signal, interactive content can be offered to more and more householders. iTV describes the phenomenon whereby users are able to actively engage with content on their TV screens. iTV adds an extra layer of functionality to IPTV beyond on-demand and linear programming services.

It can provide viewers with access to further information, communication applications, and the facility to respond to interactive programming. iTV involves the transition from one-way to two-way communications and passive to active viewing and viewer control. Applications and programming are transmitted to viewers' set-top boxes and viewers can interact by using a navigational device such as a remote control or a wireless keyboard. Information sent by viewers is directed from the set-top box to a return path to the host's network server. Although, iTV can be delivered via an analogue signal, the amount of content is restricted by bandwidth. Wider deployment of iTV has occurred because a digital signal, which is compressed and transmitted at higher speeds in binary code, can carry more iTV content and numerous channels. iTV is distributed to people's homes by terrestrial, cable, IPTV, and satellite television providers, each of these groups are consistently battling for a place in the viewers' homes. All of these providers encourage viewers to subscribe to their pay TV, broadband, and telephone services and usually, they offer a variety of packages, which are designed for different audience types and the different budgets of their audiences.

11.3 INTERACTIVE IPTV APPLICATIONS

In addition to allowing subscribers to watch linear video programming, IPTV brings Web-like interactivity to the traditional TV viewing experience. It allows service providers to differentiate their offerings from traditional television services. For consumers it allows them to view hundreds of channels, simultaneously record TV material, and access a range of interactive TV services including

- Electronic programming guide
- IP-VoD
- IPTV browsing
- IPTV e-mail
- DVR centric applications

- Walled garden portal
- Instant IPTV messaging
- IPTV-commerce
- Caller ID for TVs
- IPTV advertising
- Localized video content
- Gaming on demand
- Parental control
- IP based emergency alert systems
- IPTV program related interaction applications
- Personalized channels

11.3.1   Electronic Programming Guide

Since IPTV brings into the home many more channels and services, people need a way of navigating the myriad of choices. An application that does this is typically referred to as an electronic program guide. An EPG, also known as intelligent programme guide (IPG), is an interface application that allows IPTV subscribers to preview, select, and connect to various types of IP and interactive TV services. The EPG application is generally a standard part of an IPTV service offering and is commonly used to navigate through a growing number of channels and sources of video content available to subscribers. An EPG presents IPTV subscribers with a menu of available IPTV channels in HTML format and a remote control is typically used to navigate this menu. Owing to the two-way nature of IPTV networks, it is possible to include several full days of programming information for every channel and detailed descriptions of every programme within each channel. In addition, subscribers are able to search by genre, programme title, channel, and even by time. Once a channel is selected, the video content is downloaded over the broadband network for immediate viewing. The EPG application is usually displayed on the TV screen in tabular or grid format. This type of design is generally easy to read and understand. It is however possible to change the look and feel of the EPG interface to meet the branding requirements of different IPTV service providers. The communication process between the client EPG application running on the IP set-top box and the server application uses standard Internet protocols. If the IPTV transmission network is using multicasting, then the program descriptions retrieved by the IP set-top box also contain port numbers and related multicast IP addresses.

11.3.1.1 Introduction to IPTV EPGs   Also known as an event service guide, an IPTV EPG displays program lineup and episode information for current and future broadcast programming choices. It displays the current channel contents in a window, and allows IPTV subscribers to select a particular broadcast channel. Subscribers use the arrow keys on their remote controls to highlight and select various options. The functions of a standard IPTV EPG can include the following:

- Displaying weekly schedules of IPTV multicast channels. Note that the channels accessible on the weekly channel are defined by the end-user access rights.
- Automatic video recording.
- Alerting subscribers about the arrival of new e-mail.
- Reminding viewers when a selected program (or a program covering a selected topic) is about to be shown.
- Restricting access to TV channels that are deemed inappropriate.
- Allowing subscribers to find programs with a particular theme on a specified time or date.
- Controlling disk storage devices in IP set-top boxes.
- Previewing particular programs and personalizing TV viewing.

Some advanced IPTV systems even allow subscribers to programme the EPG to make it look the way they want it.

A picture-in-guide is a variation of the standard IPTV EPG application that displays a selection of multicast channels within a small window (stamp like images) on the TV screen. The end user is able to use the arrows on their remote control to move from one window to the next. Each time a window is selected, the IPTVCD plays the sound associated with the particular channel. In addition to sound, the set-top will also display channel information including title, beginning, and ending times. By simply pressing the OK or Enter button on their remote control, the window extends to the full length and width of the television screen, the end user can sit back and watch the channel in comfort. From a network engineering perspective, multiple streams are required to support the provision of a picture-in-guide interactive TV application. This is problematic for service providers that operate networks with low bandwidth capacities. Note the delivery of this interactive IPTV application also requires picture-in-guide functionality to be built into the encoders at the IPTV data center.

11.3.1.2 IPTV EPG End-to-End Network Architecture   Figure 11.2 describes the building blocks that comprise an end-to-end IPTV EPG solution:

The underlying technology platform has three major components, namely, a metadata generator, an EPG application server located at the IPTV data center, and a client EPG application resident on the IPTVCD. We explore these technical components in the following subsections.

Metadata Generator   The EPG needs schedule data to allow the viewer receive information about IPTV broadcast and on-demand services. The technical and industry term for this data is metadata. The metadata generator is at the heart of an end-to-end IPTV EPG system and allows IPTV service providers to acquire, edit, generate, and play out EPG schedule data over the network. IPTV metadata defines data about data and is typically formatted in XML. It combines video content

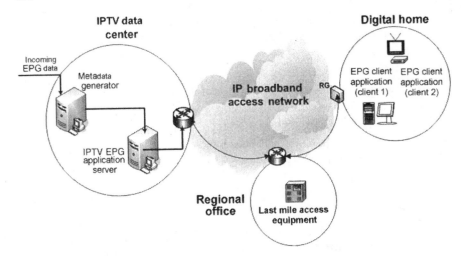

FIGURE 11.2 End-to-end IPTV EPG system

information from both the content provider and network operator. The types of metadata provided by an IPTV system can include some of the following items:

- List of channels available for the various tiered packages
- IPTV channel name
- IPTV channel description
- IPTV channel logo
- IPTV channel provider
- IPTV channel provider's Web site
- Program title
- Program start and finish times
- Program language options
- Detailed description of program contents
- Parental control details and rating standards
- Content aspect ratios
- An indicator of whether captions are available or not
- A description of any embedded advertisements
- Compression techniques used on both audio and video content
- Prices and access conditions for different items of IPTV content
- Scheduled distribution time for delivery across the IP broadband network
- Description of protocols and mechanism used to deliver the content
- Caching details
- Preview duration for IP-VoD assets
- Recording rights

- Applicability of content to particular types of IPTVCDs
- Viewing profiles of IPTV end-users

As with many technologies used by end-to-end IPTV systems, a number of meta-data industry initiatives and standards have emerged over the years. The two most successful formats used by the television world are MPEG-7 and TV Anytime.

MPEG-7 MPEG-7 is an ISO/IEC metadata standard officially called ISO/IEC 15938. Also known as "Multimedia Content Description Interface," MPEG-7 defines a set of core technologies used to describe both real-time and non-real-time AV content. At its highest level the MPEG-7 specification is split into 11 parts. Table 11.1 summarizes these 11 parts.

In the context of an IPTV networking environment, the types of MPEG-7 metadata information used to describe a video asset can include the following:

TABLE 11.1   Structure of the MPEG-7 Specification

Part Title	Brief Description
Systems	This part describes the architectural framework required to prepare MPEG-7 meta-descriptions.
Description Definition Language (DDL)	As the name suggests this part provides details about DDL. This XML based language allows IPTV service providers to create various types of descriptors.
Visual	This defines descriptors that cover various types of visual features associated with an IPTV content asset. Color and motion are examples of descriptors covered in this part of the specification.
Audio	This defines descriptors specific to audio centric content.
Multimedia description schemes	This defines descriptors that are applied to multimedia components.
Reference software	This part provides software simulation platforms for the various MPEG-7 descriptors.
Conformance	This part defines tests and guidelines to ensure the metadata conforms to the MPEG-7 standard.
Extraction and use of MPEG-7 descriptions	This part is effectively a technical report that was created for informational purposes and defines the extraction features of MPEG-7.
Profiles	This part defines a set of MPEG-7 profiles.
Schema definition	This part defines the schema based on DDL.
Profile schemas	This final part specifies the collection mechanism of profile schemas.

- A high level description of what is depicted in the video asset;
- Coloring scheme details;
- Coding and storage format of the asset;
- Parental rating;
- The context in which the content was produced.

MPEG-7 codes this information in a manner that allows search tools to efficiently identify requested items of content. Note that although MPEG-7 is applicable to a wide variety of media types, it is particularly suitable to content encoded using MPEG-4, this is due to the fact that MPEG-4 encodes a TV scene as objects and MPEG-7 allows IPTV service providers to attach meta-descriptions to each object.

TV-ANYTIME The TV-Anytime is an association drawn from various industry sectors ranging from Internet Service Providers to consumer electronic manufacturers and content providers. The organization was set up in 1999, and it had achieved its stated objectives of publishing a number of specifications by the middle of 2005. The specifications covered four key areas including

(1) Business models
(2) System, transport interfaces, and content referencing
(3) Metadata
(4) Rights management and protection

The next section gives a brief overview of TV-Anytime's metadata specification.

TV-Anytime Metadata Specification Overview The overall system model as described in the TV-Anytime Extended Metadata Schema S3 document is illustrated in Fig. 11.3.
The key features of the metadata architectural diagram are

(1) *Content creation*—This represents the creation of the video content data. Some basic metadata is gathered at this stage in the process.
(2) *Content publishing*—In the context of an IPTV network, this involves multicasting or unicasting video content to end users.
(3) *Metadata editing*—This involves the creation of metadata, which is suitable for consumption by the end user.
(4) *Metadata aggregation*—This involves the aggregation of metadata from a number of sources at the IPTV data center.
(5) *Metadata publishing*—This involves publishing the data associated with the content. This is typically facilitated through the EPG.
(6) *Content selection*—This typically involves the selection of content by the end user.
(7) *Location resolution*—This identifies the location of the video asset.

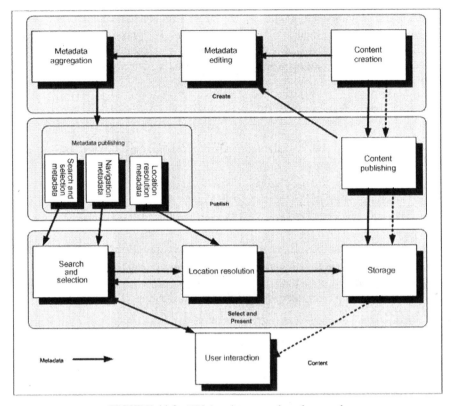

FIGURE 11.3   TV-Anytime metadata data mode

(8) *Storage*—This defines the storage requirements for the video asset.

(9) *Search and Selection*—This facilitates the searching and related selection of metadata.

Similar to MPEG-7, TV-Anytime has decided to use XML for formatting and representing the metadata. In addition to using XML, the TV-Anytime specification has also defined a number of attributes that may be used in an IPTV networking environment to describe the following four multimedia components:

- Audio
- Video
- Still images
- Games

In addition to defining various attribute types, the TV-Anytime specification also allows service providers to target video content based on a number of different IPTVCD types and network characteristics. Additionally, TV-Anytime has been incorporated into a number of international standards. For example in 2006, the

DVB published an addendum called Carriage of Broadband Content Guide (BCG) information over IP to their DVB-IPI specification. The purpose of the addendum was to define the signaling and carriage of TV-Anytime metadata over an always-on bidirectional IP network. The following are the major characteristics of the BCG addendum.

- Support is provided for both unicast and multicast IPTV communication systems.
- The TV-Anytime metadata is carried inside DVB defined units called containers.
- DVB protocols are used to transport multicast based BCG meta-data whereas HTTP is used over unicast connections.
- The specification defines a TV-Anytime profile for DVB based networks.

It is critical that metadata is kept current and is secured. Ensuring the currency of the metadata is typically achieved through the use of a notification protocol that informs the metadata generator that newer versions are available for download. Securing metadata is normally facilitated through existing IT systems.

IPTV EPG Application Server   The metadata generator interfaces with the EPG server to deliver program information to the EPG client application in a suitable format. The application server consists of software programs, a HTTP server, and a centralized database that contains the event data of all the broadcast programs on offer. It has four main functions:

(1) *Formatting*—The channel listing are generally formatted as Web pages.
(2) *Authorisation*—The server in conjunction with the conditional access and digital rights management system authorizes access to particular channels.
(3) *Provision of IP multicast addresses*—The EPG server provides IP multicast addresses which are used by the router to stream the IP video channel across the core and access IP networks.
(4) *Caching*—The IPTV EPG server also provides caching functionality, which speeds up the EPG load time for end-users.

Client EPG Application   The level of functionality provided by the EPG client application can include some or all of the following features:

- Allows end users to customize list of favorite multicast channels and on-demand titles.
- Allows end users to change the look and feel of the screen layout.
- Provides end users with an efficient mechanism of navigating through the various IPTV services and channels that are on offer from the network operator.

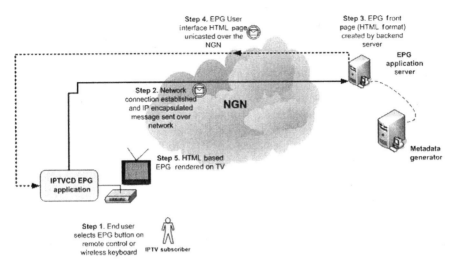

FIGURE 11.4   Launching an EPG via an IPTVCD connected to an IP broadband network

The steps involved in launching an EPG application are illustrated in Fig. 11.4 and explained in following sections:

Step 1—The EPG key is pressed and the client application processes the command.

Step 2—Launching of an EPG over an IPTV network utilizes a browser based client-server networking model. Thus, the next step in the process involves the establishment of a network connection between the EPG browser application and backend server.

Step 3—The command is received in the form of IP packets, authenticated, and a Web page is generated by the server. The Web page will contain channel information requested by the end user. This information will either be cached at the IPTV EPG server or requested from the metadata generator.

Step 4—The IPTV EPG server sends the results of the end-user request to the client EPG browser application.

Step 5—Results are received, rendered, and the EPG page is layed out on the TV display for use by the end user.

The EPG network architecture described above relies on the ability of the network and backend servers to respond immediately to EPG instructions. On some occasions IPTV end users may experience delays due to congestion or other issues that slow down the processing of EPG requests. Where this occurs on a regular basis, service providers have an option of using an architecture that involves processing the EPG locally on the IPTVCD. Under this approach the basic EPG application normally resides in the IPTVCD's storage system and accesses current meta-data from the IP network. Local storage of the EPG application speeds up response times for the end-user because it launches and operates immediately once the IPTVCD is powered on.

11.3.2 IP Based Video on Demand (VoD)

IPTV is often provided in conjunction with VoD. As the name implies VoD is a service that allows subscribers to select and download video content over an IP network. The content typically includes a library of films, music videos, and recorded television programs. The deployment of a VoD system falls into two broad categories:

(1) *Downloadable*—the VoD content is delivered to the IP set-top box and the content is viewed once the downloading is finished.

(2) *Streaming*—The IPTV set-top box receives the content via an IP stream.

Note that the networking infrastructure and technologies required to support IP-VoD applications are described in Chapter 9.

11.3.3 IPTV Browsing

Many IPTV network operators provide viewers with access to the Web. This can be in the form of either TV-based Web browsing (WebTV) or Walled gardens. TV based Web browsing is similar to browsing on a PC but it is TV based. Since many Web pages are not tailored for viewing on TV, some network operators provide an embedded browser on their IP set-top boxes that change a Web page automatically. Such software can be provided as part of a middleware platform. The concept of using a television set to bring Web connectivity has been commercially available for over a decade but due to a whole raft of reasons consumers have continued to reject the use of this interactive TV application. With advancements in IPTV and Web technologies the popularity of accessing Web content via TV applications is expected to grow in the coming years.

Web browsing is a key component of an IPTV system. In Fig. 11.5 the structure of a generic infrastructure used by IPTV operators to deliver high speed Web access to their subscribers is illustrated.

As shown, an end-to-end system includes: the IPTVCD Web browser, a HTTP Web server, caching servers, and a high-speed backhaul connection to the Internet.

11.3.3.1 IPTVCD Browsers Browsers built for IPTVCDs need the same robust functionality found in desktops, but with access to a fraction of the hardware resources. Therefore, they are optimized to run within resource constrained platforms such as set-top boxes, mobile phones, and Internet-enabled appliances. The browser software runs on top of the underlying real-time OS and middleware platforms and typically provides the following functionality:

Displaying Web pages—Putting a Web page on a television screen is very different to displaying the same page on a computer monitor. It must be

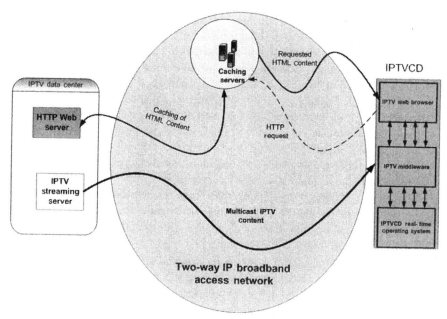

FIGURE 11.5 Technical building blocks of an IPTV browsing system

remembered that the television has been in existence for over a hundred years. Therefore, IPTVCDs employ sophisticated software browsers to ensure that Web pages are easily viewed on a TV screen. IPTVCD browsers are able to squeeze the content of a Web page on to a television screen, while limiting the affects of viewing content on a low resolution device. This process is called transcoding and involves repurposing Web content into a simpler format that is suitable for display on a standard television. Transcoding can be divided into two broad categories: HTML and Image Transcoding. The transcoding of HTML content includes checking the code for any errors. An example of an HTML error would be the absence of a </html> tag on a standard Web page. Transcoding filters these errors and produces a version of HTML that can be easily rendered by the IPTVCD browser. In addition to checking for errors, the HTML transcoding element is also capable of changing the layout to a TV centric format. For example, the default font size for a Web browser is normally 10 points. If this font size were to be displayed on a television screen, then it would be very difficult to read. Consequently, the transcoding engine will automatically increase the size of the font to 20 or 22 points. This approach means that less text can be displayed on a TV screen than in a comparable area of a computer monitor. However, the text is legible for subscribers that are viewing the page from a distance of 4–5 ft away from the television display. In addition to automatically adjusting font sizes, transcoding takes care of adjusting the size of a page to fit within the width and height of a TV screen. Image transcoding is responsible for converting the many image formats used on the Internet to a format that is suitable for an IPTV environment. Flat-color GIFs are processed more readily by IPTVCDs than JPEGs or PNGs. So, for instance, when an IPTV subscriber makes a request for a document on the public Internet, which contains JPEG images, the image transcoder is able to convert the images into a GIF format. In addition to translating the file format, some advanced browsers are also capable of formatting and scaling graphics for a television screen. This could involve adjusting color maps, resizing cells in tables, or sizing images on Web pages to fit the resolution of the TV screen. It is worth noting that parsing and laying out HTML programming code requires a good deal of processing power. Lightweight IPTVCD clients are unable to perform all of this processing, so in some IPTV deployments a transcoding server preprocesses Web content before it is transmitted.

Interfacing and interpreting commands from the IPTV subscriber—For Web browsing systems on the television to be a success they need to be very easy to use. With this in mind, IPTVCD browsers currently use navigation controls on the television screen that are accessible from a remote control or a wireless keyboard. An IPTVCD browser user interface typically consists of a standard HTML page, which is available locally or dynamically generated by the backend server. Subscribers use the arrow keys (up, down, left and right) on their remote control or wireless keyboard to highlight buttons and menu options on the TV screen. A colored box usually indicates a selected option. Pressing any one of the navigation buttons moves the colored box to the nearest selectable option. If this selectable option is a hyperlink and the subscriber also selects the enter key, then the browser will display the Web site or content associated with that particular link.

Local language translations—This feature supports the deployment of IPTV services in culturally and geographically disparate regions of the world.

Personalization—IPTVCD browsers can also support a Web personalization technology known as "cookies". A cookie is best described as a message that is transmitted from a server to the browser and is stored locally as a text file. The main purpose of these cookies is to identify IPTV subscriber habits and customize enhanced TV pages for them.

Customizable—The browsers user interface is typically customizable to include specific logos and other branding elements.

Support for content based on Internet standards gives IPTV service providers the ability to reuse existing Web content and services on their high speed IP based broadband networks.

As shown in Fig. 11.6 an IPTVCD browser consists of a presentation engine that processes and renders a number of different types of open Internet and TV centric content formats, including:

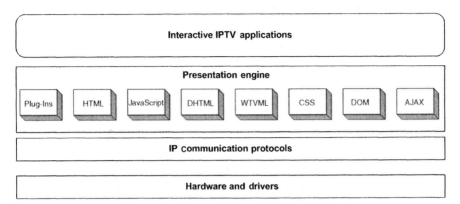

FIGURE 11.6 IPTVCDs supported browser formats

HyperText Markup Language (HTML) Browser support for HTML allows IPTVCDs to process and display a wide range of content that has been developed for the World Wide Web. HTML is a simple text-based language that gives interactive TV developers the ability to publish IP based content incorporating headings, text, tables, lists, photos, video clips, and illustrations. An overview of how an IP browser processes a Web page is described in the following steps:

(1) Once a Web page is received from the IP network, the browser will parse the HTML code and produce a vendor specific data representation.
(2) The various HTML components (images, tables, text, etc.) are then formatted and automatically optimized for easy viewing on a standard television screen.
(3) The page is sent to the appropriate output port.

The various digital TV and IPTV standard bodies have added a number of TV orientated extensions to the HTML language to allow television and Web content to coexist on the same page. Some of the additions to the core HTML language can be classified into the following groups of features.

Picture in picture—This feature supports the display and scaling of live video broadcast within a Web page.

Overlays—This feature supports the display of a Web page over a TV channel.

Windowing—By definition, windowing allows IPTV subscribers to use their remote controls to resize, close, or hide a video stream.

Transparency—Image file formats that support this technology are able to make certain designated pixels wholly or partially transparent, so that the background color or texture shows through. HTML tags associated with this feature allow IPTV designers to control how Web content integrates with a standard television screen. The normal technique used by designers is to position the television screen in the background and overlay the picture with Web content. Using transparency tags, designers are then able to show through particular sections of the live video.

Positioning of components—The use of tables and specialized HTML tags offers IPTV content designers a great deal of control over the exact positioning of multimedia and text components on a television screen.

JavaScript   JavaScript is an open, cross-platform, object-based scripting language created in order to extend the power of HTML documents. The JavaScript component allows IPTV service providers and content creators to deliver powerful interactive TV services to consumers.

CSS   Cascading Style Sheets (CSS) is an extension to the HTML language that gives IPTV content developers advanced control over how pages look and feel. The W3 consortium has actively promoted the use of style sheets on the Web since its inception in 1994 and has produced three main recommendations (CSS1, CSS2, and CSS-TV) that have been integrated with the latest releases of IPTV browsers. The CSS-TV specification is particularly relevant to IPTV professionals because it defines a set of guidelines for styling Web content for viewing on television sets. CSS constitutes a complex topic and a detailed discussion of all of its components is beyond the scope of this book. Accordingly, we suggest that you visit http://www. w3.org/ to view further details on the technology.

DHTML   Dynamic HTML (DHTML) is a term used to describe the combination of HTML, style sheets and scripts that allow documents to be animated. DHTML is not a scripting language like JavaScript but an API for HTML. It allows an IPTV content author to take advantage of the processing power of an IPTVCD to create pages that can be modified on the client without having to access the server for each change. It also provides TV content authors with enhanced creative control so that they can manipulate any page element on a television screen at any time.

WTVML   WTVML is a markup language based on XML and is used to format interactive TV content for European TV audiences. The WTVML Specification uses Internet technologies and has been standardized by ETSI.

Document Object Model DOM   DOM is a W3C recommendation that describes how objects in a Web page—text, images, headers, and hyperlinks—are represented. DOM is used by IPTV content creators to dynamically change the appearance of Web pages after they have been downloaded to a subscriber's IPTVCD browser. The DOM code is embedded within a Web document and is interpreted by the browser along with the HTML code on the page. The various DOM related specifications produced over the years by W3C are categorized into different levels. As of 2006, there existed three different levels, namely, level 1, level 2, and level 3. The number of levels supported by IPTV browsers varies between vendors.

AJAX   Asynchronous JavaScript and XML (AJAX) is a technology platform used to create Web based IPTV applications. One of the main benefits of using AJAX is it improves responsiveness for end-users accessing TV-HTML pages. To accomplish this, the AJAX platform passes small data amounts in the background, thus negating the need for full HTML page reloads when a command is received from an end user.

Plug-Ins   Additional support for animations and multimedia components may be incorporated into the browser through the use of plug-ins. A plug-in is a software module that adds a specific feature or service. For example, a new Adobe plug-in could enable an IPTVCD browser to render Flash based games, animations, movies, and interactive presentations.

Owing to constrained memory restrictions, it can sometimes be difficult to design an IPTVCD browser that will support all of the television and Internet based standards described above. Incorporating W3C technologies into IPTVCD browsers benefits both end users and the IPTV service provider communities. For end users, it allows them to view a whole range of content available on the public Internet. For service providers, it allows them to repurpose existing Web content and quickly create interactive TV applications.

At the time of publication at least three separate organizations had undertaken initiatives in the area of defining content and hyperlinking specifications that are particularly relevant to IPTV systems. The following sections provide further detail on the contributions these groups are making to the area of IPTV browser technologies.

W3C   HTML is the de facto language for publishing documents on the Web. It is however been slowly replaced by a newer and more powerful version of the language—EXtensible HyperText Markup Language (XHTML). Version one of the language became a W3C recommendation in 2000. In 2007, the XHTML2 Working Group was formed by the W3C organization to finalize and promote the development of next generation XHTML technologies. The use of XHTML by IPTV browsers allows interactive IPTV developers to publish content that easily renders in display screens of varying resolutions and sizes ranging from old television sets to modern day flat screen display panels. Most IPTVCD browsers will include support for IPTV content rendered in this format.

CEA   In addition to supporting W3C content delivery HTML standards, suppliers of IPTV browsers are also starting to add support for a new Consumer Electronics Association standard called CEA-2014. CEA-2014, also known as Web4CE allow

IPTV subscribers to browse the Internet and remotely control UPnP based home networking devices.

ISMA   ISMA has also become active in the IPTV browser sector with its publication of its hyperlinked video (HV) specification in 2006. Some characteristics of the HV specification include the following:

- It is an open specification that is compatible with existing IPTV browser standards.
- It uses URLs to trigger different types of actions that are complementary to the primary IPTV stream.
- It recommends a particular URL syntax and structure for use across IPTV networks. The URL is formatted as a text string and can include the following types of detail information:

  - Timestamp details;
  - Whether the URL is to be sent over the IPTV network at a particular instance in time or sent in response to a user request;
  - The target address of the URL;
  - A parameter that defines whether or not the URL is displayed in the IPTVCD browser window.

In addition to the above features, the specification also allows for the delivery of URLs over the RTP protocol.

11.3.3.2 A Web Server   A Web server, also known as an HTTP server, is basically a file server. A Web server's main function within an IPTV network is to listen for and respond to HTTP requests from IPTVCD browser clients. When a request is received, the server opens a connection to the browser and sends the requested file or Web page. After servicing the request, the server returns to its listening state, waiting for the next HTTP request. HTTP is extremely rapid. According to Tim Berners-Lee, the developer of HTTP, the request is made and the response is given in a cycle of 100 ms. The delays experienced are usually the result of congestion on the IP based network. In addition to responding to requests from browsers, Web servers are also responsible for completing other tasks including

- Logging activity on the IPTV network
- Protecting Web pages from unauthorized users
- Sending requests to peer Web servers on the public Internet

The hardware requirements for a Web server will vary according to the level of interaction between the IP set-top box and the Web server.

11.3.3.3 Caching Servers When using a resource constrained device such as a set-top box to browse the Web, it makes sense to incorporate optimized caching within the overall system architecture. The caching servers are normally located at the IPTV service provider's regional office at the edge of the core IP backbone network. The caching technique employed by IPTV systems involves the storage of popular Web pages on the server. The various caching servers used by an IPTV browsing application use standardized communication protocols to "talk to each other," thus ensuring that local Web pages are kept current. Caching servers fall into two broad categories: software that runs on a standard server or a dedicated hardware platform.

Note that in addition to the above technologies, a high speed and resilient connection to the public Internet is also required for IPTV browsing services.

11.3.4 IPTV E-Mail

From the earliest days of society people have used various systems and technologies to communicate with each other. With the recent explosion of the Internet, the most popular means of communicating between computers is electronic mail (e-mail). With the advent of IPTV, e-mail is now poised to enter the living room. Figure 11.7 shows how an end to end IPTV e-mail solution might look like.

As illustrated the major components of a fully built end-to-end IPTV e-mail system include the client software resident in the IPTVCD, a suite of communication protocols, servers, and a firewall located at the IPTV data center.

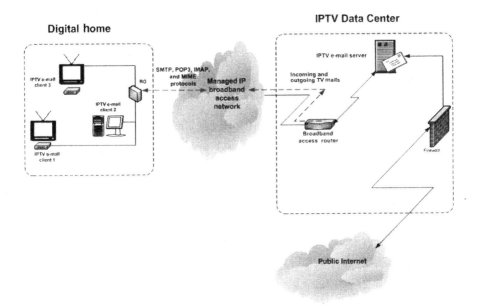

FIGURE 11.7 End-to-end IPTV e-mail solution

11.3.4.1 IPTVCD E-Mail Clients   The IPTVCD with its interactive capability is seen by many as a natural environment for e-mail in the home. The client e-mail application is normally integrated with the IPTVCD middleware and uses the TCP/IP protocol to communicate with the IPTV data center. Modern IPTV e-mail applications fulfill a wide range of functions and let subscribers do more than just send or receive mails. For instance, the attachment of files to e-mails is supported. This might sound like a trivial feature but in the world of IPTV and thin IPTVCD clients the implementation of attaching files poses some technical challenges. For instance, IP set-top boxes do not have the processing power or storage capacity necessary to run popular programs that are needed to open file attachments. Therefore, the IPTV e-mail client needs to convert the attachment into HTML format and then use the browser to display the file. Other features supported by IPTV e-mail clients include personal address books and support for encryption.

11.3.4.2 Networking and Communication Protocols   E-mail messages are always transmitted in American Standard Code for Information Interchange (ASCII) format. ASCII files are text-only files. The transfer of e-mail messages in an IPTV environment are normally based on Internet mail standards such as SMTP, MIME, IMAP4, and POP3.

   SMTP—stands for Simple Message Transfer Protocol. It is used to transfer IPTV mails from the mail server located at the data center to another e-mail server on the Internet, and is designed for reliable and efficient mail transfer. SMTP forms part of the TCP/IP protocol stack, and is used to specify how e-mails should be passed from transmitting to receiving hosts.

   MIME—stands for Multipurpose Internet Mail Extensions. MIME is an Internet standard that lets IPTV e-mail clients automatically encode and decode binary files from their original format into a format that is viewable on a television screen. It specifies how non-ASCII messages should be formatted so that they can be sent over the public Internet. Examples of non-ASCII messages include graphic, audio, and video files. This protocol allows IPTV e-mail systems to support much more than just text messages. There are many predefined MIME formats, including JPEG file graphics, and PostScript files.

   POP3—stands for Post Office Protocol version 3. It is used by IPTVCD e-mail clients to access messages stored on the mail server.

   IMAP—stands for Internet Message Access Protocol, the latest version of this protocol, IMAP4, has similar functionality to POP3 'but does support additional features, which are useful for IPTV end users. E-mail searching using keywords while the e-mails remain on the server is one example of an IMAP feature that is not supported by POP3.

11.3.4.3 E-Mail Servers   Every IPTV subscriber that avails of the e-mail service receives a unique e-mail address and a mailbox. The mailboxes are stored and managed on mail server gateways at the IPTV data center. These servers are connected to the Internet and also interface with the OBSS for billing purposes. The mail server gateway is a software program that runs on a mail server and links the IPTV network through a firewall to the public Internet.

11.3.5 Digital Video Recording

A hard disk needs to be incorporated into the IP set-top box to support DVR functionality; the ability to select programs, digitally record and store live TV content on a local hard disk. Chapter 5 on IPTVCDs provides a more detailed insight.

11.3.6 Walled Garden Portal

A walled garden portal is best defined as a Web portal or quasi-Web environment specifically developed for a TV environment. Within a walled garden, a viewer has access to a variety of content such as horoscopes, news, recipes, sports, weather, and internet applications such as e-mail and chat. In addition to providing content, some of the more advanced portals provide the following functionalities:

- Registration services for various types of IPTV content services.
- Allowing end users to purchase IPTV content services.
- Allowing service providers to run promotional campaigns.
- Providing content navigation services, which are similar to those provided by the IPTV EPG.

Since content is owned and controlled by the platform provider, only IPTV subscribers have access to a walled garden via their TV. Internet users do not have access. In some cases, IPTV subscribers have access to the public Internet from the walled garden. Each page within a walled garden is designed for TV.

11.3.6.1 Designing Pages for an IPTV Based Walled Garden Portal IPTV service providers and content creators must first consider the differences between a TV and a PC, the circumstances in which they are used and the audience using the IPTV portal. A graphical example of a viewer watching TV is presented in Fig. 11.8 and described in Table 11.2.

A graphical example of the setting in which a PC user typically uses a PC is presented in Fig. 11.9 and described in Table 11.3.

As shown in the tables and diagrams, a number of considerations need to be taken into account when repurposing Web content for an IPTV walled garden portal. The following list provides a set of guidelines for professionals who are involved in developing a portal site for a new IPTV service.

Keep Video Picture in Background When designing a Web page for a TV audience, it is preferable to have the TV picture as the background to maintain

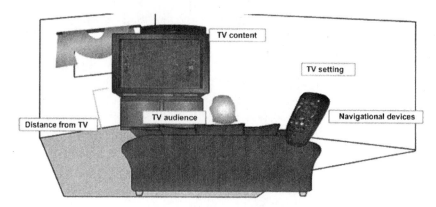

FIGURE 11.8   A typical setting in which a person views a TV

consistency and to enliven the Web page. IPTV content creators should provide a choice of semitransparent or opaque backgrounds to the Web page. It is also possible to use audio (like music and voice introductions), video, and animations to energize the material.

Keep Content Concise   Considering the medium and viewers' goals, content in a Web page should not be too long and complex. Content should be provided in easily absorbed chunks and having a large quantity of menus or hyperlinks should be avoided. Since a TV viewer will have less patience than a PC user, ensure fast download and response times.

Choice of Navigation   Given that it is difficult to use a PC keyboard while sitting on a soft chair, the navigational device most used for Interactive IPTV applications is a remote control. A remote control should be light and easy to hold in one hand. The buttons should be easy to see and to press, a maximum of 30 buttons is optimal. Most remote controls have directional buttons and a select button. There should be a clear relationship between the screens and the remote; this is often achieved by linking color-coded onscreen buttons with equivalent remote control function keys. IPTV providers often supply a wireless keyboard with a remote control so that a viewer can make the most of features such as e-mail, chat, t-commerce, and informational elements. A wireless keyboard contains most of the buttons that are found on a PC keyboard. They also include iTV-specific buttons that vary according to brand. Seeing as, not all IPTV viewers have a wireless keyboard or if they do, it is cumbersome to use in a TV setting, it is best to assist navigation by

- Keeping navigation simple
- Avoiding scroll bars or allowing limited vertical scrolling
- Presenting choices in lists
- Minimizing demands on the viewer to type

TABLE 11.2   Description of the Various Characteristics of TV Viewing

Characteristic	Description
TV audience	The TV audience • is culturally, intellectually, and technically diverse; • age profile encompasses all ages—from the very young to the very old; • uses the TV principally for entertainment purposes; • does not expect demanding interaction; • has, on average, a low level of computer skills.
Distance from TV	A TV viewer views the TV at a distance of approximately three to four and a half metres (10–15 ft). Everything displayed on screen should be large enough for viewers to see from a 5 m distance. Font size should be at least 18 point and screen elements should be arranged well spatially, to create a cleaner and clearer screen.
TV content	A TV viewer is most used to the seamless display of content in motion with audio, that is, standard television. Viewers have become more and more familiar with text in the form of program guides in large font. In Europe, many viewers have become accustomed to using teletext that is overlaid on the TV picture. It is available with opaque, semitransparent, and transparent settings.
Navigational devices	A TV viewer uses a remote control and/or a wireless keyboard as navigational devices. The remote control is most prevalent because it is a handheld device and relatively inexpensive. The keyboard can be cumbersome to use while sitting on a soft chair but is useful for text input. The context of viewers' television experience involves going up and down through broadcast channels.
TV setting	Generally, viewers watch TV in the "living" room or area of their home. Typically, this is a comfortable area with soft seating and lighting to create an environment conductive to relaxation.

- Arranging elements in grids for remote-control navigation (like EPG grids)
- Linking functions to remote-control keys
- Avoid multilayered navigational trees
- Remove nonessential links and options
- Use pull-down menus instead of navigation links
- Have site navigation at the top with sub navigation at the bottom
- Add back and next links to each page

Because Web pages are built for a mouse-controlled onscreen cursor, it is not possible to design for navigation in a TV Web page in the same way.

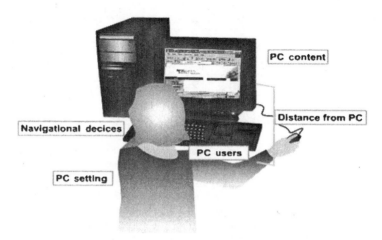

FIGURE 11.9   A typical setting in which a person uses a PC

Understand the limited Memory and Speed of IPTVCDs Particularly IP Set-top boxes   Set-top boxes range in CPU power, operating systems, and memory capacities. However, a set-top box on average will have 8 MBs of RAM and a processing speed of 400 MHz while most PCs will have at least 1 GB of RAM, a

TABLE 11.3   Description of the Various Characteristics Associated with Using a PC

Characteristic	Description
PC users	PC users • use the PC mainly for information and communication (often work-oriented); • expect lots of text, complexities, and problems; • often expect to wait for something to download; • have moderate to high computer-skill levels.
Distance from PC	The PC user is usually about half a metre (2 ft) from his or her PC. Along with high resolution and a picture that adheres to the edges of the computer monitor, the screen allows for a lot of detail. The standard text size is 11 point.
PC content	PC users are used to static, clear, and often colorful display of content in Web pages. There are few restrictions to color or color combinations. PC Web pages generally consist of menus, sub, and drop-down menus; lots of text in 11 point font, vertical and horizontal scroll bars, numerous hyperlinks, search features, crisp graphics, and audio and video requiring plug-ins
Navigational devices	Usually have a keyboard and mouse as navigational tools or similar equivalents.
PC setting	Many PC users are accustomed to viewing Web pages in a focused setting such as in a home or work office while sitting on a "hard" chair, at a desk (often both are ergonomically designed).

100 GB hard-drive, and a processing speed of 4 GHz. Viewers will not be any less impatient with loading times if aware of these differences. Loading and response times should be short.

Understand the Characteristics of Analog TV   The resolution and the safe area affects how a Web page looks on a TV screen.

Resolution—Owing to higher resolutions, the picture display on a computer monitor is much crisper and more stable (e.g., color) than an analog TV display. Resolution is generally measured in terms of pixels. The more pixels there are in a display, the better the resolution. The manner in, and speed at, which a picture display is transmitted to the screen are also determining factors.

TV safe area—Content on a full screen might be obscured because of overscanning or the casing of a TV-set. That is why it is important to determine the safe area of a screen. The safe area of a screen is the area where the audience is guaranteed to see content (with few exceptions) despite the brand or style of TV set. TV signals usually overscan the edges of the display so that the TV picture appears without "fringes." In addition, the amount of the display area covered by a TV's casing depends on the TV set.

Minimize Download Times   To minimize download times

(1) reduce amount of data to be downloaded—If there are fewer elements on the screen that need to be downloaded, the overall download time will be reduced.

(2) design pages to fit within the safe area of the TV screen—If pages fit the safe area of the TV screen, the IP set-top browser does not need to spend time resizing large images and tables.

(3) anticipate the IPTV viewer's next page—While the viewer is looking at the current page the next one can be downloading.

(4) insert a fast-downloading introductory scene in animations—While the introductory scene is playing, the remainder of the animation can be downloading.

(5) use television or a color as a background—It is best to avoid putting background images other than television or a plain color as a background on the Web page.

Identify the Types of Signals Used by Target Televisions   There are two primary types of video signals used by analog televisions: composite and S-video input. In composite video, red, green, blue (RGB), and sync information are combined into one signal going into the TV; in order to transmit this picture, the TV must separate the signal into its four components, which results in some distortions and a degraded picture quality. S-Video provides an improved mechanism for carrying signals resulting in cleaner and sharper pictures. Content developers need to be aware of these differences when customizing applications for an IPTV platform.

Avoid Color Problems   TV sets with a composite signal usually generate visual artifacts. Artifacts are distortions of display, content, or sound caused by a limitation or malfunction in the technology used. Artifacts can be minor or extreme. Such artifacts do not appear on computer monitors. If the chrominance difference between two colors is great, an artifact known as "chroma crawl" is created on a composite video signal. Wherever two contrasting colors meet, dots appear to crawl. Sometimes they can form a shape and move across or down the screen. This can make text and images appear unclear or distorted. It is possible to avoid chroma crawl by using colors of the same hue but with varying luminance.

Minimizing Color Problems   By eliminating chroma crawl IPTV content creators restrict their choice in colors. However in many cases, it is sufficient to just minimize the effect. Minimizing chroma crawl means reducing the chrominance distance between adjacent colors. This is possible by replacing vivid colors with pastel relatives, for example. In addition to chroma crawl there are other TV-generated color problems to consider:

- *Hot colors*—To reduce the overvibrant appearance of colors, reduce saturation by 80 to 85%.
- *Bleeding*—Bleeding can occur with either a composite or a S-video signal. Adjoining colors appear to bleed into each other. Bleeding can be remedied in the same way as chroma crawl.
- *Moiré effect*—Moiré patterns occur with interlaced scanning when contrasting colors (like, black and white in a herringbone pattern) appear in close proximity. Moiré effects vary depending on colors. The area around the pattern can produce a shimmering effect or it could produce a red glow, for example. It is possible to avoid Moiré effects by avoiding detailed patterns.
- *Blooming*—In interlaced scanning, brighter scan lines are wider than darker ones. Therefore, if there is a large number of alternating bright and dark scan lines grouped together the vertical sides appear to wave in and out. Blooming is highlighted when you have vertical lines, like those in a table.
- *Flickering*—Flickering is caused because TV displays are interlaced. Without a flicker filter, a thin horizontal line just one-pixel thick would appear to flicker on and off. Horizontal lines should be at least 2 pixels in width. Similarly, horizontal edges between two contrasting colors would pulsate.

There is a gradual move across the world from old analog TV sets toward digital and high definition televisions, which deliver better picture, sound, and overall stability. Eventually there will be global deployment of these. When this happens the development of Web pages for IPTV Web portals will become more straightforward. Until that happens, content creators will need to comply with the various guidelines outlined above. From a market demand perspective, TV viewers are more accustomed to watching TV than viewing a Web page on its screens. However, this author is more optimistic than ever that the concept of walled gardens will eventually catch on and firmly believes that IPTV will act as a catalyst for increasing the popularity of this application.

11.3.7 Instant IPTV Messaging

This application uses "presence technology" to allow subscribers to use their televisions to participate in interactive chat forums. The network service provider decides on the types of chat forums that are available to their customer base. From a technical perspective the IPTV operator needs to install a powerful community chat server in the headend. Chat servers are easily customizable and seamlessly integrate with advertisement banners. The client IPTV application has very low memory and processor requirements and makes it very suitable for use in IPTVCDs. Most IPTV chat programs offer one-to-one and one-to-many communication channels. New versions of IPTV messaging software applications allow interactive TV designers to split the TV screen into two frames. The first frame includes the forum, while the second frame displays the related TV programme.

11.3.8 IPTV-Commerce

IPTV-commerce supports business activities and enables viewers to purchase goods through a TV using a remote control instead of a keyboard and an IPTVCD. The term IPTV-commerce can refer to online shopping, instant shopping, online betting, and home banking. In the case of instant shopping, end-users can purchase a product featured on an advertisement without having to leave the channel they are watching. The ability to facilitate the provision of e-commerce via an IPTV platform helps to enhance the overall revenue stream of service providers. Security measures that apply to PC based e-commerce also apply to IPTV-commerce. This is an ideal application for people who are not comfortable with using their PC to buy online.

11.3.9 IPTV Social Networking

The popularity of Internet based social networking sites has grown rapidly over the past couple of years. IPTV facilitates the expansion of the phenomena of social networking onto the television. Implementing a social networking environment on a TV set requires specialized software on both the IPTVCD and the backend servers. Typical features supported by this type of interactive IPTV application range from buddy lists that allow IPTV subscribers to view what their friends are watching on television to real-time chatting while watching the same program.

11.3.10 Caller ID for TVs

This IPTV application allows consumers who have subscribed to a caller ID service through their telephone company to view caller information on the TV screen. In other words, when the phone rings a pop-up window appears on the television to notify the end user of the name and phone number of the incoming caller. Advanced versions of this IP application allow the simultaneous pop-up of caller ID windows on multiple PC's and TV screens around the house. A high level overview of the two main components required to deploy a caller ID service on an IPTV network are shown in Fig. 11.10 and explained in the following sections.

  An application server—This server is normally located in the IPTV data center and supports a wide range of telecommunication protocols. When a call arrives, the signal is routed to this server and a communication session is established with the IP set-top box. Additional functions of this application server range, from communicating with the back office provisioning and management systems to managing data transfer across the network.

  A client IPTVCD software application—the display of caller ID details on a TV requires the installation or activation of a specialized piece of software on the IPTVCD. This software module integrates with the IPTVCD operating system and is typically branded by the IPTV service provider. This application is responsible for ensuring that the name and number of the incoming caller appears on the subscriber's TV screen.

The caller ID application can also be used to display notifications of other types of incoming messaging services including voicemails, and text messages. The caller ID for TVs is an extremely popular application amongst IPTV subscribers

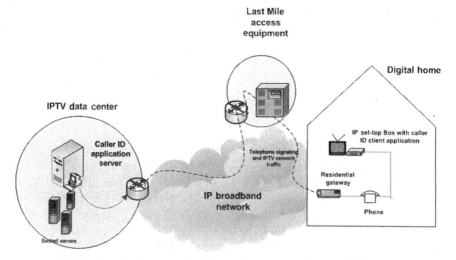

FIGURE 11.10 Components of an end-to-end caller ID IPTV application

and is one practical example of how converged services will have an impact on people's lives in the years ahead.

11.3.11   IPTV Advertising

The use of interactive TV as a medium to deliver advertising is rapidly emerging into a mainstream application. The fact that it allows advertisers and service providers to engage consumers with a compelling suite of advertisements is considered to be one of the main drivers of this type of IPTV application. IPTV is capable of supporting a range of different IPTV ad formats, including telescoped long form adverts and integrated links that are embedded directly into the content. In addition to the various ad formats, IPTV allows operators to place adverts into both the standard broadcast programming streams and the increasingly popular method of delivering content on-demand. Not only does IPTV advertising provide advertisers with a one-to-one connection with consumers but it also represents a significant revenue opportunity for telecommunication operators deploying IPTV services. The advent of new targeted advertising technologies that run across IP networks are now starting to appear in different IPTV deployments across the world. The two key benefits of these new IP based systems compared to traditional advertising systems that operate over RF based networks are

(1) *Two-way capabilities*—Owing to the fact that all of the IPTV deployments are operating over bidirectional broadband networks means that these new advertising systems are able to collect and monitor in real-time viewer reactions to advertisements. This feature allows advertisers to immediately gauge the effectiveness of a particular marketing campaign.

(2) *Individual IPTVCDs can be addressed directly*—The deployment of targeted ad insertion techniques on an IPTV system enables service providers, content programmers, and advertisers to target their advertising messages to individual IPTV viewers. This is a major benefit over current advertising systems, which are limited to targeting commercial advertisements to specific geographical areas.

The functionality provided by this application allows subscribers to retain control of their video content viewing experience while ensuring that they only receive adverts on products and services that align with their interests in an opt-in basis. From an advertiser's perspective, the availability of accurate user data helps to increase the success of their advertising campaigns by allowing them to focus on viewer's interests. Service providers also benefit from this level of functionality because it helps to create a new television advertising revenue stream. The six major hardware and software components of a fully built end-to-end addressable advertising system are depicted in Fig. 11.11 and described in the following sections.

Content reception—In an IPTV advertising system, the television channel is typically received from off-air or from a satellite feed and forwarded to the

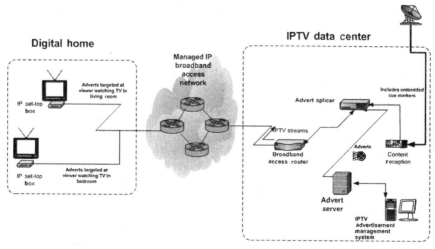

FIGURE 11.11   IPTV targeted advertising system

splicer. Special advert markers are sometimes included in this stream,
which pinpoint the sections of the video stream that are allocated for ad
inserts.

Advert server—Advertisements are typically pre-recorded and stored digitally
on this special purpose video server located at the IPTV provider's data
center. The server streams these advertisements to the splicer.

Advert splicer—The purpose of this piece of hardware is to insert adverts
retrieved from the advert server into the IPTV streams.

Advertisement management system—This is a piece of software that runs on a
PC or server. It ensures that the correct adverts are inserted into the correct
IPTV streams at the specified time slots. Some of the more advanced systems
support the creation of specific marketing campaigns based parameters such
as programming type, geography, and demographic data.

Broadband access router—This router streams the IP video streams with the
embedded digital adverts on to the core and access IP networks.

IP set-top box—The targeting of adverts over an IPTV network generally starts
with the subscriber who uses their IPTV interface to create a personal profile
that outlines the types of adverts they would like to view. This profile
information and related viewing statistics is then used by advertising
companies to reach IPTV subscribers who have an interest in a specific
category of products or services.

The deployment of an IPTV targeted advertising system is rarely implemented
in isolation and is usually integrated into an IPTV data center in conjunction with a
number of other third-party systems.

11.3.12   Localized Video Content

The deployment of localized video content is seen by many telecom operators as a differentiator for IPTV. Most of the content delivered by the media nowadays has a countrywide or international focus, with a limited emphasis on providing content that is local to particular geographical communities. With IPTV, telecom operators now have the ability to offer local information to their subscribers. The local content IPTV application typically allows subscribers to retrieve local weather, schedules for garbage collection, school announcements, results of sporting events, and traffic videos on their televisions.

11.3.13   Gaming on Demand (GoD)

IPTV technologies also allow telecom operators to deliver interactive TV gaming applications to their customers. The addition of online gaming applications to a portfolio of IPTV services can not only help to generate additional revenue streams for telecom operators but also reduce customer churn. In addition to subscriptions, IPTV providers can also place adverts in IP based games that connect advertisers with end users. The GoD application allows IPTV subscribers to

- play high quality premium games from the comfort of their homes;
- play with other gamers connected to the Internet;
- participate in text, phone, or emoticon messaging during game play.

The two major components of an IP based on-demand gaming system are depicted in Fig. 11.12.

Gaming servers—A client-server computing architecture is generally used to deploy this type of IPTV application. The servers are located at the IPTV data center and typically integrate seamlessly into existing VoD infrastructures. A server may be involved in two types of tasks with the deployment of online games: store the various game titles and interpreting instructions from the IP set-top box.

Online gaming client applications—IPTV gaming applications run locally or remotely, dependent on the hardware capabilities of the IPTVCD. A remote control or specialized joystick is typically used to play the games. The UDP/IP protocol layers of the IPTVCM are used to carry the gaming content both upstream and downstream to the client IPTVCD.

Although games deployed by IPTV systems are a far cry from the advanced games that are played on dedicated game consoles, the level of sophistication used by IPTV games has improved dramatically in recent times and this trend is expected to continue into the future.

IPTV Data Center

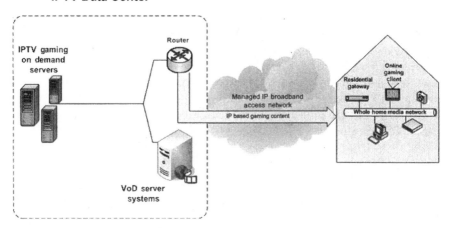

FIGURE 11.12   IP based gaming on demand system

11.3.14   Parental Control

IPTV networking platforms also include filtering systems that allow telecom operators and subscribers to restrict access to certain on-demand titles or broadcasting channels that contain inappropriate content. Channels or VoD titles requested by an IPTV subscriber are checked against a database of objectionable content. If the server finds that the material is rated, it will not allow that video asset to be passed on to the person making the request. The application itself uses PIN codes to enforce content filtering regulations.

11.3.15   An IPTV Based EAS

Digital TV systems based in the United States will typically interoperate with a national system called the Emergency Alert Systems (EAS). This system was established in the late 1990s by the FCC and alerts people of danger. Till now, TV and radio were the two primary carriers of these messages but now IPTV systems have started to also provide this level of functionality to subscribers. As shown in Fig. 11.13, implementation involves the installation of a standalone FCC Type-Certified EAS receiver and character generator at the IPTV data center.

Once this device is live on the network, it is possible to automatically send text and digitized audio based EAS alert messages to IPTV end users.

11.3.16   Personalized Channels

The ability of IP based technologies to allow subscribers to create their own TV programs and to broadcast this on personalized channels is yet another application that helps to differentiate IPTV from traditional RF based TV services.

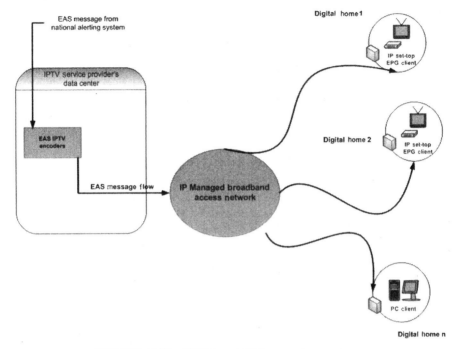

FIGURE 11.13 IPTV based EAS network components

11.4 IPTV PROGRAM RELATED INTERACTION APPLICATIONS

Program related based iTV applications allow IPTV subscribers to interact with the incoming video stream while continuing to view the video program. A poll is a typical example of a program related iTV application. There are approximately five different approaches to providing interactivity between the video stream and the IPTV subscriber. These approaches are discussed below.

11.4.1 Datacasting

Datacasting refers to the provision by IPTV service providers of extra information as an enhancement to the television video stream. This information can be either related to programming or unrelated to it. Data related to programming would need to be synchronized whereas the unrelated data may not need to be. Datacasting can be a useful means for IPTV providers to generate revenue. Information pushed to the viewers by the broadcaster can arrive at their TV sets in the form of a trigger or a ticker.

11.4.2 Interactive Programming

Interaction in programming can feature in different ways. For example, IPTV subscribers can be prompted to give their opinion in the form of voting, respond to live shows via e-mail or SMS, connect to a related chat room, play programme-related quizzes, purchase programme-related offerings, or access programme-related data. Most viewers have found themselves shouting answers to quiz-show questions. With iTV, viewers have the opportunity to participate in a real-time quiz show. Broadcasters can offer prizes to at-home viewers and benchmark them with other viewers.

11.4.3 Picture in Picture

Picture in picture can be another feature of interactive programming. It describes the feature that enables an IPTV end user to view a broadcast within a window on a TV screen while watching another broadcast or interacting with an iTV interface. Picture in picture can be a feature of an interactive sports program. A viewer could select to watch several matches or games simultaneously. Alternatively, they could watch one match or game from several different angles simultaneously.

11.4.4 Individualized TV

Individualized TV, multiangle-camera events, and synchronized television are more specific kinds of interactive programming. They require a lot of design, programming, and support work from the service provider. Individualized television describes an iTV-enhanced program where the viewer controls camera angles, selects further information, commercials, and can purchase products. Data is collected at the network-suppliers end and further information is customized according to viewer preferences and sent to the viewer.

11.4.5 TV-Synchronized Applications

TV-synchronized applications enable broadcasters or content providers to synchronize interactive, related content with a program. Typically, it refers to synchronizing dynamic content in a Web page with a program or synchronizing other interactive information. Triggers can be used to inform the user that such interactive content is available. The content that is pushed by the broadcaster can vary from related data to polling and quizzes. Other types of interactive entertainment can include.

- Magazines
- Music selection
- Gaming

Note that each of these TV-synchronized applications is generally accessible via the EPG.

11.5   DEPLOYING INTERACTIVE IPTV APPLICATIONS

The mechanism used to deploy an iTV application will often depend on the hardware capabilities of IPTVCD. There are three common mechanisms for executing interactive IPTV applications.

11.5.1   Embedded Applications

A resident application is a program or a number of programs that are built into the memory of the set-top box. This helps to increase the responsiveness of the application and minimizes the need to interact with the network. Typical embedded applications include

- Basic navigation
- Emergency alert system
- Parental control
- Internet browsing
- Managing configuration settings

These applications are generally proprietary and written for specific hardware platforms. In addition to the proprietary nature of these application types, resident programs are limited in their capabilities due to inadequate computing and storage resources available on the local hardware platform.

11.5.2   Network-Centric Applications

In this case, the program code is downloaded from the network into RAM in response to a request from the TV viewer and executed on the digital set-top box. The files associated with the iTV application are periodically transmitted and the set-top box will wait until the files become available in the stream. The method used by the set-top box to interact with the network depends on the location of the iTV application. There are two possibilities:

In a data carousel—The creation of data carousels is one popular mechanism of broadcasting interactive TV application source code and data to a digital set-top box. In a traditional TV transmission system, the data and code are typically sent from the headend using the DSM-CC protocol. DSM-CC offers, among other things, a mechanism to broadcast objects (like files and directories) in a carousel-like fashion. Once broadcasted over the network in a cyclical pattern, the digital set-top box uses details available in the SI tables to identify and locate specific interactive TV applications. Compared to other download protocols, DSM-CC download is designed for lightweight and fast operation in order to meet the needs of devices like set-top boxes that contain limited memory. DSM-CC is physical layer independent.

In the OOB channel—The OOB channel is separate from the video delivery channel. This allows providers to use this channel for carrying interactive video applications. The set-top box is instructed to tune to the OOB channel when a subscriber requests an iTV service. The interactive TV application files are then retrieved and loaded into memory for execution. Standard IP based protocols are often used by this channel.

Note that network centric iTV applications are generally deployed on broadcast networks that do not support a return channel.

11.5.3 Server-Centric iTV Applications

Under this mechanism, the interaction between IPTVCDs and the backend servers follow the classic client-server computing model. Under this model, the processing tasks involved in running an interactive IPTV application are intelligently divided between a client application running on the set-top box and one or more application servers located at the service provider's IPTV data center. The bulk of the backend processing tasks, such as accessing databases and retrieving rich multimedia applications are performed by the application servers. The client program running in the IPTVCD acts as the front end to the TV viewer and the application server responds to various types of requests. Note communications between the iTV application servers and the IPTVCDs require a high capacity network connection.

11.6 ACCESSIBILITY TO IPTV SERVICES

Similar to traditional TV viewing, the delivery of IPTV services need to be accessible to people with special needs and disabilities. Although accessibility is not a specific interactive IPTV application, its functionality needs to be available in all IPTV related services. Thus, IPTV network administrators need to consider the technical implications of supporting interactive IPTV accessibility features across their systems.

SUMMARY

Telecom operators are deploying two-way digital IP based networks that have the ability to support sophisticated interactive TV applications. Advances in technology since the 1950s when the concept of iTV originated, have been great and widespread, from the improvements in how television is broadcast and delivered, to advances in hardware, networking, and software.

IPTV service providers are adding interactive TV applications to their existing portfolio of services in order to differentiate themselves from their cable and satellite competitors. The two-way capabilities of IP based networks offers providers an ideal platform for delivering interactive TV services to consumers. As shown in Table 11.4, IPTVCDs can provide consumers with access to a number of different services.

TABLE 11.4   Popular Interactive IPTV Applications

IPTV Application	Description
EPG	To aid in program selection, advanced IPTVCDs feature a fully integrated EPG that informs subscribers of the vast programming available. The digital EPG is the gateway to a portal of interactive digital services such as banking, shopping, games, Internet applications and on-demand television. An EPG needs to be imaginative and compelling to win new customers and retain a company's existing subscriber base. Its primary function is to provide the viewer with an overview of the programming currently available, as well as the ability to browse upcoming television programs. Basic and advanced EPGs are important keys to the future of IPTV. A basic EPG informs a viewer what is on now and what is on next. Advanced EPGs provide viewers with detailed information on programs, channels and even actors. Other features include the ability of subscribers to personalize the EPG to meet their own needs.
On-demand video content	End users are able to use their IPTVCDs to access video content on their own schedule rather than at a predetermined time.
IPTV browsing	While the viewing of Internet TV never took off in the early days of deploying digital TV services, Web browsing has now become an integral part of deploying IPTV services. An IP set-top box browser is an essential component for the successful delivery of next generation, Web-based TV services. Its main function is to adjust existing Web pages for presentation on a TV where necessary. Most of the browsers currently available for deployment of IPTV networks feature support for the latest television and W3C standards. The use of international standards allows service providers to repurpose existing Web content for delivery over an IPTV networking environment.
IPTV e-mail	An advanced digital IPTVCD with a keyboard can serve as a method for a subscriber to access e-mail.
Digital video recording	The addition of a hard disk to an IPTVCD allows consumers to record live broadcast video content and digitally store for viewing at a later time.
Walled garden portal	Walled gardens are designed to emulate a Web portal. The interface to the portal is typically branded with the coloring schemes and logos of the network operator hosting the IPTV portal.
Instant IPTV messaging	The use of IM as a mode of communication has grown in recent time. As a result IM has been integrated with a number of commercially available IPTV networking platforms.

(Continued)

TABLE 11.4   (*Continued*)

IPTV Application	Description
IPTV-commerce	IPTV enables the delivery of electronic commerce applications on a TV.
On-screen caller ID	This interactive IPTV application combines video content with telephony services.
Targeted interactive TV advertising	This application allows service providers to generate additional revenue streams by inserting adverts into their IP based video channels. It also helps advertisers to pinpoint consumers with appropriate advertising messages and match products to individual IPTV end users.
Gaming on demand	The combination of the TV medium with the two-way capabilities of broadband is an ideal platform for gaming on demand applications.
Parental control	Parental control in the context of an IPTV networking environment allows parents to manage the types of on-demand and multicast content viewed by their children.
IP based emergency alert systems	This application is primarily used in the United States to notify residents of emergency situations such as severe weather storms and terrorist attacks.
IPTV program related interaction	This application involves the presentation of additional information on a screen while the viewer is watching the video content.

Even though IPTV technology is able to support all of the applications in Table 11.4, the initial deployments of IPTV have tended to focus on interactive TV services such as EPGs and caller ID over TV services.

iTV applications can be broadly classified into three categories: embedded, network, and server centric. Table 11.5 provides an overview of these systems.

TABLE 11.5   iTV Applications Category Overview

Type	Description
Embedded	An embedded IPTV iTV application is a program or a number of programs that is built into the memory of the IPTVCD.
Network centric	This type of application is widely used on networks that do not support a return path. The interactive application is broadcast on a specific channel and the set-top box tunes to this channel, loads into memory, and executes. DSM-CC, and OOB channels are the two most common methods of delivering multimedia software applications over a broadband network.
Server centric	A typical server centric iTV system includes one or more application servers connected to a large number of set-top box clients.

12

IPTV NETWORK ADMINISTRATION

The delivery of TV services over an IP based network comes with its own set of technical and commercial challenges. One of the first challenges arises during the operation of an IPTV network on a day-to-day basis, where the service provider needs to be able to manage video traffic and the various network components which are used to make up an IP network. To enable such control, IPTV providers need a network management system (NMS) that monitors and identifies problems that could affect the delivery of television services to their customers. Another challenge faced by providers is to simplify the provisioning of new subscribers and services onto the network. The installation of an IPTV service is relatively complex and puts a strain on service provider resources. Therefore, many operators attempt to automate the installation and provisioning of IPTV services in an effort to reduce time taken to carry out installations.

Moving beyond the management and provisioning of services, IPTV network operators need to also ensure that their customers receive a TV experience that compares favorably with other services that are on offer from cable and satellite pay TV operators. At its most basic level, IPTV operators need to ensure that their customers receive adequate responsiveness to channel change requests. This is quite a basic requirement from a subscriber's perspective; however, the implementation of this functionality, particularly for a large IPTV network, can be problematic for network operators. To help meet these challenges, this chapter outlines a number of engineering and operational functions that are critical to the success of IPTV deployments.

Next Generation IPTV Services and Technologies, By Gerard O'Driscoll
Copyright © 2008 John Wiley & Sons, Inc.

12.1   AN INTRODUCTION TO IPTV NETWORK ADMINISTRATION

Network administration defines the technologies and procedures used for monitoring and controlling a network. It encompasses a variety of tasks:

- Supporting the IPTV networking management system
- Managing installations, service problems, and terminations
- Network testing and monitoring
- Ensuring service availability and managing redundancy
- IP address space management
- Routine IT and network administrative tasks
- Managing IPTV QoS requirements
- Monitoring the IPTV subscriber experience
- Remotely managing in-home digital consumer devices
- Scheduling and managing delivery of software updates to IPTVCDs
- Troubleshooting IPTV problems
- IPTV and business continuity planning

We explore these various IPTV network administration tasks in the following subsections.

12.2   SUPPORTING THE IPTV NETWORKING MANAGEMENT SYSTEM

Today's IPTV delivery networks offer enormous potential for operators to generate a range of new revenue streams. The management of an end-to-end IPTV system is however a daunting task! Occasionally, these systems can develop problems that have catastrophic effects for the IPTV service provider. To minimize the risks associated with network problems, operators use network management systems to proactively monitor the end-to-end IPTV networking infrastructure. These tools are used to ensure that the IPTV end-to-end infrastructure achieves an uptime availability in excess of 99.999% and ensures a high quality video signal for consumers. Owing to the variety of technologies and devices associated with an end-to-end IPTV system, a typical IPTV network operations center will contain a number of different systems that manage different parts of the network. The types of functions performed by these IPTV management systems can include

Maintaining 24/7 network visibility—An NMS includes graphical network maps that display status information on a number of mission critical network elements:

- IPTV data center specialized video equipment and servers
- IP core network devices
- IP access network devices
- IPTVCDs
- Home networks

Network optimization—An NMS helps in the process of optimizing network resources that support an increasing mix of IP based applications.

Help support personnel to resolve problems—An NMS collects error messages from the various components that make up an end-to-end IPTV system and allows support personnel to quickly identify and resolve problems that may arise during the day-to-day running of an IPTV network.

Trouble ticketing—the ability of an NMS to support trouble tickets is a useful tool for coordinating problem-solving and repair activities.

Reporting—An NMS gathers statistics about the various network components on a regular basis and generally includes a reporting system that allows IPTV network managers to:

- Track and evaluate IPTV service outages;
- Check the health of the various network components;
- Identify potential problems on the network through the analysis of trend data;
- View network utilization over specific periods of time.

Graphical visualization of network components—Intuitive interfaces typically provide a graphical diagram or map of the availability state of the various elements that make up the end-to-end IPTV network.

Managing headend server faults—IPTV Servers play an extremely important strategic role within the broadband TV network. When a server goes down, multiple IPTVCDs are unable to function properly resulting in loss of revenue for the IPTV operator. Constant monitoring of the servers in real time is required to reduce downtime. The NMS supports this monitoring through the reporting of alarms and problems to support personnel.

Configuration management—The NMS will store items of configuration information about each device connected to the IPTV network in a database. The types of information can range from an IP address to the version of middleware operating on each IP set-top box. The NMS allows engineering staff to modify these parameters.

Bandwidth management—IPTV is a bandwidth hungry application. Therefore, special attention needs to be given to monitoring the amount of bandwidth consumed by IP video based applications. An NMS allows service providers to keep a tight control on network bandwidth usage and optimize their networks for IPTV based applications.

Traffic prioritization—IPTV is a delay sensitive application. To improve the quality of the service delivered to end users an NMS allows service providers to prioritize IP based video content over less sensitive applications such as Internet browsing and peer-to-peer application traffic. Some of the advanced NMSs also support the prioritization of network traffic during certain parts of the day (peak versus off-peak times).

Managing network logs—This function as the name implies is responsible for recording and storing logs of events that occur during the live operation of an IPTV network. The type of logging information stored varies between NMSs and can include the following:

- IPTV end user login and logout details;
- Details of system and IPTVCD configuration changes;
- Network device operational details.

The challenge for IPTV system administrators is to correlate the data from the various NMSs in order to gain an accurate overall view of network performance. Additionally, administrators need to also be able to support the variety of hardware platforms that run the NMS software packages in the IPTV data center. An IPTV management system normally uses Simple Network Management Protocol (SNMP) to control and monitor devices connected to the network.

12.2.1   Using SNMP To Manage an IPTV Network

SNMP operates over the TCP/IP communication protocol stack. Most if not all of the IPTV network components are compliant with the SNMP standard. The purpose of the protocol is to allow managers monitor and retrieve status information about various types of networking components including routers, network switches, and servers.

SNMP is an application layer protocol, which is used by the management software to retrieve data from the various network devices. The types of data gathered from an IPTV network falls into three distinct categories:

(1) *Alerts or alarms*—Alerts will inform the engineering personnel at the IPTV data center about unusual activity on the network. These alerts are used by technical staff to identify the source of a service failure.

(2) *Current status*—Data on the current status of the network is collected on a regular basis to determine whether the various IPTV components are operational or not.

(3) *Statistics*—Network statistics such as network traffic levels are collected by the management system on a regular basis. IP based networks can perform

well or badly, and it is important to be able to recognize when performance is below par. The statistical data gathered by the management system is used by technical personnel to optimize IPTV network performance. An optimized network improves the delivery speed for IPTV and VoD services.

A typical NMS comprises of an application and a database stored on a server and uses serial connections, closed contacts, or ethernet to communicate with the following IPTV network devices:

- IPTV servers
- IPTV data center video processing equipment
- IPTVCDs
- IP Routers
- DSLAMs
- CMTSs

Each of the devices is preloaded with a software module called an SNMP management agent, whose main function is to provide information about the IPTV device to the server application. This information is stored within an embedded database resident in the device called an Management Information Base (MIB). The information in a MIB is standardized so that various types of network management systems can use it. A simplified block diagram depicting the interoperability of an IPTV NMS with an SNMP agent is depicted in Fig. 12.1.

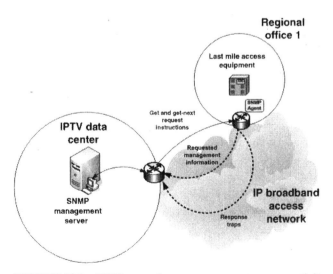

FIGURE 12.1   IPTV network management system connectivity

In the example illustrated above, communication between the server and agents is based on a number of commands. Once the *get* and *get next* instructions are received from the IPTV NMS, the agent performs the instruction as requested and sends back the required information. In addition to gathering and returning requested information, the SNMP agent software is also responsible for sending alerts without receiving specific instructions from the NMS. Using alerts called "traps" an SNMP agent can inform the NMS of events or faults that occur at the router. This allows IPTV administrators to proactively manage network devices.

12.3   MANAGING INSTALLATION, SERVICE PROBLEMS, AND TERMINATIONS

Because of the complexity and choice of products available on IPTV networks, the installation, management of service problems, and in some occasions the termination of services can be a particularly challenging and expensive process for network operators around the globe. The following sections give a brief overview of the business processes and work flows adopted by a number of service providers to deal with these functions.

12.3.1   Installing and Provisioning New IPTV Services

The addition of a new IPTV subscriber to an IP broadband network generally involves the following:

(1) Subsequent to a successful sales and marketing campaign, a potential customer calls or e-mails the service provider to request a subscription to a particular IPTV service.

(2) The order is taken by a customer service representative (CSR), details are added to the OBSS, and an installation date is agreed with the new customer.

(3) The details are automatically sent to the operations department who in turn sends a field service technician to the customer's home to install the IP set-top box and in some cases an RG.

(4) At the subscriber's home the technician un-boxes the IP set-top box and installs the device. Most IPTV installs will also require some in-home cabling because the modem and the TV are rarely colocated together.

(5) Once installation is done, the customer needs to be authorized to view the requested services. This can involve the technician calling a customer care assistant and requesting authorization.

(6) Once authorization is complete the installer carries out some or all of the following tests:

- Verifying IP connectivity with the servers at the IPTV data center
- Checking video signal levels
- Testing incoming stream quality.

- Ensuring that the addition of the new IPTV service has not had any adverse affects on existing IP based services.

(7) Once tested the video channels are viewed by the technician and the subscriber to ensure they are operating correctly and within predefined quality levels.

(8) An installation update record is completed either manually or online and contains some or all of the following information items:

- Subscriber details
- Tracking details
- Provisioning activities completed
- Time that the installation was completed
- Specific data that was used to provision the IPTVCD

Note that on rare occasions the installation may not go according to plan. As a result, the installer will typically fill-out an error report that describes the reasons behind the failed install and a date for the next attempt is agreed with the subscriber.

(9) Once the IPTV service has been installed to the satisfaction of the subscriber, an invoice or bill is generated.

The process described above can add significantly to the work load of field service technicians who are already involved on a daily basis with the installation of phone services and high speed data connections. In an effort to simplify the above subscription process, some IPTV providers implement automated provisioning systems for preplanning installations and adhering to a particular workflow. For instance, it may be possible to identify any network issues prior to dispatching a technician to the subscriber's premises. The sections below describe the functionality of modern day self-provisioning systems.

12.3.2 Automating the Provision of IPTV Services

The implementation of an automated provisioning system needs to be addressed from two different perspectives. To start with, consumers require a straightforward procedure in place that allows them to activate and deactivate IPTV services at their own convenience. To meet this need, many of the IPTV deployments include a subscriber self-management and provisioning application, which enables subscribers to use their TV interface to interactively:

- Manage profiles and subscription details;
- Check and pay their bill;
- View the types of IPTV services available and associated costs;
- Upgrade their programming package.

Note that a personal identification number is typically used to log onto the self-service interactive TV portal site.

Service providers also need to make sure that they have the infrastructure in place to support the provisioning system. Once a selection is made to activate or deactivate a particular product, the instruction is sent back to the backend systems. At the backend a number of tasks take place:

(1) The billing system is notified of the service upgrade. The mechanism of billing for IPTV services has a number of requirements that are not typically used when billing for telephone and broadband access services. For instance, the system needs to support a pay-per-use model for subscribers who decide to purchase on-demand content. Additionally, these systems need to support the charging for services in real-time, which again is a new feature not supported by billing systems used in traditional pay TV operations.

(2) Once the billing system has been alerted the subscriber can be charged for the new service.

(3) Some of the modern systems allow the transmission of a "getting started e-mail." If this facility is not available, then a courtesy call is made to inform the end user of the status of their upgrade request.

12.3.3 Processing and Managing Service Problem Requests

The types of problems occurring on an IPTV network that impact the end-users Quality of Experience (QoE) may include but are not limited to the following:

- Full or partial multicast channel outages at the subscriber's household;
- Problems with viewing IP VoD assets;
- EPG not operating correctly;
- Interference experienced while making a VoIP call;
- Picture distortion and synchronization issues;
- Power outages;
- A reduction in the quality of the service as perceived by the subscriber;
- Difficulties with network communications inside the home;
- Issues with breaches of DRM and CA security;
- Problems with ingesting source content at the IPTV data center.

Once any of these problems have been reported by the subscriber or by the internal monitoring systems at the data center, it is critical that they are resolved immediately. The resolution process typically starts with the generation of a trouble ticket that contains the following items of information:

- An identification number for the report.
- The date and time that the problem was detected.

- Level of severity associated with the problem. For instance, a network link breakdown that provides services to hundreds of subscribers receives a higher severity rating when compared to a problem, which is isolated to a single IPTV subscriber.
- Target timescale to restore service to normal operation.

The ticket is typically channeled to another department—operations or engineering and is subsequently used to diagnose and resolve the problem. On resolution the trouble ticket is updated with the following items of information:

- Identified cause of the service affecting problem
- Date and time that normal operational service was restored
- Equipment and time required to resolve the problem

Once the trouble ticket is completed the network engineers and managers implement appropriate escalation procedures to ensure that a reoccurrence of the same problem at a future date is unlikely.

12.3.4   Terminating IPTV Services

In addition to defining clear procedures for installing and provisioning new IPTV services, it is also critical that service providers have systems to support the termination of services. A flow chart outlining some of the possible steps associated with this activity is shown in Fig. 12.2 and explained in the following sections.

(1) The subscriber calls the CSR department and requests termination of the service.
(2) A termination request record is generated by the OBSS and sent to the operations department for execution.
(3) The termination is implemented either remotely or via a truck roll. This will often depend on whether the subscriber plans to terminate the entire suite of triple-play services or only access to the IPTV channels.
(4) The IPTVCD needs to be recovered.
(5) The database record on the OBSS is updated to reflect the fact that the service has been shutdown effective from date of termination.
(6) Various items of documentation are generated by the OBSS including a final billing statement.

12.4   NETWORK TESTING AND MONITORING

An IPTV data center system is made up of many complex components. As these components handle more and more services, network problems must be quickly detected and resolved. To maximize system uptime and ensure the services

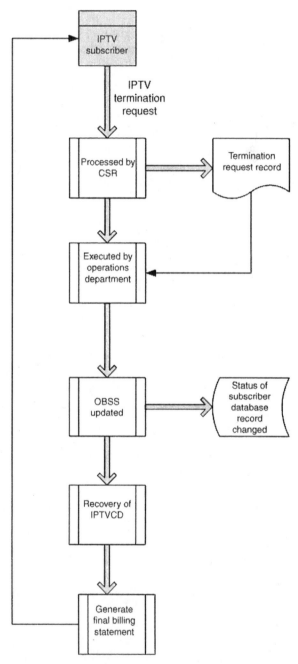

FIGURE 12.2   Termination of IPTV services flow chart

delivered to customers are of a very high quality, regular testing and monitoring needs to be performed on the IP based networking infrastructure. Close monitoring of an IPTV network has a number of benefits:

- It allows network managers to identify network traffic patterns that indicate that an outage or degradation in image quality is imminent. Early identification of a potential issue reduces the possibility of a major outage or sharp degradation in video quality occurring.
- It helps to immediately pinpoint a network outage.
- It gives the engineering team an estimate of the extent of a network problem when it occurs. The severity level of the problem will then dictate the resources that are used to address the problem. For instance, if the problem is a slight degradation in video signal quality, which results in pixilation for a second every hour then the strategy for fixing the problem will be different to a situation where the IP-VoD server fails at the IPTV data center.
- It can help to reduce subscriber churn because IPTV end users will not tolerate frequent outages that go undetected and take too long to fix.

To ensure that IPTV end users are experiencing a high quality viewing experience, it is possible to conduct a number of tests on the following IPTV network components:

The last mile equipment—A suite of tests is typically conducted to verify that the network hardware and software components installed closest to the IPTV end user are performing as expected and are optimized for delivery of both IPTV unicast and multicast services. Typical tests carried out on these components include:

- Checking the rate of data transfer across the last mile loop.
- Evaluation of errors and performance issues that might occur at the IP, transport and video packetizing layers of the IPTVCM.
- Verifying that the IGMP messages are being handled and processed in a prompt manner.

IP Network core—Tests carried out on the core network are typically conducted to verify that all hardware and software components used are performing as expected and are optimized for delivery of IPTV services. The parameters of a network configuration can often change, which can have an affect on the delivery of IPTV streams. Thus, it is important that IPTV network managers periodically test performance levels of streams when changes occur on the network. Changes can range from the addition of a switch to the network to a modification of a devices firmware. Other tests carried out on the core IP network segment can include the following:

- Checking to see if the network can support the delivery of multiple multicast channels.
- Determine if any packet loss is occurring.
- Measurement of latency introduced by the IP core network.

IPTV data center equipment—The various pieces of equipment installed at the IPTV data center need to be fully tested to ensure that it is capable of delivering multiple IPTV services to a large number of IPTVCDs in a secure and efficient manner. A test plan needs to also be regularly implemented on content that is supplied by external third-party providers to ensure that quality levels comply with agreed parameters.

IPTVCD—It is essential to know that the various IPTVCDs used to provide end users with access to the service are capable of handling a range of different types of IP services. The types of tests carried out on IPTVCDs can vary from verifying that the channel requested by the viewer's remote control is the actual channel received to ensuring that the channel change time is low when a large number of subscribers are connected to the IPTV streaming servers. Other more specific tests can include the following:

- Measurement of latencies associated with IGMP joins.
- Measurement of latencies associated with IGMP leaves.
- Measuring network jitter.

Note that these IPTVCD tests can be simulated or implemented using an embedded piece of client software.

The implementation of a comprehensive testing program helps service providers to optimize their networks for delivery of IPTV content.

12.5 MANAGING REDUNDANCY AND ENSURING SERVICE AVAILABILITY

IPTV has high availability and reliability requirements. To prevent periods of service disruption, the network and IPTV data center equipment need to support a rapid switchover from a device that fails to a separate backup device. The time taken for the switchover should be in the milliseconds to ensure that the fault is not service affecting. Additional levels of hardware and software redundancy are sometimes employed by service providers to minimize the affect that outages have on IPTV end users.

Figure 12.3 illustrates a network design that supports an advanced level of redundancy at the IPTV data center as well as the network itself. As depicted the network comprises a number of redundancy features.

- A symmetric ring topology is used to carry the IPTV traffic at the core of the network. So when a link goes down anywhere in the ring the IPTV traffic gets rerouted around the ring in the opposite direction.

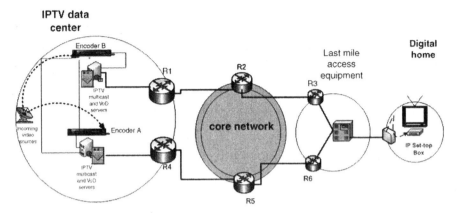

FIGURE 12.3 Network design that supports redundancy

- Various video infrastructure components are duplicated. In this example the VoD servers, the IPTV multicast streaming servers, and the encoders are all duplicated.
- A redundant distribution router is also installed to ensure that a router is available to take over the load if one of the devices fails.
- Each encoder features a dual-IP output.
- Some redundant links are also available between the distribution router and the edge routers who are serving IPTV to a large number of end users.

It is important to note that many of the individual IPTV network components already come with in-built redundancy features, such as dual processors and multiple failover hard disks.

The levels of redundancy built into the design of an IPTV network will often depend on the types of IPTV services that are deployed across the network. The delivery of live broadcast content for instance has an extremely high availability requirement, whereas IP-VoD has a lesser availability requirement because subscribers can rewind back to the point of disruption when the network problem has been resolved. The level and sophistication of the redundant and backup routing paths is typically the responsibility of the engineering team managing the IPTV networking infrastructure.

The scheduling of preventative maintenance on different segments of the network is another technique commonly used to maximize uptime of an IPTV networking infrastructure. Sophisticated maintenance systems are often employed by service providers to track daily and weekly maintenance activities. In addition to tracking maintenance activities, these systems also maintain details of spares that are available to engineering staff and employ advanced reporting systems that allow IPTV network managers to analyze and plan future maintenance activities.

12.6   IP ADDRESS SPACE MANAGEMENT

The allocation of IP addresses will always align with how the network is designed to carry services. For instance, in a triple-play environment it is possible to configure a different IP subnet for each service—voice, Internet access, and IPTV. It is also possible to design an IP addressing scheme that supports the sharing of common backend infrastructural components such as DNS management, OBSS, and DHCP servers for all three services. Once the IP addressing scheme is configured on the shared backend components, it is then possible to configure a separate IP subnet or pool of addresses for network resources, which are specific to the delivery of each triple-play service. For instance, the VoD server and the IP set-top box at the subscriber's home are both configured as part of the IPTV service subnet. The allocation of IP addresses on an IPTV network is the responsibility of DHCP servers. DHCP offers IPTV network managers many advantages:

- By using DHCP, all the necessary configuration details for the IPTVCD are fixed without the end user having to know about the details.
- DHCP offers centralized management of IP addresses.
- The setting up of a DHCP server to automatically allocate addresses to devices on an IPTV network is generally straightforward.
- It is possible to specify a range of available IP addresses for specific geographical areas.
- It is possible to integrate the DHCP server with the OBSS to support the provisioning of new IPTV services.

12.7   ROUTINE IT AND NETWORK ADMINISTRATIVE TASKS

The expectations of IPTV subscribers are very high. To meet these demands, the IPTV backend servers and networking infrastructure need to run smoothly. Thus, support personnel need to perform a variety of maintenance and administration tasks in the following areas.

12.7.1   Backups and Restoration

One of the most important jobs of managing backend IPTV servers is to maintain backups. Having a proper backup system allows IPTV service providers to restore critical content and data in the event of a disaster at the data center.

12.7.2   Monitor Disk Space Usage

The storage of compressed video places huge storage demands on IPTV servers. The storage space needs to be monitored closely; otherwise, the hard disk space on the IPTV content servers will become depleted.

12.7.3 Networking Infrastructure Optimization

The unpredictable traffic volumes and high bandwidth requirements associated with transporting content within the data center can compromise the performance of mission-critical IPTV servers and middleware applications. By optimizing the performance of the networking infrastructure, the servers will be able to interact with IPTVCDs in a speedy and efficient manner, which in turn keeps customer satisfaction high.

12.7.4 Securing Internal IT Systems

Maintaining the security of the IPTV data center is critical. In addition to enforcement of a security policy, IPTV network managers need to identify any unknown vulnerabilities or weaknesses in their networking infrastructure. Viruses, spyware, and malware all pose a serious threat to the integrity of an end-to-end IPTV infrastructure.

12.7.5 Application Management

Each IPTV application will generally include a software tool, which is used for administration purposes.

12.8 MANAGING IPTV QOS REQUIREMENTS

Delivering a video signal that is comparable or improved on the video signals delivered by existing pay TV providers is one of the main prerequisites for delivering IPTV services. For a service provider to achieve this level of viewing experience, IPTV network managers need to ensure that the network supports a QoS system. Deploying a QoS system ensures a high quality customer experience. The QoS for an IPTV network defines the network resources and parameters required to ensure that the IPTV streams travel from the source streaming server to the destination client IPTVCD in an unhindered manner. In other words, the quality of the stream does not degrade while in transit across the network. Before describing a QoS infrastructure, it is first helpful to understand how IP video packets are routed across the network. The following section gives a brief overview of the different traffic patterns of IP Video traffic and how these varying patterns are processed by intermediate devices such as network routers and switches.

12.8.1 Routing and Queuing of IPTV Packets

When an IP packet gets transmitted onto the IPTV distribution network, it needs to pass through a number of network components such as routers, switches, DSLAMs, and other types of devices that are specific to the transport technology used on the network. The "passing through" of a packet, typically means that the packet enters the device on a specific interface, gets stored in a memory buffer is copied out of the buffer, processed by the network device, and sent onward to the next network device on the routing path via a particular output interface. Of these activities, the order in which this packet gets processed is sometimes referred to as "queuing." The queuing of packetized data is an issue that has direct impact on the levels of packet loss and jitter. Since IPTV network traffic is particularly sensitive to packet loss and jitter the issue of queuing needs to be examined in closer detail. At its most basic level, the queuing of IPTV packets involves a router or a switch storing packets of IP data for a very short period, while the network device is busy processing other IP packets. Once these IP packets are processed the network device selects the next packet in the buffer queue. The number of queues supported by routers will vary according to the number of services that are supported across the network. The approach to selecting the next packet in the queue often depends on a QoS priority system, where certain packet types are selected for immediate processing, where others are left to wait in the queue. So in a triple-play environment, IP packets that are tagged as containing voice or video content will often receive a higher priority rating in the queue and are therefore processed before other types of less delay sensitive IP traffic, such as packets that contain Web or e-mail content.

The process of queuing packets can be illustrated by the following example. Suppose a broadcast IP streaming server is sending an H.264 compressed IPTV multicast channel onto a distribution network at a constant bit rate (CBR) of 2.5 Mbps. Some of the potential network traffic patterns between the streaming server and the RG located at an IPTV end users home are illustrated in Fig. 12.4.

The first scenario illustrates the situation when network conditions are ideal and the packets are separated by a time difference of 1.5 ms. In this environment the intrapacket gap stays the same as it traverses each of the network hops on route to the RG. This scenario is unlikely to occur across a live network. In the second scenario, two of the IPTV packets have been injected onto the distribution network with no time gap and the third packet is 3.5 ms behind the other pair of packets. Although no queues are occurring, the injection of another a pair of packets back-to-back results in the formation of a mini-queue. The third scenario is more representative of what occurs across a network that is used to transport triple-play IP services. In this final scenario, the VoIP, IPTV, and Web servers have simultaneously injected IP packets into the network. Once on the network the intrapacket delays of all three streams are more or less the same as they arrive on three different router interfaces at the same time. The router is not able to process three streams simultaneously so a queue is formed and packets need to wait a finite period of time in memory before being forwarded to the correct output port. In this example the queue overflows and a packet storing Internet access content is discarded. Meanwhile, the VoIP and the IPTV traffic receive preferential treatment in the queue and the intrapacket time gap remains the same at the output of the router.

This is a very simplistic model of the types of network patterns that could develop across an IPTV network. Given the sheer variety in the network traffic patterns that differ from component to component and between traffic types, it is

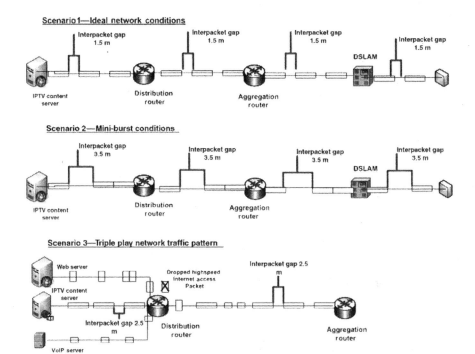

FIGURE 12.4 Potential IPTV traffic patterns

critical that a QoS system is applied to the networking infrastructure to ensure that packet queuing is properly managed. Methods called Differentiated Services Architecture (DSA) and MPLS-DiffServ are becoming increasingly common as a means of improving the delivery performance of both time-sensitive unicast and multicast IPTV streams.

12.8.1.1 Understanding DSA IP networks were originally designed for a best effort approach to carrying data. This by inference means that the correct sequence or even the actual delivery of an IP packet is not guaranteed. To further complicate the delivery of IPTV traffic, the mechanism used by IP networks to control network traffic during times of congestion involves discarding IP packets. Finally, IP treats all network traffic the same, so equal priority of passage through the network is given to both real-time traffic and delay-tolerant traffic. This would result in losses and delays in the arrival of IPTV packets at the IPTVCD. To overcome packet delays, most of the IPTV networks deployed today provide enough bandwidth capacity to ensure that a maximum QoS level is met and queuing rarely occurs. There are however some networks that do not have sufficient network resources to support the addition of IPTV services to their current mix of VoIP and high speed Internet access products. As a result, a QoS mechanism is required to prioritize IPTV packets over other types of IP packets during transmission across the network. This occurs through the classification of IP traffic into different groups. The provision of QoS over an IPTV network is generally implemented in compliance with an IETF standard called differentiated services architecture (DSA). The differentiated services architecture, also known as DiffServ, is used by some IPTV providers to manage and guarantee a particular level of QoS for subscribers. Implementing the DSA basically means that IPTV traffic is given higher priority over other types of IP based traffic. The specification for DiffServ was published in 1998 by the IETF and can be found in RFC 2475. The DSA comprises of a number of network devices that are all configured with similar QoS policies and are combined into a logical unit called a domain. DSA requires that any device connected to a physical link that may experience network congestion must be able to implement DiffServ functionality. Thus, routers, network switches, DSLAM, (CMTSs in the case of cable broadband networks), and residential gateways typically handle the implementation of DiffServ on an IPTV network. As previously described, DiffServ allows IPTV service providers to give precedence to some types of IP traffic. This has the affect of dividing the network traffic into different classes. For instance, in a triple-play environment real-time multicast traffic, which requires an uninterrupted flow of data might get precedence over other kinds of traffic, which can handle instances of interrupted data flows. This prioritization process will only generally get activated in times of network congestion (Fig. 12.5).

Consider a simple example of an IP-VoD stream and a multicast stream simultaneously arriving at the interface of R1 and destined for the same output port during a period of network congestion. If no QoS mechanism was applied to the router, then during times of congestion the multicast packets get stuck in the queue resulting in dropped packets and delays will be introduced into the delivery of a "live" multicast channel. This is a major issue for multicast TV as these types of streams are very sensitive to dropped packets. By applying DiffServ in the second sample configuration, the router is able to prioritize and move the multicast IPTV packets ahead, while leaving the IP-VoD packets in the queue until the congestion has cleared. In this case an IP-VoD packet gets dropped. Although the quality of the IP-VoD service is affected, the end user can however issue a command to rewind the video title to a point prior to where the signal quality started to degrade.

DiffServ operates at the IP layer of the IPTVCM and does not interact with the lower layers of the protocol stack. DiffServ enabled routers on an IPTV network classify IPTV packets by marking them with a six-bit field, known as the Differentiated Services Code Point (DSCP). The DSCP replaces the type of service (ToS) field of a standard IP header. The purpose of this field is to instruct routers and network switches along the transmission path on how to deal with the packet while it is waiting in a queue for forwarding. In other words, it defines the behavior of that packet at a router from a QoS perspective. By using the six bits the number of possible "forwarding treatment" approaches that can be applied to any IPTV packet is 64. The different "forwarding treatment" options are also called per-hop behavior or PHBs. Each PHB provides a particular level of service. Of the 64 possible options the IETF has defined a set of 14 PHB values that are summarized in Table 12.1.

To summarize, the DiffServ architecture is a useful mechanism for network service providers to enhance the performance of their IPTV networks. DiffServ

Sample network scenario with no diffserv

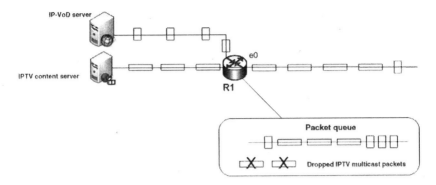

Sample network scenario with diffserv

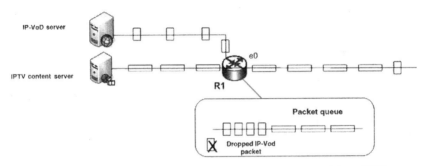

FIGURE 12.5 Simplistic example illustrating the benefit of applying DiffServ on congested links

typically operates in combination with other networking technologies such as MPLS.

12.8.1.2 MPLS-DiffServ QoS   As described in Chapter 2, MPLS is a traffic engineering system that boosts the efficiency of IP routing networks. By combining the many benefits of MPLS with the QoS guarantees of DiffServ, network operators are able to deploy services that require strict performance guarantees such as IPTV. The mechanisms used by MPLS-DiffServ QoS systems are defined in RFC 3270.

To summarize this section it helps to enforce QoS within the network, which ultimately translates into an improved QoE for end users. It is important to also note that effective implementation of an end-to-end QoS system requires adequate network resources.

12.9 MONITORING THE IPTV SUBSCRIBER EXPERIENCE

Using IP instead of traditional RF transport technologies to carry video over a network comes with its own set of unique challenges. Consumers who move over to

TABLE 12.1   Standard Set of PHBs as Defined by the IETF

PHB Type	Number Assigned	Description
Best effort (BE)	1	The IP traffic receives no special treatment at the router. This PHB is typically applied to internet access traffic. The implementation of this PHB occurs at the routers and guarantees a minimum amount of capacity to ensure that the Internet access traffic traverses the network. Full details of this PHB are available in RFC 2474.
Expedited forwarding (EF)	1	IP packet tags that are assigned this PHB experience minimal delays when been processed by a router. It offers IPTVCDs a connection to the server with low levels of latency, and jitter. Because low latency and jitter levels are particularly critical to a voice service, this PHB is generally used to support the deployment of VoIP services. Specialized algorithms are used by the routers to implement this PHB. Further details of this PHB are available in RFC 3246.
Assured forwarding (AF)	12	Latency and jitter are not such a big an issue for IPTV traffic as they are for voice services. However, IPTV does have a strong requirement of minimal packet loss. This is due to the fact that modern compression algorithms are able to "pack" a large amount of video into an IPTV packet. Thus, lost or corrupted packets have a serious impact on image quality displayed on the IPTV end users screen. An AF PHB is used to guarantee a certain amount of bandwidth to particular video streams. Guaranteeing bandwidth reduces the likelihood of lost packets, which in turn reduces the probability of noticeable degradations occurring in the quality of the video signal. Note the AF PHB is typically used by multicast IPTV and IP VoD type services. Other features of the AF PHB include the combination of four levels of forwarding assurance with three levels of packet drop probabilities. This yields an overall figure of 12 AFs. Full details of this PHB are available in RFC 2597.

IPTV services have high expectations and are quick to point out any glitches that are present on the system. IPTV end users have a number of basic requirements:

- They are used to viewing high quality video content, which has a high availability from the existing cable, satellite, and even terrestrial providers. There is no tolerance for poor quality video signals.
- They require access to a broad portfolio of video services ranging from multichannel TV to real-time access to video on demand titles.
- The system must be easy to use and accessible to all family members.

Therefore, ensuring that the subscriber is completely satisfied with a new IPTV service is the number one challenge for service providers. One of the core requirements of achieving high satisfaction levels is to implement a QoE measurement system that will closely monitor and benchmark the perception that an IPTV end user has of the service. The term QoE refers to the experience associated with watching an IPTV service. This not only relates to picture quality but can also cover other areas such as responsiveness and usability. This contrasts with QoS where the measurements are purely based on networking parameters such as jitter, packet loss, and delays. Before discussing the various QoE measurement models, it is first helpful to understand the factors that affect the delivery of IPTV Services. The following section gives a brief overview of the factors that negatively impact the quality of the viewer's experience.

12.9.1 Factors That Affect the Delivery of IPTV Services

The visual and audible quality of IPTV services has to be as good, if not better, than traditional RF based TV services. There are 11 key areas that can impact IPTV quality, namely, poor quality source content, encoding used at the IPTV data center, GOP length, packet corruption, packets arriving out of order, dropped IP packets, latency, video jitter, competition with other triple-play services, incorrect configuration parameters, and server congestion. These problems are discussed in the following sections.

12.9.1.1 Poor Quality Source Content   Poor quality source content reduces the efficiency of the coding process and has a negative impact on the quality of the content delivered to the end user.

12.9.1.2 Encoding   The mechanism used to encode the source video material can have a major impact on the viewing quality of the signal delivered to the IPTV end user. IPTV service providers typically use MPEG-2, H.264/AVC, or VC-1 encoding systems to compress content prior to forwarding over the IP networking infrastructure. All of these systems can compress content at various rates. Highly compressed video signals for instance will reduce the amount of network bandwidth required to carry the content across the network; however, the perceived picture quality may not be of a very high standard. The types of image distortion introduced by encoding can include

Tiling—IPTV video tiling refers to the rendering of a video image that includes sections of the picture that are in the wrong locations when compared to the positioning contained in the source video image.

Distortion of video blocks—A block is a portion of an MPEG video image that consists of a matrix of pixels. During encoding an algorithm is applied to this matrix that in some cases can distort the quality of the video block when this occurs the block will not accurately represent the content of the source. Because a number of blocks are required to build a complete picture, the visual affect of video block distortion is limited if a small number of blocks are involved. However, it does become an issue if the encoding process starts to distort the video quality of a large number of blocks.

Quantization noise—During the digitization of a video signal some noise errors can be introduced as part of the process. Large amounts of quanti-zation noise is evidenced by a "snowy" affect that is visible in the video sequence, when played back at the IPTVCD.

Image jerkiness—An IPTV video stream consists of a sequence of images that are shown on a TV or a display panel at a fast rate. To the end user the rapid change in image displays goes unnoticed and the stream is interpreted as "motion." If however the rate at which the sequence of pictures slows down to a particular rate then the stream is perceived as a series of still or jumpy images rather than a smooth motion of video content. Jerkiness is typically introduced by encoders who do not compress the content into sufficiently high enough frame rates. This may be a necessity in some cases, where bandwidth availability is an issue and the content needs to be encoded at a very low frame rate.

Retention of screen objects—This distortion occurs when a particular object or section of the screen remains when the video sequence has changed and moved forward.

Having outlined the various distortions that could occur at the encoding phase it becomes obvious that in some situations IPTV managers need to make the trade-off between compression levels and network resources available to them.

12.9.1.3 GOP Length In addition to issues relating to compression, the structure of the GOPs used in an IPTV stream also has a bearing on video quality. For instance, shorter GOPs provide for higher quality levels but suffer from a reduced compression ratio. It is the responsibility of the IPTV engineering departments to optimize GOP lengths in order to maximize signal quality for IPTV end users.

12.9.1.4 Corrupted Packets Corruption of packets is another factor that contributes to the distortion of IPTV signals. Corruption generally occurs during the transmission process and involves the modification of the original IPTV payload or header data contained within the overall packet. Impulse electromagnetic noise is one of the main causes of packet corruption within an IPTV environment. Error techniques at the lower layers of the IPTVCM are generally used to handle corrupted IPTV packets. As such the probability of corrupted packets reaching the IPTVCD decoder is typically quite low.

12.9.1.5 Packets Arriving Out of Order   Packets arriving in the wrong order and misrouted IP packets at the IPTVCD are due to network impairments. Packet buffering is one technique that is often used to address the issue of packets arriving out of sequence. Buffering involves the allocation of a certain amount of IPTVCD memory, usually DRAM, to temporarily store incoming IP packets. If packet difficulties are occurring on the access network, the buffer will store a certain amount of data. The storage capacity of the buffer can range from microseconds to tens of seconds and is specific to each deployment. It is worth bearing in mind that buffers are also used by cable, satellite, and terrestrial set-top boxes; however, the buffer size of these devices is somewhat smaller because the arrival of MPEG packets is somewhat more predictable than packets of IP data. Once the misrouted or out-of-sync IP packets arrive, they are reinserted into the IPTV bit stream in the correct order. It is also worth noting that buffers can also be configured on the servers located at the IPTV data center to support the retransmission of IP packets. This whole process is dynamically completed in the background and reduces the likelihood of lip-sync and video blocking problems. Network protocols such as UDP can also add to the problem of packets arriving out of order because it does not support a mechanism for detecting out-of-order packets. Control protocols such as RTP and RTCP can however be used to identify issues associated with IP video packets arriving out of order.

12.9.1.6 Dropped and Lost IP Packets   Some packet loss is okay for applications such as Web browsing and e-mail applications and generally goes undetected by users. However, dropped packets over an IPTV network can degrade the overall viewing experience due to the highly compressed nature of the video content. Compression technologies like VC-1 and H.264/AVC are extremely sensitive to lost packets. For instance, packet loss during the streaming of an IP channel can result in graininess on the TV screen, frozen picture frames, and in some cases the viewer will see momentarily blanks of a second or more on the screen. The percentage of the image affected by the packet loss will largely depend on the number of pixels and the type of information contained inside the packet. In general, a small amount of packet loss can affect large sections of the IPTV video or audio stream.

   In very severe cases the connection between the video server and the IPTVCD will end. This is particularly true for IPTV content, which is transported by UDP because the packet is never resent and the content is gone. TCP is of no major benefit either because the time taken to request a new copy of the packet would probably result in a buffer underflow and associated freezing of the picture. This type of performance is not tolerated by consumers and the repercussions of this are

TABLE 12.2 Typical Sources of Packet Losses

Source of Packet Loss	Explanation
Electrical interference	Packet loss may occur due to electrical interference being introduced at various points of an end-to-end IPTV networking infrastructure. For example, IPTVCDs connected to the home network run off residential powerline outlets. These electrical sources are subject to various impairments including brownouts and surges in the power levels. This can introduce bit error problems, which can result in packet losses, while the device is operational.
signal-to-noise (SNR)	The SNR is another factor that has an affect on the number of IPTV packets that are lost or dropped during the transmission of an IPTV unicast or multicast stream. On FTTH networks this value is quite high, which means that the packet loss is very low. This is due to the fact that optical transmission techniques are used. The opposite is true for some DSL networks where the SNR is quite low, resulting in a larger percentage of packet losses. This is due to a number of factors ranging from corrosion of cable to long local loop distances.
Congestion on the network	Congestion is one of the most obvious sources of packet loss. It is normal for most IPTV networking infrastructures to experience some element of congestion during certain periods. As long as the time period of congestion is short and is not extended, then packet losses are typically quite low for well engineered networks.

all too obvious. The reasons behind packet loss on a network can range from the IPTVCD buffer been overwhelmed with traffic to signal problems on the network to large geographically dispersed networks that consist of numerous components where errors could be introduced. Table 12.2 summarizes some of the factors that can cause packet loss across an IPTV network.

Service providers generally use one or a mix of network management techniques to reduce the affects of dropped packets on an IPTV service. The following sections give a brief overview of five of the most popular techniques.

Using Retransmissions   As described in Chapter 3, one of the main functions of the IPTVCM transport layer is to ensure that video content is reliably delivered to its destination. The reliable delivery of content across a network in some cases requires the use of some type of feedback mechanism of detecting packets that have not arrived at their destination. There are three feedback variants that are relevant to IPTV networks, TCP, RTP, and RUTP retransmissions techniques.

USING ARQ WITH TCP   Although TCP is not particularly suitable to the delivery of live IPTV video content, its retransmission techniques are useful for unicasting content across a network. When used by a TCP connection, a suite of protocols called

Automatic Repeat Request (ARQ) are used to identify lost packets. Once the status of a packet is identified as lost it is subsequently retransmitted. The identification of lost packets involves the use of an acknowledgement system. The most basic system involves the used of an ARQ protocol called Idle RQ or Stop-and-Go. The principle of this system is that the IPTVCD acknowledges each packet. Once the source or IPTV delivery server receives the acknowledgement, also known as an ACK, it transmits the next packet across the network. If an acknowledgement is not received within a particular time period, then the source will assume that the packet is either lost or discarded and subsequently retransmits the packet. If the server detects that a large number of packets have been lost, it will assume that congestion is occurring and will reduce the packet transmission rate. Although the Idle RQ protocol is relatively straightforward, it is unusable within the context of an IPTV environment because of the long delays associated with waiting for acknowledgements. As a result, Idle RQ is rarely used for the delivery of real-time traffic; however, some implementations do use another type of ARQ protocol called Continuous RQ for particular types of nonreal-time IPTV applications. In a Continuous RQ protocol the server is allowed to send packets before receiving an acknowledgement from the previous packet sent. This is more appropriate to applications such as IPTV services that need to send multiple packets at a high rate. Although some ARQ error control schemes are unsuitable for multicast IPTV networking environments because retransmitting lost packets to multiple IPTVCDs consumes large quantities of bandwidth, the ITU and IETF are both developing versions of ARQ that address this issue. Note that the TCP retransmission technique is advantageous to use where network conditions are poor and in situations where subscribers are downloading video content to their IPTVCD.

USING RTP TO RECOVER FROM PACKET LOSS In addition to providing quality monitoring information, the RTP and RTCP protocols also provide support for recovering lost packets. Like the recovery technique used by TCP, the RTP mechanism also requires feedback from the receiving IPTVCD to identify lost packets. Under the RTP recovery mechanism, the IPTVCD needs to first detect any packet losses. This is achieved by identifying a gap or multiple gaps in the RTP packet numbering sequence. Once a missing packet has been identified, the details are sent back to the server via a negative acknowledgement (NACK) message. The format of an RTCP NACK is shown in Fig. 12.6 and described in Table 12.3.

A unicast RTCP connection is used to carry the NACK between the IPTVCD and the server back at the IPTV data center. Note that it is possible to configure the error feedback system to support multiple NACKs inside a single RTCP packet. Aggregating NACKs does cut down bandwidth requirements; however, it will increase the buffer requirements of the IPTVCD because the sending of NACKs is postponed until the RTCP packet has reached a certain limit. Once the NACK is received by the server, the missing packet is resent to the IPTVCD. It is critical that the missing packet arrives at the IPTVCD before its scheduled decoding time commences.

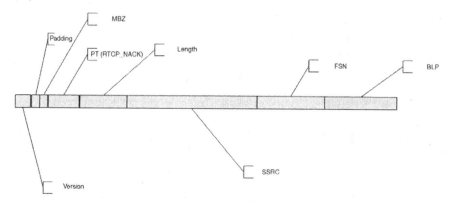

FIGURE 12.6   Format of an RTCP NACK packet

The immediacy of receiving information about lost packets is obviously critical in a multicast IPTV environment. As a result, the IETF are creating extensions to the RTCP feedback mechanism called Audio–Visual Profile with Feedback (AVPF). The purpose of this development is to improve the efficiencies and speed of feedback based recovery techniques for applications such as IPTV.

Note also that both retransmission techniques described above involve a certain amount of latency because of the round trip delay introduced from the following activities:

(1) Time taken to send an ACK or NACK from the IPTVCD to the server.

(2) Time taken for the server to process the request and respond to the request.

TABLE 12.3   Structure of a RTCP NACK Packet

Field Name	Description of Functionality
Version (V)	This 2-bit field identifies the version of RTP used in the RTCP NACK packet.
Padding (P)	This defines whether or not padding octets are present in the RTCP NACK packet.
MBZ	Reserved for future use
PT	This packet type field is set to the RTCP_NACK identifier 193.
Length	As the name suggests this field defines the length of the RTCP_NACK packet.
SSRC	The purpose of this 32 bit field is to identify the synchronization source on the IPTV network. This field is often used in conjunction with the sequence number field to rectify problems that may arise in the IPTV stream from time to time.
First Sequence Number (FSN):	The purpose of this 32 bit field is to identify the first sequence number.
Bitmask of following lost packets (BLP)	This field is set to a value of one if the corresponding IPTV packet has been lost. Otherwise the field is set at zero.

(3) Time taken to send the missing packet back to the IPTVCD.

As a result, the network needs to be engineered to cope with these delays when delivering IPTV services. Additionally, retransmission packet loss mitigation systems require additional bandwidth to carry the various ACK and NACK messages back to the source server. A NACK message for instance is typically between 50 and 60 bytes in length. On a large IPTV network with high packet loss levels the number of NACKs generated can result in high levels of bandwidth consumption. From an engineering perspective this can have an affect on the design of the network. For instance, the network path for ACKS and NACKs should be kept to a minimum. As a result, caching is typically used at regional offices to reduce the number of retransmission requests routed back to the server at the IPTV data center. Additionally, standard security mechanisms are also required to minimize the risks posed by hackers to the IPTV service provider's internal systems.

USING RUDP   Reliable UDP (RUDP) is another retransmission technique, which may be used by IPTV providers to detect video packets that have not arrived at IPTVCDs deployed in the field. Although RUDP was originally designed to transport telephone signals over an IP network, its flexibility provides support for the delivery of IP video content. RUDP is layered above layer 4 of the IPTVCM and just below the optional RTP layer. The basic method of RUDP operation is to send multiple copies of the same packet to the client device. This increases the likelihood that one of the packets will arrive at the desired destination and any additional packets are discarded. Although the bandwidth inefficiencies of RUDP are not desirable within an IPTV networking environment, some of its congestion control features can be effective in some instances.

Delay Masking Methods   Another technique used to tackle packet losses on an IPTV network is to conceal or mask the affects of packet losses from the IPTV end user. Delay masking methods are based on the principle of replacing lost packets or picture frames with estimated units. In other words, rather than requesting a retransmission of the original packet(s) the IPTVCD esti-mates the contents of the lost packet and replaces with a packet that has a high likelihood of been very similar to the original. The level of sophistication used by delay masking technologies varies between implementations. The number of delays due to packet loss that can be masked obviously has a finite limit before the IPTV end user notices the QoS problems with the incoming stream. As a result, this technique is generally used in conjunction with other packet loss mitigation systems such as retransmission systems and forward error correction (FEC).

Using Forward Error Correction (FEC) and Interleaving   An error control technique called FEC is often used in combination with interleaving to eliminate the affect of dropped IP packets and improve the end users TV viewing experience. FEC operates on the principle that an algorithm is used to generate error correction data, which allows an IPTVCD to detect lost packets without requesting a retransmission of data from the source server. The FEC error control technique reduces the number of retransmissions required to support IPTV services; however, the additional redundant data added to each packet increases the overall bandwidth requirements of the network. The use of FEC is typically applied at both the physical and upper layers of the IPTVCM. FEC techniques used at the physical layer provide error correction to individual bits using the Reed-Solomon schemes. This scheme was originally designed in the 1960s and works by adding extra bits to each IPTV packet before sending across the network. At the receiving end of the connection, the extra bits are used by the IPTVCD decoder to identify errors and recover the original bits that have been lost. FEC at the upper layers is typically applied to full packets and is based on the principle of sending additional "repair" packets in addition to the actual packets that are part of the baseline IPTV stream. These additional packets are used to replace original packets, which have been lost whilst traversing the IPTV distribution network. The amount of redundant packets sent to aid in the recreation of lost packets is defined by a parameter called "framing." The level of framing applied to a networking infrastructure is defined by the IPTV network manager. Full details of how FEC is used in an IP multicast environment is available in RFC 3452.

INTERLEAVING is a mechanism that is used to sends bits on to the network at different time intervals. This mechanism spreads the bits across a number of FEC blocks that means that in the event of a lost packet the impact is spread across several blocks.

Implement an ACM   For some IPTV operators it may not be feasible to continue to increase bandwidth to meet every possible peak requirement that may occur. Consider an example of a service provider who uses MPEG-2 to encode its entire program lineup of 120 SD channels and 20 HD channels. The bandwidth capacity required to support transmitting all the channels at the one time would be approximately 780 Mbps.

   Although this capacity may exist in the service providers core network, it would be unreasonable to expect this amount of capacity to be available across the downstream links used to connect the core backbone with regional DSLAMs or CMTSs. As a result an admission control mechanism (ACM) may be used to prevent the networking infrastructure from becoming overloaded. An ACM works by examining each IGMP request or RTSP message and determining if sufficient network bandwidth and resources are available to fulfill the request. If the network is incapable of handling the request and is overloaded, then the IPTVCD request is rejected by the network. As a result of the failed request, a blank screen is displayed on the TV or display panel. Therefore, it is critical that ACM limits are set at levels that minimize the probability of problems occurring when IPTV end users request a channel change. In the context of an IPTV environment, these levels define the number of streams that are authorized to traverse a particular link. Statistical analysis and operational data are the two most common methods used to determine stream limits in different parts the network.

Although an ACM reduces the possibility of congestion occurring on the network and ensures that capacity usage stays within a particular limit, it does impact the latency associated with channel changing. In addition to reducing the possibility of network congestion, an ACM can also be used to detect and minimize the impact that network link failures have on IPTV viewers.

Engineer the Bandwidth Requirements Correctly   A network that has enough bandwidth to carry IPTV traffic loads during peak demand times will dramatically reduce the probability of congestion occurring on the network. Unless well dimensioned and engineered the core network in particular is an area that remains susceptible to congestion because all traffic types get aggregated at this point. Congestion subsequently leads to packet loss and ultimately leads to a reduction in the QoE for the IPTV subscriber. Techniques such as defining a class of service and bandwidth allocation for each of the IP services are approaches used by service providers to reduce the possibility of congestion occurring in the core network.

It is important to note that each of the above packet loss mitigation systems each have their own strengths and weaknesses within the context of different networking conditions. Thus, the applicability of these various mechanisms will vary between networks. In fact, the use of different packet loss mitigation systems for different segments of an end-to-end networking infrastructure is a common occurrence.

12.9.1.7 Latency   Latency is a parameter that is used to measure the time taken for IP packets to travel from the IPTV server to the IPTVCD. In any IPTV deployment some delay or latency is expected because the video content has to be decoded by the IPTVCD. The latency figure does however vary from network to network. Low latency levels are critical to delivering good quality video content to IPTV end users. If the latency value is too high, then end users will start to see impairments such as corrupted images, picture blocking, and frozen frames on their TV displays. The causes of latency can vary from not having enough bandwidth to deliver IPTV channels to high bandwidth IP services increasing their usage levels beyond particular limits. Allocating additional bandwidth for the video traffic is one of the most common yet effective approaches for solving issues such as packet loss or transmission delay.

12.9.1.8 Video Jitter   IPTV services are particularly sensitive to delays caused by overloaded servers, routing, network congestion, and queuing as the IP video packets traverse the network. The quality of a video signal very much depends on the delivery of a lossless IP stream at a constant bit rate. The decoder in the IPTVCD requires a steady and dependable incoming IP stream. This is achieved through a sophisticated synchronization and clocking process that occurs between the decoder in the IPTVCD and the encoder at the IPTV data center. Any time variation associated with packets arriving too early or too late result in a behavior called jitter. Figure 12.7 depicts the impact that jitter can have on an IPTV stream.

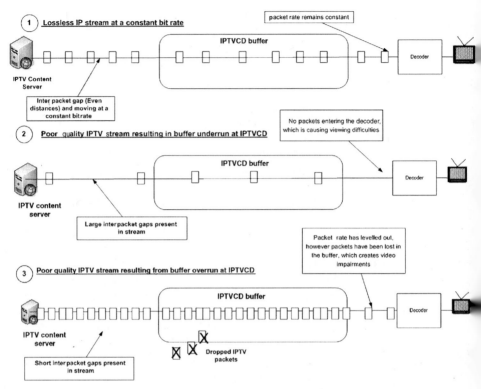

FIGURE 12.7   Different jitter patterns across an IPTV network

(1) This first stream pattern is considered to be ideal and will deliver a high quality TV signal to the end user. As shown, the packets are evenly spaced and the bit rate remains constant throughout the network path.

(2) As illustrated in the second pattern, picture viewing problems also occur if the rate of IPTV data flow is too slow. This is because the buffer does not fill fast enough and the feed to the decoder does not happen at a constant rate.

(3) In the third pattern illustrated the stream of IP packets is flowing too quickly. Thus, the IPTVCD buffer overflows and packets are lost. Once this starts to occur then jitter effects such as flicker on the TV screen and lockouts to the IPTVCD become obvious to the end user. This affect can be minimized or offset by using large memory buffers in the IPTVCD or increasing the bandwidth capacity on the broadband network. Although the use of larger buffers help to minimize jitter, it does however introduce delays because it takes extra time to fill the buffer full of packets. The size of the buffer varies between different types of IPTVCDs. For IPTV set-top boxes for instance the buffer sizes can store anything between 5 and 40 s of video before forwarding to the decoder.

The jitter on an efficient IPTV network is generally measured in terms of milliseconds.

12.9.1.9 Competition with Other Triple-Play Services Video is one element of a triple play of IP services on offer from many operators nowadays. Simultaneous demands by voice, video, and high speed data applications for networking resources can cause quality issues for IPTV services. To resolve these issues, IPTV providers can prioritize the delivery of video traffic to prevent delays and distortion of the IPTV stream.

12.9.1.10 Incorrect Configuration Parameters There are a number of configurations that need to be implemented when designing a network to support the transmission of IPTV services. The I and general frame rate are two characteristics that require a special mention due to their potential impact on QoS.

I Frame Frequency Rate The video transmission parameters chosen can have a bearing on the quality of the signal. If, for example, an IPTV stream uses a large number of P and B frames, then the amount of bandwidth required to transport the stream is reduced. Although the bandwidth requirements are reduced this type of configuration could affect the quality of the signal if an error occurs on the network. So if data is lost, then the IPTVCD must wait until the next I frame becomes available in the IPTV video sequence. If a large number of P and B frames are contained in the stream, then the waiting time period increases, which results in problems such as picture freezing. It is possible to mitigate against this issue by increasing the rate that I frames are sent across the network. Increasing the frequency of I frames in the incoming stream allows the IPTVCD to improve the time taken to recover from problems such as packet losses. Note, as a rule of thumb the insertion of an I frame every 10–15 frames should provide adequate performance in terms of dealing with error corruption.

General Frame Rate Reducing the frame rate of the video content within the IPTV stream reduces the bandwidth requirements. However, this configuration approach could result in video signal degradation, particularly for fast changing motion scenes.

12.9.1.11 Server Congestion IP-VoD and multicast IPTV streaming servers can also have a negative impact on QoS if the hardware is not sized correctly to meet the demands of end users at peak times. The main symptom of server congestion is frozen frames occurring during the viewing of a VoD title or a multicast channel.

12.9.2 Introduction to IPTV Video Signal QoS and QoE Measurement Systems

One of the objectives of QoE is to ensure that the main characteristics of the original source signal are recreated at the IPTVCD. Achieving this objective requires a strong understanding of how to detect and prevent problems that may occur in the delivery of IPTV services. As such QoE measurement models and metrics are often used to identify factors that could directly impact the end-users viewing experience. There are three primary models used by IPTV quality measurement systems to identify the presence of IPTV stream impairments, namely, full reference, zero reference, and partial reference.

12.9.2.1 Full Reference (FR) Based QoE Measurement Systems   This system makes a copy of the stream at the IPTVCD and compares it with a reference signal obtained from the source of the video content. Note the size of the signal varies between measurement equipment but is typically uncompressed and quite large. This measurement will determine the level of distortion and degradation that occurred during the encoding and transfer of the original video content across the network. Fig. 12.8 illustrates a basic reference topology used to compare the original signal with a signal post transmission across the IP distribution network.

The measurement system then uses an algorithm to compare both video signals in terms of a broad range of distortion parameters:

- Frame delay differences
- Blurriness caused by coding impairments
- Chrominance and brightness levels
- Image jerkiness
- Image blocking

The FR approach provides a very accurate assessment of video quality received at an IPTVCD because it compares both signals at a very detailed level including individual pixels.

12.9.2.2 Zero Reference (ZR) Based QoE Measurement Systems   There are also systems available that do not require a reference video signal to score the video quality of an IPTV stream. Instead, a sample compressed signal is obtained from the IPTVCD in real-time but not from the source. ZR systems are particularly

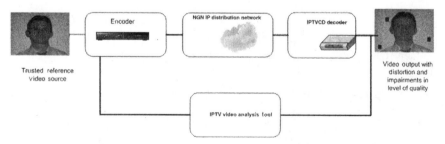

FIGURE 12.8   Full Reference (FR) based QoE measurement system

suited to measuring real-time IPTV streams because they analyze fewer factors compared to FR systems. A ZR measurement system focuses on analyzing video frame headers and monitoring a wide variety of key metrics including the number of PCR errors and of impaired I frames contained in the IPTV stream.

12.9.2.3 Partial Reference (PR) Based QoE Measurement Systems Similar to the approach used by FR systems, the PR measurement equipment is designed to take a sample at the source and also at the destination IPTVCD, compare the signals, and output a metric. PR systems require less computational complexities when compared to FR systems and use a smaller reference IPTV stream sample when comparing both signals.

Of the three types of models FR based systems are considered to be the most accurate for measuring IP video quality.

12.9.3 IPTV QoS and QoE Metrics

IPTV QoS and QoE metrics fall into two broad categories: subjective and objective metrics.

12.9.3.1 Subjective Centric Metrics This approach uses a group of participants to grade and assess picture quality. The environment and people used for subjective testing varies between service providers. For instance, some service providers may simply use experienced engineers to assess picture quality at the IPTV data center whereas other providers may conduct tests in a purpose built testing facility with a number of qualified video quality specialists. Typical steps associated with identifying a subjective metric for an IPTV service include the following:

(1) Identify a series of video sequences for the test.
(2) Select a number of configuration parameters.
(3) Set up the testing environment in compliance with the desired configuration parameters.
(4) Gather a number of people to take part in the test.
(5) Implement the test and analyze the results.

Formal testing environments generally follow an opinion based system defined by the ITU called Mean Opinion Score (MOS). This rating system is explained in the following section.

MOS   A QoE system needs to also take human factors associated with IPTV end users into account. The MOS system allows a sample number of people to assign a numerical value between 1 and 5 to the perceived quality of the viewing experience (see Table 12.4).

The MOS rating is subsequently calculated by averaging the results. There are four different MOS variants:

TABLE 12.4   MOS Scores Used to Measure IPTV Quality Levels

Perception of IPTV Channel	MOS Score
Excellent	5
Good	4
Fair	3
Poor	2
Bad	1

MOS-V—This metric scores the viewing quality of an IPTV stream.

MOA-A—The audio version of MOS is used to score the audio portion of an incoming IPTV stream

MOS-AV— As the name implies this metric is used to rate the overall A/V quality of an IPTV stream.

MOS-C—This metric is used to score the experience associated with interacting with the IPTV stream. Channel changing and usage of the EPG are examples of interactions, which are measured using this metric.

Industry initiatives and specifications such as the TR-126 from the DSL Forum recommend the use of MOS as a mechanism for determining video services QoE.

12.9.3.2   Objective Centric Metrics   Under this approach test equipment is used to measure video signal quality in terms of comparing original picture frames with the compressed versions and measuring the signal quality degradation. Commonly used objective centric IPTV QoS and QoE metrics include PSNR, MPQM, VQEG, and MDI. These IPTV QoE metrics are explored in the following subsections.

PSNR   In addition to identifying packet delivery problems and associated issues, it is also necessary to measure and analyze the actual quality of the content inside the packets after encoding and transmission across the network. A metric called peak signal-to-noise ratio (PSNR) is sometimes used to examine and rate the quality of this content as perceived by end users. It expresses the ratio between the power of a video signal and power generated by electromagnetic noise in terms of decibels. This information is used to identify impaired video frames and generate a score that represents the quality experienced by the end user of the IPTV service.

MPQM   MPQM (Moving Pictures Quality Metric) is used to assess the quality of MPEG compressed video streams. It includes technologies that replicate the experience of a human observer and rates the IPTV stream on a scale from 1 to 5. The quality rating scale for MPQM is described in Table 12.5.

**TABLE 12.5   MPQM Video Quality Rating
Scale**

Rating	Description
1	Excellent
2	Good
3	Fair
4	Poor
5	Bad

Note, MPQM considers both network and video compression issues when rating a particular IPTV stream.

ITU-T J.144   ITU-T J.144 provides a set of guidelines on objective centric video quality assessment techniques.

MDI   A measurement metric called the Media Delivery Index (MDI) is an industry standard defined in RFC 4445. It is used to measure the quality levels at various points on the network, namely, the data center, at various network routing and switching components, and at the IPTVCD. MDI is a scoring mechanism that indicates video quality levels and also identifies network components that are affecting the end-users QoE. This is achieved by measuring instances of jitter levels and packet loss occurring on different points of the IPTV network. So, for instance, if the MDI value at the input on a router is low and the value at the output port is high, then there is a strong indication that the router is having some difficulties that are adversely affecting stream quality. MDI is also used to diagnose network problems. The MDI metric is displayed as two values, namely a Delay Factor (DF) and a Media Loss Rate (MLR), which are separated by a colon.

A DELAY FACTOR   This value measures the amount of buffering time required to accommodate the jitter present in the network. The jitter value is derived by measuring the latency with regard to the rate of the stream. In other words, if the output rate of the streaming server is 2.4 Mbps and the stream is arriving at the IPTVCD at the exact same rate of 2.4 Mbps, then the jitter value in these ideal circumstances is zero. When the jitter value is zero, the IPTVCD decoder is processing the packets at a constant rate. A smooth graph illustrating the decoding of IPTV packets at a constant rate over a 100 ms time period is shown in Fig. 12.9.

A zero jitter value is a rare or nonexistent occurrence because each component on the stream path will contribute some element of jitter. Thus, a nonzero video jitter value occurs when the bit rate is affected by various network factors and variations of packet arrival times at the decoder. Thus buffers are used to accommodate the underflow or overflow of the stream. The graph in Fig. 12.10 illustrates the presence of jitter. In this graph the packets are flowing at a varying pace, which increases the risk of packet losses occurring at the IPTVCD.

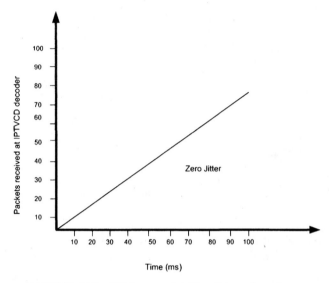

FIGURE 12.9   IPTV network with a jitter value of zero

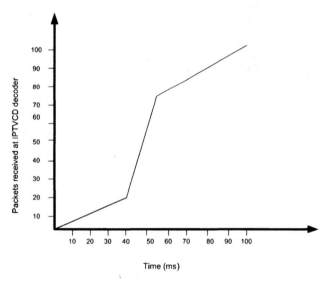

FIGURE 12.10   IPTV network with jitter present

It is also important to note that jitter does not actually damage the video content inside the IPTV packets. Thus, if the video quality is good at source then the content decoded at the IPTVCD will have the same quality levels. QoS test and measurement tools are generally used to calculate the DF value for a particular network device and involves the following steps:

- Once a packet arrives then measurement of the time interval between receipt in the buffer and the rate at which packets are taken out of the buffer is taken.

- For a particular time period the maximum and minimum intervals or deltas for these variations are measured.
- The result is divided by the IPTV stream rate.

Once the DF value is calculated, the IPTV administrator knows the amount of buffer space or delay required at each network component to accommodate the video jitter present on the network. So, for instance, if a particular IPTV stream has a DF value of 40 ms then any IPTVCD that needs to decode this stream will require a buffer, which will store at least 40 ms of the stream. Note that a high DF value is also a strong indicator of congestion occurring on parts of the network.

A MEDIA LOSS RATE   This second MDI metric defines the number of IPTV packets lost or dropped per second. In addition to lost packets, the MLR also provides administrators with a metric for the number of packets that are out of sequence arriving at the IPTVCD decoder. Any MLR value, which is above zero, can have an affect on the IPTV end users QoE.

The actual MDI combines both of these values and expresses the final QoS metric as DF:MLR. A much more accurate analysis of the systems QoS and queue performance levels begins, to emerge when the MDI values for the various elements of the IPTV infrastructure are combined together. Fig. 12.11 briefly illustrates the difference in MDI values calculated on a live IPTV multicast stream at different points along the networking path.

As shown, an MDI metric is taken at each output port along the network route from the data center to the RG. The MDI for the broadcast channel 1 at the server

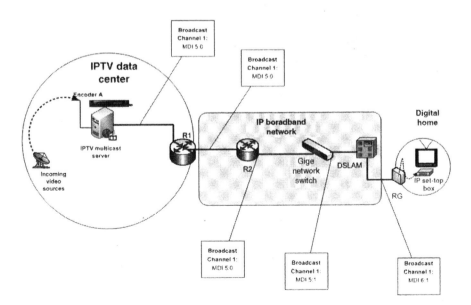

FIGURE 12.11   Sample MDI calculations of live MDI calculations

output is 5:0. Thus, server buffering needs to accommodate 5 ms of jitter delay. The packet loss remains at an acceptable level of zero. Moving further downstream, the MDI remains unchanged at R1 and R2; however, an increase in packet loss occurs at the network switch. This indicates that the network switch may have a problem dealing with the varying traffic patterns and loads. The solutions to this problem could range from upgrading the switch to increasing the bandwidth capacity of the incoming or outgoing links. This packet loss issue has a cascading affect and is also present on the link between the DSLAM and the RG.

The DF value or "burstiness" of the stream does however increase to a value of "6" on the final hop to the RG. This indicates that the DSLAM has introduced some minor jitter, which needs to be examined in greater detail. The most likely cause of an increase in the DF value is long queuing delays at the DSLAM. Thus, the use of MDI not only provides administrators with a QoS measurement parameter but also helps to isolate the network segment causing issues for the various IPTV streams. Once issues are identified fixes and adjustments can be made to improve the overall QoS of the end-to-end networking infrastructure.

12.9.4 Tools Required for Measuring and Analyzing QoS and QoE

Specialized test and measurement tools allow service providers carry out QoE and QoS tests to identify causes of poor digital image quality. A fully rounded end-to-end test and measurement system generally consists of a number of probes and a centralized server based system. Fig. 12.12 shows how these components might be deployed on a DSL based network.

The role of both components are discussed in the following sections.

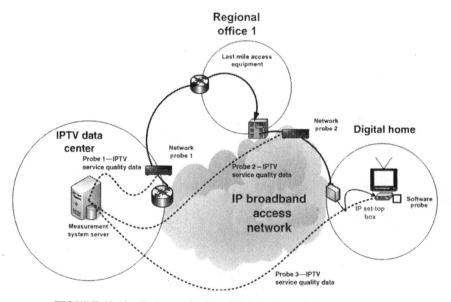

FIGURE 12.12 End-to-end test and measurement system components

12.9.4.1   Network Probes   The purpose of a probe is to monitor and continuously test key parameters of multiple standard definition, high definition, and on-demand IPTV channels. Results of the analysis are stored in a database that is subsequently forwarded to the server located at the IPTV data center. The analysis of an IPTV stream generally involves an in-depth examination of both video packet and network protocol layers. Both layers throw up different types of problems including

- *Network layer*—Out of order packets, packet loss, and jitter.
- *Video packet layer*—Problems with the video streams.

As shown in Fig. 12.12, the two hardware probes are set up at different ingresses and egresses points in the networking infrastructure. Typical locations for probes in DSL based networks are at the IPTV data center, at critical routers used to build the core network and at the edge of the network, namely, switches and routers in the regional office. In this example, a nonintrusive software probe is also gathering data in real time at the IPTVCD. These types of probes require a small amount of memory and processing resources to operate. It is important to note that portable probes may also be used instead of software probes to monitor QoS and QoE at the end-users home. Probes are often configured to send a notification in the form of an e-mail or text message to the network administrator once a particular predetermined network threshold level has been crossed that indicates IPTV quality is threatened or the signal quality levels have fallen below an acceptable level. A sample structure for these types of notification messages is shown in Fig. 12.13.

12.9.4.2   Server Application   The purpose of the server application is to correlate service quality real-time data such as alarms and threshold levels received from the various network probes. This information helps IPTV administrators to pinpoint causes of poor image quality. It presents results from various probes in an easy to understand interface giving the network administrator a clear indication of the IPTV end-user's viewing experience. In addition to alarm and performance details some probes can send back statistical information on IGMP Join and Leave transactions to the centralized server unit. Some of the advanced measurement systems also link in with existing NMSs. In addition to servers and probes, test devices called MPEG

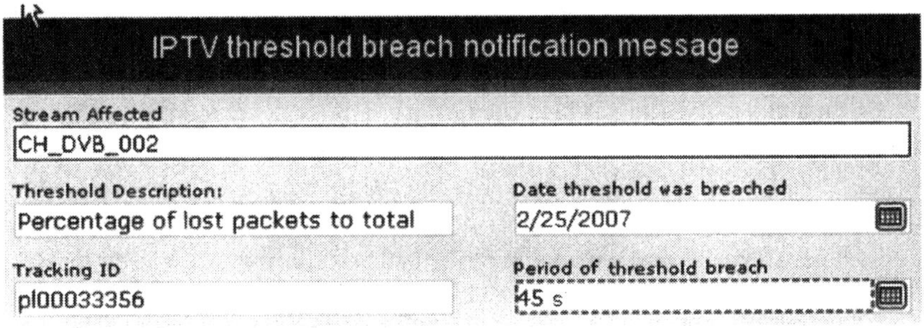

FIGURE 12.13   Sample threshold breach notification message

analyzers are also used by IPTV engineering departments to measure various QoS and QoE parameters.

12.9.4.3   Video Analyzers   Video analyzers (VAs) are electronic devices used by IPTV engineers to measure video quality. Additionally, they are used to identify and debug IPTV stream impairments that affect the QoE for end users. Features of a VA can include the ability to

- capture MPEG-2, H.264/AVC or VC-1 IPTV encoded streams in real time;
- record parts of a live IPTV stream to hard disk for offline analysis;
- monitor stream parameters such as frame and bit rates;
- analyze IPTV streams to identify causes of problems that may be affecting the signal QoS levels.

Using the visual diagnostic capabilities of VAs, IPTV engineering professionals are able to quickly locate intermittent issues that contribute to the degradation of IPTV end users QoE and QoS.

12.9.5   Multicast TV and IP-VoD Test Programs

Once an IPTV network has been deployed it is critical for the IPTV network manager to test and assess end users QoE. This is achieved by regularly checking both channel changing and VoD components performance levels across the IP network. In addition to carrying out tests on the live network, many engineering departments have set up test networks in their labs that allow the emulation of a number of different configurations. These lab based networks may be used to determine the effects that IPTV end users will experience under circumstances such as network faults and increased traffic loads. To identify any potential QoE problems that may occur when a new service is commercialized it is important that IPTV network managers define a number of QoE test programmes. Specialized test and measurement equipment are typically used to execute these test programmes. A description of the more popular QoE programmes is provided in the following sections.

12.9.5.1   A Channel Changing Test Program   One of the key challenges associated with deploying an end-to-end IPTV infrastructure is making sure that the channel-changing performance has been optimized. To meet this challenge, IPTV network managers typically plan a series of tests that are carried out prior to deployment on a production network and comprise of the following activities.

Define IPTVCD Performance Metrics   Performance metrics for the IPTVCD need to be defined. Typical performance metrics gathered during a test need to include IGMP join and leave latencies, channel change delay at the IPTVCD, and duration of the channel overlap. These metrics are illustrated in Fig. 12.14.

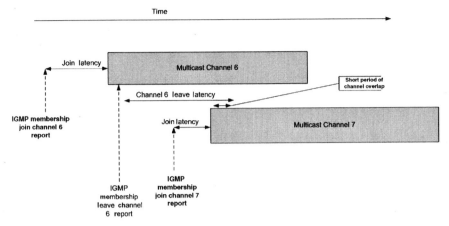

FIGURE 12.14   Channel changing metrics used to analyze performance at the IPTVCD

Compile a Test Plan   Creating a comprehensive test plan is the first step of the testing program and needs to include the following:

A definition of different network loading conditions—A description of different network loading scenarios needs to be complied. Each scenario should include parameters such as the number of multicast streams to be sent across the test network and the number of simulated end users accessing these streams. Some of the advanced test plans also determine the affect of additional traffic such as VoIP and high speed data services on the IPTV channel changing performance levels.

A description of the different channel changing patterns—In addition to defining the numbers of end users, the test plan needs to also include details about the different channel changing and usage patterns for each IPTV end user. The user channel change behavior in particular can vary enormously and has a direct impact on the network resources responsible for delivering the video service. For instance, some IPTV end users will sequentially zap through channels at a rapid pace when sitting down initially to watch TV, while others will only engage in channel surfing during commercial breaks.

IGMP configuration parameters—IGMP is the industry standard used to support multicast IPTV services. As described in Chapter 3, different versions of IGMP include support for different features that affect channel changing performance levels. The test plan needs to include a provision to test each of these parameters and measure the trade-offs that need to be made before deploying across the live network.

In addition to defining the various inputs to the test network, the plan should also define baseline parameters that constitute an acceptable level of QoE for end users of the service.

Configure the Test Network   The IGMP protocol details and load conditions outlined in the test plan are used to configure the test network.

Run the Tests and Analyze Results   Run the test to determine channel change time delays that are experienced by IPTV end users. The output from the test will be a series of results, which include:

- Jitter levels
- Packet latency
- Average interpacket arrival times
- Join latency
- Leave latency

Once the test results are gathered and analyzed it is advisable to run a number of iterations of the initial test. This is achieved by using different load requirements and channel changing behavior configurations.

Tune the Network   Once the network has been emulated and analyzed for various network conditions, the various network resources and protocols need to be tuned to ensure that end users receive acceptable channel changing delays when accessing their multicast IPTV services.

12.9.5.2 *IP-VoD Server Testing*   VoD servers are a key element of an end-to-end IPTV networking infrastructure. As a result, it is critical that IPTV network managers implement a test programme that measures the performance of an IP-VoD server under varying load conditions. Video performance metrics such as the number of concurrent connections and jitter levels are gathered during the test to determine performance capabilities of the VoD server when operational in a live networking environment. Similar to the tests carried out on measuring channel changing delays, a test plan is also created to measure performance characteristics of IP VoD servers. The types of input parameters used to stress test an IP VoD server include the following:

- Number of IP-VoD client IPTVCDs.
- Number of VCR type instructions issued by each IP-VoD client over a specified period of time.
- RTSP protocol parameters.
- Video stream configurations (constant bit rate or mixed bit rate streams).

Once the network is configured with the parameters included in the test plan a number of key server metrics are gathered after the test(s) are run including

- CPU performance levels
- Memory usage
- Disk access speeds
- Streaming rates of VoD content at network interfaces
- Number of concurrent unicast streams supported before packet loss and jitter occurs

The performance results of this program are typically used to determine the QoE that end users are likely to receive when the server is deployed in a production environment.

12.9.5.3   Video Content Testing   As compression rates continue to improve, it is critical that IPTV network administrators are able to test the quality of the actual video signal delivered to end users. Advanced test and measurement equipment is generally used to identify and analyze the structure of raw video streams. These tests can produce a range of statistical information on the quality of the video signal contained within the stream. For instance, if the content contained in the stream is MPEG compressed, then the following fields are examined to determine if there are any QoE issues:

- *The synchronization byte*—If this byte is out of order, then the IPTVCD will be unable to process the video packet.
- *The PCR*—This field indicates the level of stream jitter generated by the transport networking infrastructure.
- *The continuity counter*—This field helps to identify missing packets that could adversely impact on the end-users QoE.
- *PID values*—Incorrect PIDs will cause confusion for the IPTVCD decoder and degrade the end-users QoE.

Once a comprehensive engineering testing program is in place, service providers are able to guarantee an acceptable level of QoE for new and existing IPTV subscribers.

12.9.6   QoE and Service Level Agreements (SLAs)

An SLA is a contractual agreement between two parties that defines a certain level of service. In the context of an IPTV networking environment an SLA defines various performance levels that must be adhered to during normal operations. SLAs come in three variants.

12.9.6.1 IPTV Subscriber SLAs   Although not in widespread use at the time of authoring the level of support for the introduction of SLAs between paying subscribers and their service provider is expected to increase over time. The purpose of these SLAs is to make sure that the service provider is delivering services at a particular quality level. In addition to video quality levels, these types of SLAs can also include other types of performance parameters such as minimum service outage levels and commitments on the time taken to repair IPTVCDs.

12.9.6.2 Internal SLAs   It is good business practice for service providers to establish an internal SLA that ensures that certain quality expectations are met. The actual implementation of these types of internal SLAs involves the measurement of QoS and QoE parameters at various segments of the network.

12.9.6.3 Third-Party Content Provider SLAs   As discussed previously, service providers will obtain most of their VoD assets and real-time broadcast streams from third-party content suppliers. As a result, it is critical that the service provider has a strong and up-to-date knowledge of the incoming content quality. To ensure that quality levels remain high, an SLA is entered into between the network provider and the various source content suppliers. A third-party centric SLA defines a set of agreed QoS indicators and performance levels that are mutually agreed between both parties. The implementation of a third-party content provider SLA occurs during the ingesting of the incoming source content. At this point the content quality and associated metadata is measured and compared to the reference parameters defined in the SLA. If impairments such as packet loss or jitter are detected, then a signal degradation report is normally generated and e-mailed to the content provider. The structure of a sample signal degradation report is shown in Fig. 12.15.

Once a violation of any SLA type is detected, the issue needs to be addressed by the content supplier immediately. Otherwise the QoE of IPTV end-users may be seriously affected. These reports are typically archived for future purposes.

12.9.7 QoE and KPIs

Also known as Key Performance Indicators, a KPI provides management at IPTV service providers with a clear "picture" of the overall quality of the service that they are responsible for delivering to end users. A KPI is a metric generated through the aggregation of quality performance results discussed in previous sections of this chapter, with quality details on the content received from third-party providers. A metric that defines the percentages of sessions that are experiencing high levels of packet losses is one example of a KPI. KPIs may be used by management for the following purposes:

(1) Determine how well the service is performing in general.
(2) Assess the quality of the IPTV content both from the perspective of the end user and the content being sourced from providers.

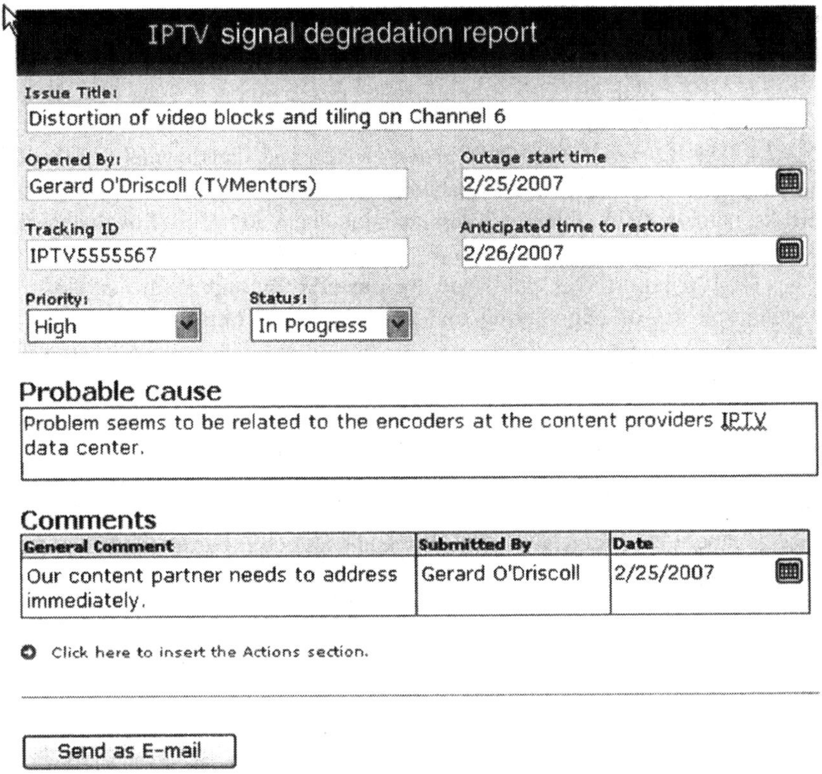

IPTV signal degradation report

Issue Title:
Distortion of video blocks and tiling on Channel 6

Opened By:	**Outage start time**
Gerard O'Driscoll (TVMentors)	2/25/2007

Tracking ID	**Anticipated time to restore**
IPTV5555567	2/26/2007

Priority: High

Status: In Progress

Probable cause

Problem seems to be related to the encoders at the content providers IPTV data center.

Comments

General Comment	Submitted By	Date
Our content partner needs to address immediately.	Gerard O'Driscoll	2/25/2007

○ Click here to insert the Actions section.

Send as E-mail

FIGURE 12.15   Sample IPTV signal degradation report

(3) Analyze any impairments that may be associated with the various IP video and audio streams.

(4) Ensure that the KPI scores are enough to satisfy the requirements set out in either the internal SLA or the SLA agreed with third-party content suppliers.

(5) Planning of long-term strategic goals with regard to upgrading networking infrastructure and ensuring that services carried over the network are delivered to end users at high QoE levels.

Some IPTV service providers have installed systems that monitor KPIs and trigger alerts when scores have reached certain minimum and maximum thresholds. A threshold breach for instance may indicate a gradual decrease in service quality. Once this occurs an escalation procedure is activated and the quality of the service is improved back to acceptable operational levels. It is key that IPTV network managers monitor KPIs on a regular basis and tune the network components when particular thresholds have been breached.

12.9.8   IPTV QoS and QoE Industry Initiatives

The four main industry efforts at the time of writing, with regard to defining a standard for IPTV QoE and QoS is being undertaken by the following organizations.

12.9.8.1   DSL Forum   The *DSL forum* TR-126 specification defines the QoE requirements of a number of applications that are transported across broadband networks, namely, IPTV, Internet access, gaming and VoIP. A brief overview of the TR-126 specification is provided below.

The TR-126 report was published by the DSL Forum at the end of 2006 and defines a set of engineering and end user QoE objective guidelines for delivering voice, data, and video services across an IP broadband networking infrastructure. Rather than attempting to cover the multitude of video services that are commercially available, the report focuses on entertainment video applications such as multicast IPTV programming and VoD.

In addition to defining the various factors that contribute to a decrease in QoE, TR-126 specifies the use of three different approaches to measuring QoE for IPTV services, namely subjectively, objectively, and indirectly. Furthermore, the report defines a set of encoding bit rate objectives for deploying popular IPTV applications (see Table 12.6).

Note, the report also recommends a set of engineering specifications for encoding digital audio streams.

12.9.8.2   ITU Affiliated Video Quality Experts Group (VQEG)   This group was created in 1997 to develop and standardize video quality measurement techniques. The group has been instrumental in standardizing the ITU-T J.144 video quality measurement technique.

12.9.8.3   CableLabs   This organization has defined a QoE framework in its DOCSIS 3.0 specification that supports the delivery of downstream IP multicast traffic over HFC networks.

12.9.8.4   ATIS Group   As mentioned in Chapter 1 the *ATIS group* has defined a framework for monitoring QoS metrics for various types of IPTV services.

TABLE 12.6   Summary of TR-126 IPTV Bit Rate Objectives

IPTV Compression Category	General SD Multicast TV Services — CBR (Mbps)	SD Premium and VoD Content — CBR (Mbps)	HDTV — CBR (Mbps)
MPEG-2 - MP@ML	2.5	3.18	15
H.264/AVC - MP@L3	1.75	2.1	15
VC-1	1.75	2.1	10

12.10   REMOTELY MANAGING IN-HOME DIGITAL CONSUMER DEVICES

The management of SD and HD IPTV services once they enter a subscriber's home comes with its own set of unique challenges, namely:

- The multitude of in-home networking technologies available to consumers can complicate the day-to-day operations of the in-home network and may affect the quality of IPTV content distributed to televisions connected at various locations.
- In addition to an IP set-top box, which is used to receive IPTV content a typical home network will also include a PC, and a Wi-Fi access point. Problems with these devices or the network itself can have an impact or completely stop the delivery of IPTV services.
- Sending a technician to resolve a home networking problem is often cost prohibitive.
- Reproducing the problem is sometimes difficult and can often delay the reinstatement of a fully operational in-home network.
- Video impairments such as jerkiness and pixelization are transient and may only occur for a fraction of a second and will disappear almost immediately. This makes it difficult for CSR personnel to detect the causes of the impairment.
- Solving and diagnosing problems with IPTV end users over the phone can be problematic. Many end users will not be able to describe in detail the video and audio impairments that are present in the IPTV service.
- Some of the performance problems such as a degradation of video quality may not get detected as an alarm situation. The service provider may only realize that there is a QoS problem when customers call in and report a fault in the system.

IPTV network administrators need to be in a position to pinpoint and fix problems with in-home digital consumer devices which will affect the quality of an IPTV viewer's experience. This is typically achieved through the use of a specialized in-home remote management server software tool. A home device management tool allows service IPTV administrators to do a number of things:

- Provide support personnel with a means of remotely monitoring the performance of IPTV subscriber's in-home networks.
- Collect performance data, which is used to pinpoint IPTV service quality problems within the home.
- Allow CSRs to check signal levels on last mile broadband connections.
- Provide administrators with a mechanism of remotely fixing a problem. This type of functionality negates, in most cases, the need to send a technician to the subscriber's home.

- Provide administrators with visibility into the various devices that are connected to the IPTV subscriber's in-home network and identify abnormal activity such as a WHMN application stopping.
- Allow administrators to remotely control IP address configurations.
- Provide a means of reporting events that occur on the home network on a daily basis.
- Gather historical statistical data and provide reporting features that allow IPTV providers to support SLAs.
- Maintain an inventory of hardware and software configurations of all IPTVCDs connected to the network. Note the availability of configuration details about the last known functional home networking environment is useful during the problem resolution process.
- Generate exception reports when uncharacteristic behavior occurs on the home network.
- Connect with other systems including the NMS and the OBSS.
- Checking IPTVCD inventory in order to determine whether devices require software or hardware upgrades.
- Installing and upgrading software in all the IPTVCDs connected to the broadband network.
- Provide software metering capabilities that enable support staff to keep track of IPTVCD software licenses.

The use of a home device management tool can often lower the operational costs associated with providing IPTV services and also improve customer service. The networking infrastructure of an end-to-end home management system architecture is shown in Fig. 12.16.

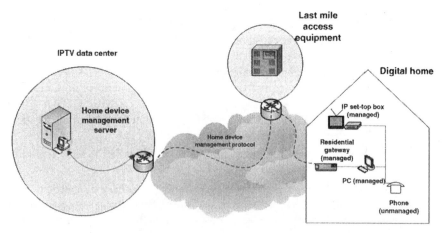

FIGURE 12.16   End-to-end home management system architecture

As shown on the diagram the digital devices in the home are defined as either managed or unmanaged in terms of remote management. A managed device includes a remote management software client whereas an unmanaged device has no support for the software required to interact at a detailed level with the backend home management server. The DSL Forum's TR-069 standard is an example of a software infrastructure that supports this type of client-server interaction. In some cases unmanaged home devices do however support other types of protocols such as DHCP and UPnP. The figure also shows the use of the home device management protocol across the IP access and core networks. The nature of IPTV systems can also cause engineering difficulties. For instance, it is quite common for TVs to be turned on and no-one watching in some households. This is fine for cable based networks, however on IP based networks an unwatched TV is eating up precious bandwidth.

12.11 SCHEDULING AND MANAGING DELIVERY OF SOFTWARE UPDATES TO IPTVCDs

The ability to upgrade software running on the various types of IPTVCDs connected to the distribution network is a vital part of maintaining and supporting an IPTV service. In addition to increasing the efficiency of maintaining thousands of IPTVCDs, an update in software can also increase customer satisfaction. Rolling out a truck to upgrade features in an IPTVCD is not cost effective. There are two methods of upgrading software. The first method is manual and involves a subscriber physically returning the IPTVCD to the service provider or else sending a technician to each home to upgrade the software. This does not make sense from an economic perspective if new versions of the software need to be frequently delivered and makes life difficult for subscribers. The second and most commonly used method of updating an IPTVCD operating system and resident software application is to securely download the software over the network. Using this approach service providers can

- download software bug fixes to IPTVCDs that are already deployed in the field;
- upgrade IPTVCDs with new releases or versions of software to improve functionality and ensure that all IPTVCDs connected to the network have the same version of software;
- improve performance by migrating to a new software standard;
- apply a digital signature to a software image to secure the download;
- download new decoding software to ensure compatibility with new IPTV data center video encoders;
- ensure that all IPTVCDs receive the same software upload at the same time.

The steps involved with a software download procedure to a group of installed and operational IPTVCDs are shown in Fig. 12.17 and explained in the following sections.

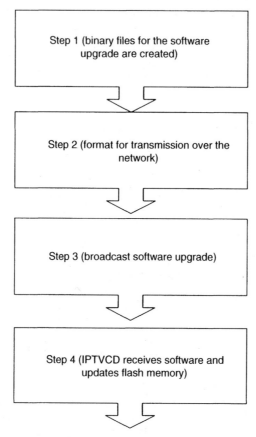

FIGURE 12.17   IPTVCD software download process

Step 1 (binary files for the software upgrade are created)—The first step of the
   software download procedure involves sourcing and compiling the binary
   files required for the upgrade.

Step 2 (format for transmission over the network)—Next the files need to be
   structured in a format or image, which is compatible with the file system used
   on the transmission network. In addition to formatting the files, some
   operators will also add a digital signature or certificate to the files to ensure
   that only valid software updates are sent to the set-top box.

Step 3 (broadcast software upgrade)—The files required for the upgrade are then
   transmitted across the network. The timing of the upgrade needs to be well
   planned and can generally takes place during off-peak hours. Trivial File
   Transfer Protocol (TFTP) is generally used to copy the files over the network.

Step 4 (IPTVCD receives software and updates flash memory)—Once the
   updated software is downloaded from the network, a software module present
   in the IPTVCD will then test the code for any errors or corruption that may

have occurred while traversing the network. Following the verification of the code, authentication is then performed to establish a trust relationship with the server back at the IPTV data center. Once verified and authenticated, the image is reassembled and the update to the IPTVCD can commence. The IPTVCD will generally report back to the data center that the software download was completed.

This update can be done in either the foreground or the background. A foreground IPTVCD software update is intrusive to the viewing experience. A message normally appears on the television or display panel notifying the viewer that a software update is available. If the IPTV end user chooses to proceed with the update the IPTVCD is taken off line while the software update occurs. Foreground updates can take as little as 2 min or as long as a half an hour, depending on the write performance of the flash memory and the amount of software code that needs to be downloaded. IPTVCDs designed for foreground updates require less flash memory and often rely on a single flash component. By contrast a background update does not interrupt the viewing experience. The viewer is not even aware an update is taking place. In this case the IPTVCD may contain two flash memory devices or partitions. The first one stores the existing code while the second is used for updates. The system switches to the flash device or partition with the new software code when the power is recycled. An onscreen display will notify the IPTV end user of the new features that have been installed. While this method of downloading software is more elegant, it is obviously more expensive. Once the software upgrade has been written to nonvolatile storage, the IPTVCD will generally reboot. Once rebooted the IPTVCD resumes normal operations using the upgraded software. The decision of which is the most suitable approach varies between IPTV service providers. Specialized software applications are also available that allow IPTV service providers to automatically distribute software upgrades and patches over their existing network infrastructures. It is also important to note that in some instances it may be necessary to downgrade the software resident in the IPTVCD and revert back to the previous software image. Although no standard exists for the download of software images to IPTVCDs, organizations such as CableLabs have developed a mechanism called the OpenCable common download specification, which is used in the United States to upgrade digital cable set-top boxes.

12.12   TROUBLESHOOTING IPTV PROBLEMS

Problems experienced by IPTV subscribers with the service are generally reported through a call to the CSR department. If problems occur on a network, it is important that they are quickly resolved. Support personnel and administrators need to be able to determine and troubleshoot problems. Troubleshooting an IPTV network is similar to troubleshooting any other type of IP based application. First and foremost, the problem needs to be isolated and will depend on the nature and extent

TABLE 12.7   IPTV Network Errors by Area

Problem Area	Common Failures and Issues
Network	Errors can occur in different parts of the network as IP streams get transported to the IPTVCD. A network sniffer is sometimes used by IPTV engineers to examine data traversing a live network. A typical protocol analyzer comprises a PC, a network card and an application that analyses problems on the network. Real-time network analysis helps detect and resolve network faults and performance problems quickly.
Backend servers	Specialized server software is typically used to troubleshoot problems that occur on servers.
IPTVCDs	A high level of reliability is expected from IPTVCDs such as IP set-top boxes and gateway devices. Therefore, IPTV service providers have developed sophisticated systems for gathering error messages from IPTVCDs and using this information to troubleshoot problems that may arise with the IPTV service. This information is generally stored in the home device management application and is typically used by IPTV engineers to identify the root cause of a particular service affecting problem for individual subscribers. Common problems include issues with buffering, hardware malfunctions, and software crashes. Once the problem area is identified the next stage is to resolve. This may involve one or more of the following actions: • Make configuration changes to the IPTVCD. • For usage problems speak directly to the end user and provide instructions on how to resolve the issue. Note e-mail may also be used as a complementary mode of communications when issuing usage instructions. • If the problem cannot be resolved over the phone than a service technician(s) needs to be mobilized to the end-users home. • Arrange equipment replacement. Note that local troubleshooting and debugging is typically done via a physical interface on the unit.
Customers in-home network	Problems such as bad in-home wiring runs, low quality cables, viruses, incorrect IP address settings, and corrupted drivers can all have an impact on the IPTV service. The remote home management system in combination with either telephone or e-mail support are generally used to correct problems that occur across a subscriber's in-home network.
External content provider	Typical issues with ingesting of incoming content range from poorly encoded video material to transmission problems. Once identified the third-party provider is notified and the problem is rectified.

of the problem. The next step is to fix the problem. This is typically achieved by making a configuration change or replacing a failed network component. Most errors on IPTV networks generally fall into one of five areas: network, backend servers, IPTVCDs, customer's in-home network and external content providers (see Table 12.7).

In addition to solving a particular problem, it is also crucial that an IPTV service provider includes processes and procedures that document or electronically record from once the problem has been detected to when it gets resolved. This is typically achieved through the creation of formalized "troubleshooting reports" (TRs) that are initiated when a failure is detected. The content of a TR depends on the location of the problem. For instance, the format of a TR created to describe a problem with a single IPTVCD (sample TR shown in Fig. 12.18) is different to a TR used to

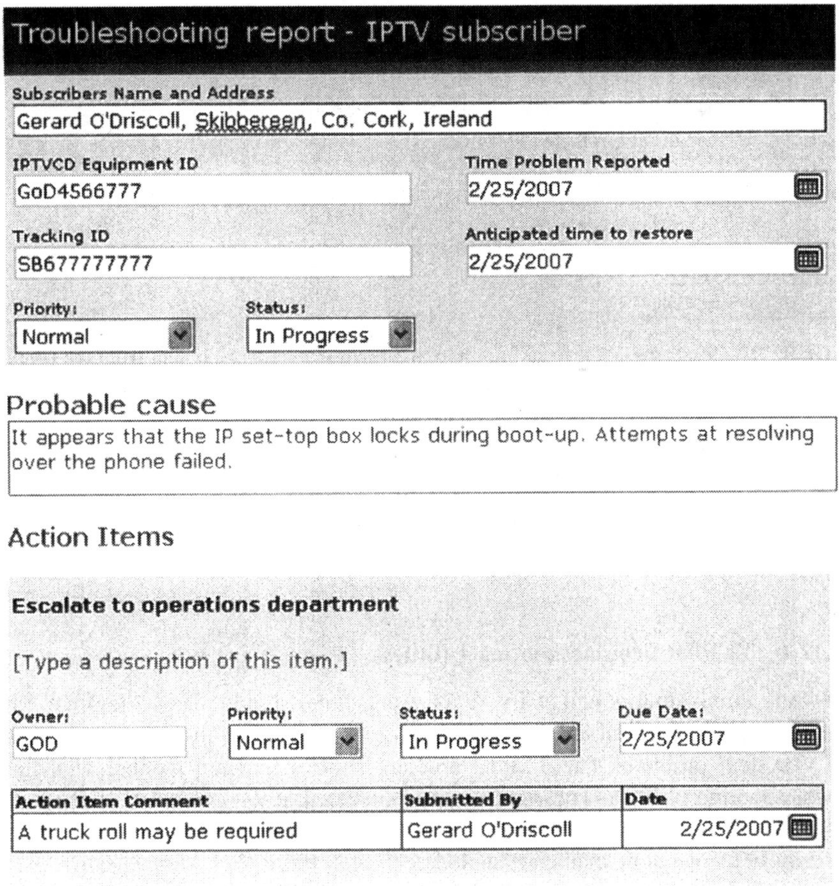

Troubleshooting report - IPTV subscriber

Subscribers Name and Address
Gerard O'Driscoll, Skibbereen, Co. Cork, Ireland

| **IPTVCD Equipment ID** | **Time Problem Reported** |
| GoD4566777 | 2/25/2007 |

| **Tracking ID** | **Anticipated time to restore** |
| SB677777777 | 2/25/2007 |

| **Priority:** | **Status:** |
| Normal | In Progress |

Probable cause
It appears that the IP set-top box locks during boot-up. Attempts at resolving over the phone failed.

Action Items

Escalate to operations department

[Type a description of this item.]

| **Owner:** | **Priority:** | **Status:** | **Due Date:** |
| GOD | Normal | In Progress | 2/25/2007 |

Action Item Comment	**Submitted By**	**Date**
A truck roll may be required	Gerard O'Driscoll	2/25/2007

FIGURE 12.18 Sample TR for a fault identified with an IP set-top box at a subscriber's home

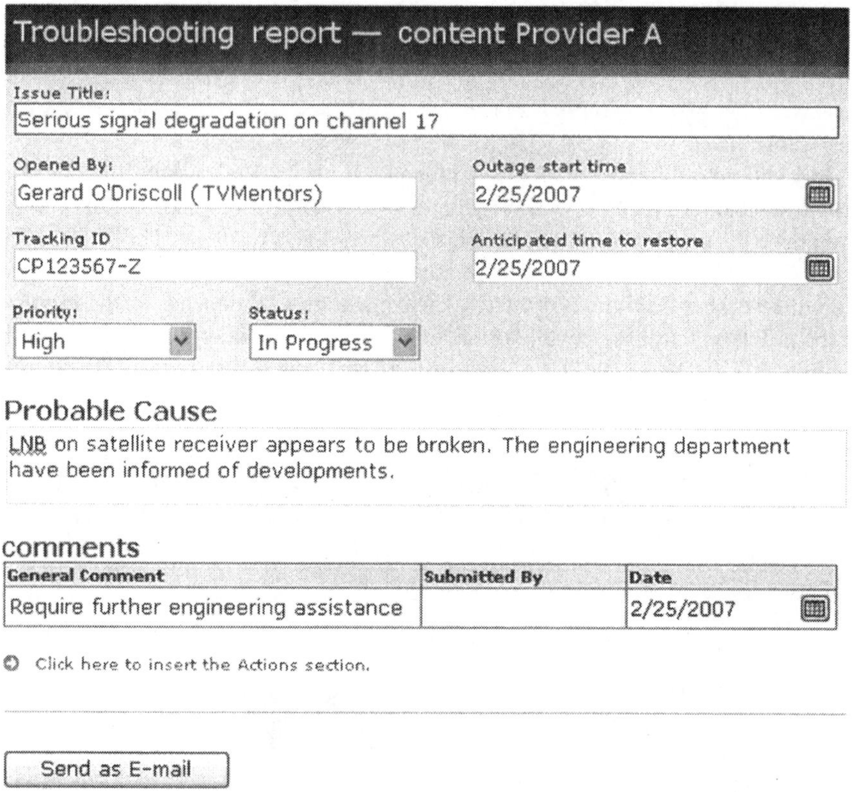

Troubleshooting report — content Provider A

Issue Title:
Serious signal degradation on channel 17

Opened By:
Gerard O'Driscoll (TVMentors)

Outage start time
2/25/2007

Tracking ID
CP123567-Z

Anticipated time to restore
2/25/2007

Priority:
High

Status:
In Progress

Probable Cause

LNB on satellite receiver appears to be broken. The engineering department
have been informed of developments.

comments

General Comment	Submitted By	Date
Require further engineering assistance		2/25/2007

◯ Click here to insert the Actions section.

Send as E-mail

FIGURE 12.19   Sample TR for a fault identified whilst ingesting content from the content
provider

document a fault that has arisen at the content providers premises (sample TR
shown in Fig. 12.19).

Once the problem associated with the specific network component has been
cleared the final step of the troubleshooting process is to change the status of TRs to
closed.

12.12.1   TCP/IP Troubleshooting Utilities

Network administrators and IPTV support personnel should also be familiar with
some basic TCP/IP troubleshooting techniques to help them in troubleshooting an
IPTV system problem. The TCP/IP protocol suite includes a number of default
troubleshooting utilities. These utilities are executed at the command prompt of a
device, which has been configured to support TCP/IP. The most popular utilities
used by IPTV network managers include:

12.12.1.1   The PING Utility   The ping command is the most basic tool used for
isolating hardware problems on an IPTV network. It enables support personnel

to verify connections to different parts of the IPTV networking infrastructure by sending Internet Control Message Packets (ICMPs) to a remote location and then waiting for response packets to be returned. If the response packets are not returned within a predefined time period, then the ping command will timeout. This command is typically used by IPTV network managers to verify that components such as routers, gateways, CMTSs, DSLAMs, and network switches are working.

12.12.1.2 The TRACERT Utility TRACERT is a TCP/IP utility that allows IPTV support personnel to display route details to any destination point on the IPTV network. The TRACERT utility traces the route taken by ICMP packets sent across the IPTV network to a specified destination point. This destination point could be equipment operating in a remote regional office or an IPTVCD connected to a subscriber's home network. It works by decrementing the "Time to Live" header value by one when the ICMP packet passes through the various routers and switches between the source and destination points. The number of decrements provides the utility with a hop count from source to destination, which is subsequently displayed onscreen. In addition to hop count, the TRACERT utility also displays round-trip timing for each hop along the routing path. This information is helpful to IPTV network administrators who need to determine routing irregularities or where network slow-downs are occurring.

12.12.1.3 PathPING Although the TRACERT provides routing details, it also identifies links that suffer from latency and packet loss. A Microsoft utility called PathPING may be used to provide IPTV network administrators with latency and packet loss statistics for each link between the source and end destination point. It works by sending multiple messages to the various routers along the network path over a specified time period and calculating the number of packets returned in response to each request message. The results of these calculations are displayed on screen. This data allows an IPTV engineer to determine if problems are occurring on certain sections of the distribution network.

12.12.1.4 Address Resolution Protocol (ARP) This utility enables IPTV engineers to match IP addresses to underlying hardware addresses.

12.12.1.5 NETSTAT This utility displays active TCP connections of a particular piece of IPTV networking equipment. Details about the routing table and the various types of protocol statistics are also outputted onscreen.

12.12.1.6 IPCONFIG As the name implies this utility provides an IPTV engineer with configuration details of a particular IPTV networking device. This utility is generally used in Windows based servers. The types of information generated by the utility include the IP address, the subnet address, and the hardware address of the machine.

Although all of the TCP/IP utilities are useful in isolating different TCP/IP problems, the most widely used commands by IPTV support departments are PING and TRACERT.

12.13   IPTV AND BUSINESS CONTINUITY PLANNING

Today's IPTV data centers need to be operating 100% of the time to grow revenues and sustain customer satisfaction. Interruptions in the delivery of IPTV services, particularly multicast channels, can damage a company's reputation and cause subscribers to move to other pay TV providers. Interruptions can be caused by a number of incidents ranging in severity from full-scale disasters that render the IPTV data center in-operable to unexpected peaks in video traffic. Therefore, the area of business continuity planning (BCP) is a critical part of administering and managing an IPTV networking infrastructure. The purpose of BCP within the context of an IPTV environment is to define a set of procedures for recovering normal services when a system failure or a major incident occurs.

In addition to operational procedures that define the process of restoring services, some thought also has to be put into designing the network to support a secondary site, which could support the delivery of critical IPTV services in the event of a major event occurring at the primary IPTV data center. There are a number of enabling technologies that are used to protect the integrity of IPTV content and optimize the availability of IPTV services. Consider the simple example shown in Fig. 12.20 of an IPTV service provider who establishes a backup data center to meet the disaster recovery and business continuity goals of the company.

As depicted, each data center is built with two physically separate infrastructures and fiber optic links are used to interconnect both centers. The availability of high speed and redundant interconnections between data centers is a key part

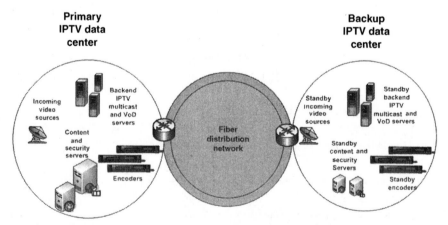

FIGURE 12.20   Example of an IPTV network architecture required to support BCP objectives

TABLE 12.8 Key Functional Areas Associated with Managing and Supporting an IPTV Networking Environment

Function Name	Description
Supporting the IPTV NMS	The operation of an end-to-end IPTV network is typically managed by an NMS. The NMS allows IPTV network administrators to monitor the performance of the underling networking infrastructure and identify any problems that may be affecting the delivery of IPTV services to end users.
Management of installs, service problems and terminations	Provisioning new subscribers and related services is one of the key tasks associated with managing the day-to-day operations of an IPTV network. This provisioning process requires an integrated backend operational and billing system to cope with these demands. In addition to new subscribers, IPTV service providers also require procedures and processes to deal with service problems and terminations.
Network testing and monitoring	Monitoring the network allows IPTV administrators to identify and resolve problems before it results in a total network failure. A test and measurement system is often used by service providers to check the health of the networking infrastructure.
Ensuring service availability and managing redundancy	The management of backup systems and redundant components are a key part of operating a network that delivers IPTV services to end users.
IP address space management	The assignment of IP addresses is a critical part of deploying and running an IPTV networking infrastructure. All devices connected to the infrastructure require an IP address to access IPTV services. DHCP servers typically provide these addresses.
Routine administrative tasks	Monitoring disk space and maintaining high security levels are examples of periodic tasks that are part of supporting a live IPTV system.
Managing IPTV QoS requirements	Because of the fact that most IPTV services operate over a private IP broadband network, it is possible to implement a QoS policy when delivering video content to paying subscribers. Implementing a QoS policy basically means that IPTV traffic is given higher priority over other types of IP based traffic. Deploying a QoS system shows up problems that may cause complaints from subscribers ensuring a high quality customer experience.

(*continued*)

TABLE 12.8   (*Continued*)

Function Name	Description
Monitoring the IPTV subscriber experience	The success of IPTV will largely come down to how end users perceive the quality of the IPTV experience. The perceived quality is determined by the following factors. *The video and audio quality*—Eliminating end-to-end delays and making sure that the IPTV data packets arrive smoothly and in the correct order at the IP set-top box interface is critical to providing viewers with a high-quality viewing experience. Bursty IP traffic and lost packets can result in serious picture artifacts and mar the user's viewing experience. *Response rates for broadcast and VoD title selection*—Zero latency responses are expected when navigating and selecting IP-VoD and multicast services. Keeping a close tab on these factors, will help prevent degradation in the quality of video services delivered to subscribers. A set of QoE metrics is typically used to measure the satisfaction levels of IPTV end users. An increased understanding by network administrators of QoE is critical in ensuring that the end-to-end IPTV system operates in an effective manner.
Remotely managing in-home digital consumer devices	Remote set-top box and home networking management tools provide IPTV network administrators with the functionality required to manage the delivery of IPTV services throughout a subscriber's home.
Scheduling and managing delivery of software updates to IPTVCDs	When deploying IPTVCDs, the software must be easily and economically upgraded. TFTP servers are typically used to upgrade IPTVCD firmware and software applications. IPTV engineering personnel are normally responsible for managing this process.
IPTV and BCP	Many of the large IPTV carriers have deployed at least two data centers to provide redundancy in the event of a failure at one of the facilities. The development of a BCP function is a key part of running an IPTV operation.
Troubleshooting IPTV problems	Despite the fact that the installation of the IPTV hardware and software components were installed according to plan, problems will still occur. Part of an IPTV engineer or technician responsibility is to troubleshoot these problems when they happen.

of replicating and mirroring IP video data between both data centers. In the event of a failure at the primary data center, standby servers and video processing equipment are used to maintain a replicated copy of the on-demand content library. Thus, if the primary IP-VoD or streaming server becomes unavailable, the standby servers at the disaster recovery site can quickly take over the servicing of requests from IPTV end users.

SUMMARY

Consumers expect and demand a high-level of service availability from their IPTV provider. To meet this challenge IPTV network operators employ a number of network administration processes to manage the many complex components that make up an end-to-end IPTV system. Table 12.8 summarizes the primary functions required to manage an end-to-end IPTV networking infrastructure.

INDEX

5C content protection, *see* DTCP

ACM (Admission Control Mechanism),
450–451, also *see* QoE
ACAP (Advanced Common Application
Platform), *see* CableLabs
ADSL (Asymmetric Digital Subscriber
Line)
ADSL2, 30
ADSL2+, 30
ADSL Reach Extended 30
Technical Overview 27–30
AES (Advanced Encryption Standard),
44
AON (Active optical networks), 26
ARIB (Association of Radio Industries and
Businesses), 9, 246
ATIS (Alliance for Telecommunications
Industry Standard), 15–16,
468
ATSC (Advanced Television Systems
Committee)
ATSC-T (ATSC Terrestrial),
216

DASE (Digital Television Application
Software Environment), 245
Overview, 8
VSB (Vestigial sideband), 217
ATM (Asynchronous Transfer Mode), 25,
31, 57
AVC (Advanced Video Coding), *see*
H.264/AVC
AVS (Audio Video Standard), 80

B-frames (Bi-directional Frames), 69, 77,
453
Blu-ray, 372–373
BPI+ (Baseline Privacy Plus), 44
Broadband TV, *see* IPTV
BSF (Broadband Services Forum), 15

CableLabs
ACAP, 245–246
CableCARD, 255–256
CableHome, 189–193
DOCSIS, *see* DOCSIS
OCAP (OpenCable Applications
Platform), 241–245

Next Generation IPTV Services and Technologies, By Gerard O'Driscoll
Copyright © 2008 John Wiley & Sons, Inc.

CableLabs (*continued*)
  OCCUR (OpenCable Unidirectional
    Receiver), 279
  OpenCable, 221
  OpenCable common download
    specification, 473
Callers ID for TVs, *see* IPTV application types
CA (Conditional Access) systems
  Algorithms, 259
  Downloadable CAS, 264-265
  ECMs (Entitlement Control Messages),
    252, 254
  EMMs (Entitlement Management
    Messages), 251, 254
  Hardware centric, 250-257
  Hybrid approach, 263
  Smart cards, 254-255
  Software centric, 257-263
  Removable security modules, 255-257
CBR (Constant Bit Rate), 71, 438
CEA (Consumer Electronics Association)
  Home networking protocols, 328
  IPTV "Principles" Initiative, 18
  TV Web browsing standard, 401
Channel Changing
  Different mechanisms, 163-166
  Improving rates, 166-169
  Sources of delays, 159-163
  Testing program, 462-464
CIF (Coral Interoperability Framework),
  278-279
CMTS (Cable Modem Termination System)
  CableHome deployment, 191
  Communication with hybrid IP
    Cable set-top boxes, 222
  Modular CMTS (M-CMTS), 37
COFDM (Coded Orthogonal Frequency
  Division Multiplexing), 28, 217-218
Compression
  Audio, 205-208
  Drawbacks, 65-66
  MPEG, 66-79
  Spatial, 68
  Temporal, 69
  VC-1, 79-80
  Video, 67-70
CRC (Cyclic Redundancy Check), 114
CRM (Customer Relationship
  Management), 121, 122

CSMA/CD (Carrier sense multiple access
  with collision detection), 292-293

DCAS, *see* CA systems
DCT (Discrete Cosine Transform), 68,
  76
Decoding
  Audio, 205-208
  Video, 205
DHCP (Dynamic Host Configuration
  Protocol)
  DOCSIS 3.0 Deployments, 42
  IP Address Management, 436
  RG IP address assignment, 180-181
  UPnP deployments, 318-321
Diffserv (Differentiated Services), 179,
  328, 440-441
Digital certificates, 238, 261-262
Digital signatures, 238, 261
Digital TV
  Benefits, 9-10
  History, 6
  Introduction, 1
DisplayPort, 375-377
DLNA (Digital Living Network Alliance)
  Guidelines, 315
  Technical architecture, 316-324
  UPnP (Universal Plug and Play),
    316-322
  UPnP AV (UPnP for Audio Video
    streaming devices), 322-324
DOCSIS (Data Over Cable Service
  Interface Specifications)
  Channel Bonding, 41
  DOCSIS 3.0, 37-45
  DOCSIS over Satellite, 47-48
  Evolution, 36-37
  IGMPv3 support, 144
DRM (Digital Rights Management),
  About, 265-268
  Analog watermarking, 276
  Copy protection, 271-276
  Digital Watermarks, 268-269
  DLNA deployments, 322
DSA (Differentiated Services Architecture),
  439-441
DSCP (Differentiated Services Code Point),
  179, 440
DSL (Digital Subscriber Line), 26-27

DSL Forum
  Introduction, 13-14
  TR-069, 186-187, 188, 471
  TR-126, 468
DSLAM (Digital Subscriber Line Access
    Multiplexer)
  Enabled for IGMP snooping, 154
  Implementing proxy functionality, 167
  Overview, 29
DSM-CC (Digital Storage
    Media-Command and Control), 237,
    419
DTCP (Digital Transmission Content
    Protection), 272, 274
DVB (Digital Video Broadcasting)
  DVB-ASI (DVB-Asynchronous Serial
    Interface), 344
  DVB-C (DVB-Cable), 221
  DVB-CI (DVB-Common Interface),
    255
  DVB-HTML (DVB-HyperText Markup
    Language), 239
  DVB-IPI (DVB Technical
    Module Ad Hoc Group on IP
    Infrastructure), 17-18, 169, 394
  DVB-J (DVB-Java), 235-239
  DVB-RCS (DVB Return Channel via
    Satellite), 47
  DVB-S (DVB-Satellite), 218
  DVB-S2 (DVB-Satellite-Second
    Generation), 218-219
  DVB-T (DVB -Terrestrial), 216
  H.264/AVC implementation, 75
  MultiCrypt, 263
  Multi-protocol encapsulation, 238
  Overview, 8-9
  SI (Service Information) tables,
    87-88, 171
  SimulCrypt, 263
DVI (Digital Visual Interface),
    373-374
DVRs (Digital Video Recorders) see IP
    set-top boxes

ECMs see CA systems
EMMs see CA systems
Encoding
  Advantages & disadvantages, 65-66
  Affect on QoE, 443-444

HDTV encoders, 371
  Real time encoders, 119-120
Encryption,
  Defined, 252
  DRM functionality, 269
  IPSec (IP security), 258-259
  Software centric CA system functionality,
    258
EPG (Electronic Program Guide)
  Defined, 388-389
  Technical architecture, 389-396
Error correction, 114
Ethernet, see GigE
ETSI (European Telecommunications
    Standards Institute) 14, 169
EuroDOCSIS, 45-46
EV-DO (Evolution-Data Optimized),
    53

FCC (Federal Communications
    Commission) 6, 7, 264, 275
FEC (Forward error correction), 449-450,
    also see QoE
Fiber Access Networks, see FTTx
Firewall, 183, 193
Flow control, 114
FSAN (Full Service Access Network)
    group, 25
FTTx
  FTTA (Fiber to the apartment), 22
  FTTC (Fiber to the curb), 22
  FTTH (Fiber to the home), 22
  FTTN (Fiber to the neighbourhood),
    21
  FTTRO (Fiber to the regional office),
    21

G.983, see PON
GENA (General Event Notification
    Architecture), 321
GigE (Gigabit Ethernet),
  Ethernet technical architecture, 288-293
  Introduction, 35
  Technical characteristics, 293
Globally Executable MHP (GEM),
    240-241
GOP (Group of pictures)
  GOP's length affect on QoE, 444
  Introduction, 70

H.264/AVC
  Benefits, 75
  Network abstraction layer, 83
  Profiles, 78
  Technical architecture, 76–78
  Video coding layer, 83
HANA (High-Definition Audio–Video
    Alliance)
  About, 324
  IEEE 1394, 324–327
HDCP (High-Bandwidth Digital Content
    Protection), 274, 275
HD–DVD, 373
HDMI (High-Definition Multimedia
    Interface), 374–375
HDTV (High definition TV)
  Classifications, 367–369
  Over IP, 369–370
  Technical architecture, 370–380
Headend, 5
HFC (Hybrid Fiber Coaxial)
  Features 32, 34
  IPTV Topology, 34–35
HGI (Home Gateway Initiative), 187–189
Home network overview, 5
HomePlug AV
  Home powerline networking
    characteristics, 298–301
  Specification overview, 302–305
HPNA (HomePNA)
  HomePNA 3.1 technical characteristics,
    309–311
  Introduction to phoneline networking,
    307–309
HSDPA (High-Speed Downlink Packet
    Access), 53

IEC (International Engineering
    Consortium), 67
IEEE 802.11n, 294–298
IEEE 802.1q, see VLANs
IEEE 802.16, see WiMAX
IEEE 802.16e, see WiMAX
IEEE 802.3, see GigE
IEEE 1394, see HANA
I-Frames (Intra Frames), 69, 162, 167–168,
    453
IGMP (Internet Group Membership
    Protocol)

  Devices, 130
  Proxy functionality, 155, 167
  Snooping, 61, 153–155
  Version 1, 132–133
  Version 2, 134–139
  Version 3, 140–147
IMS (Integrated Multimedia Subsystem), 50
Interactive IPTV applications
  About iTV (interactive TV), 387
  Deployment, 419–420
  Evolution, 382–387
  IPTV application types, 387–416
  Video centric IPTV application types,
    417–419
Internet TV 3–4, 53–56
IP (Internet Protocol)
  Addressing, 107, 109–111
  IP as a backbone technology, 57–58
  IPv4 video packets, 107, 108
  IPv6, 107, 111–113, 155–158
IPDR (Internet Protocol Detail Record
    Organization), 16, 44
IP set-top boxes
  Buffer size, 168
  Characteristics, 200–201
  DVRs and storage, 212–215, 405
  Future trends, 197, 199
  Hardware architecture, 201–212
  HDTV enabled, 372–377
  History, 195–196
  Hybrid cable, 220–222
  Hybrid satellite, 47, 218–220
  Hybrid terrestrial, 216–218
  Overview, 193–195
IPTV (Internet Protocol Television)
  Architecture overview 4–5
  Benefits 1, 2
  Definition 2
  Growth drivers 10–12
  Market data 12–13
  Standardization 13–18
IPTVCDs (IPTV Consumer Devices) 5, 124
IPTVCM (IPTV communications model)
  Data link layer, 113–114
  Defined, 64
  Encoding layer, 82–83
  HomePNA 3.1 compliance, 310
  IP layer, 106–113
  Overview, 81–82

Packetizing layer, 83–84
Physical layer, 114–115
RTP (Optional) layer, 88–95
Transport layer, 95–106
TS (Transport Stream) construction layer,
  85–88
IPTV middleware
  IPTVCD software, 232–247
  Server overview, 122–123
IPTV network management
  Day-to-day operational tasks, 436–437
  Implementing business continuity plans,
    478–481
  Network management system
    functionality, 424–426
  Management work flows, 428–431, 432
  Managing IP addresses, 436
  Managing QoE, 441–468
  Managing QoS, 437–441
  Managing service availability, 434–435
  Remote management of IPTVCDs,
    469–471
  Scheduling software updates, 471–473
  Testing and monitoring, 431, 433–434
  Troubleshooting, 473–478
IPTV Security
  CA systems, 250–265
  DRM systems, 265–282
  Intranet protection, 282–283
  Overview, 121–122, 249–250
  Protecting IP-VoD content, 364
IRD (Integrated Receiver Devices), 119
ISDB (Integrated Services Digital
    Broadcasting-Terrestrial)
  BCAS, 256–257
  ISDB-C, 219
  ISDB-S, 219
  ISDB-T, 217
ISMA (Internet Streaming Media Alliance)
  Channel changing initiative, 173
  DRM industry initiative, 281–282
  Hyperlinked video specification, 402
  Overview 16
ISO (International Organization
    Standardization), 67
ITU (International Telecommunication
    Union)
  HomePNA standardization, 309
  ITU-T FG IPTV, 2, 15, 16

ITU-T J.144, 457
IPTV QoS and QoE metric, 455

KDC (Key distribution center), 193

LCD (Liquid crystal display), 379–380

MAC (Media Access Control)
  Ethernet implementation, 292
  HomePlug AV implementation, 303–304
  IEEE 802.11n implementation, 296
  Introduction, 113
  MAC based VLANs
  MoCA implementation, 313
  UPA-DHS implementation, 306–307
MDI (Media Delivery Index), 457, and also
    see QoE
Media Servers, 224
Metadata
  EPG metadata management, 389–391
  Industry initiatives, 391–394
  VoD metadata management, 346
Metro Ethernet, 58–59
MHP (Multimedia Home Platform)
  Association with MHP, 239–240
  GEM, 240–241
  History, 233–234
  Overview, 233
  Technical architecture, 235–239
MIB (Management Information Base), 192,
    427
MIMO (Multiple-input multiple-output), 297
MoCA (Multimedia over Coax Alliance)
  Inside the MoCA IPTVCM layers,
    313–314
  Specification characteristics, 311–313
Modulators 35
MOS (Mean Opinion Score), 455–456 and
    also see QoE
MPEG (Moving Picture Experts Group)
  About MPEG Compression, 66–67
  Blocks and macroblocks, 68, 77
  Elementary streams, 82–83
  Frame types, 69–70
  Overview 14
  Packetized Elementary Streams, see PES
    packets
  Slices, 69
  Transport Streams, see TS packets

MPEG-1, 67
MPEG-2
 Audio, 206–207
 Decoding, 205–208
 Overview, 67–72
 Profiles and levels, 71
MPEG-4, 72–79
MPEG-4 Part 10, *see* H.264/AVC
MPEG-7, 391–392
MPEG-21, 67, 277
MPLS (Multi Protocol Label Switching),
 57–58, 441
MPQM (Moving Picture Quality Metric),
 456–457 and also *see* QoE
Multicasting
 Any source multicast, 141
 Distribution shared trees, 149–150
 Distribution source trees, 147–149
 DOCSIS 3.0 Deployments 42
 Groups and Addressing, 131–132
 Introduction, 127–129
 IPTV Multicasting 129–130
 MLD (Multicast Listener Discovery),
  156–158
 PIM (Protocol Independent Multicast),
  150–152
 Protocols, 132–147
 RPF (Reverse Path Forwarding), 152–153
 Source specific multicast–141
Municipal wireless networks, 51–53

NAL (Network Abstraction Layer) units
 Introduction, 83
 RTP integration, 91
NAT (Network Address Translation), 181
NCTA (National Cable and Television
  Association), 264
NTP (Network Time Protocol), 123
NTSC (National Television System
  Committee), 6

OBSS (Operational and Business Support
  System), 120–121
OFDM (Orthogonal Frequency Division
  Multiplexing), 28, 49, 313, 303,
  327
OLT (Optical line termination) 22
ONT (Optical network terminal), 22, 23
Open IPTV Forum 14

OSGi (Open Services Gateway Initiative),
 186

PDP (Plasma display panel), 378–379
PES (Packetized Elementary Stream)
  packets, 83–86
P-frames (Predictive frames), 69, 453
PKI (Public Key Infrastructure), 260–261,
  262, 270
PON (Passive optical network)
 BPON (Broadband PON), 24–25
 EPON (Ethernet PON), 25
 GPON (Gigabit PON), 25–26
 Overview, 22–23
PPPoE (Point-to-point protocol over
  Ethernet), 180, 359
PSI (Program specific information) tables,
  87–88, 169
PSNR (Peak signal-to-noise ratio), 456 and
  see QoE
PVCs (Private Virtual Circuits), 178, 179,
  180

QAM (Quadrature amplitude Modulation),
  222, 314
QoE (Quality of Experience)
 Factors that affect QoE, 443–453
 Industry initiatives, 468
 Key performance indicators, 466–467
 Measurement systems, 453–455
 Metrics, 455–460
 Service level agreements, 465–466
 Testing tools and programs 460–465
QoS (Quality of Service)
 CableHome implementation, 192
 Enforcement on an RG, 179
 HomePNA QoS guarantees, 310
 IPTV QoS management, 437–441
 Speeding up channel changes, 169
 WHMN implementation, 328–329
QPSK (Quadrature-Phase-Shift-Keying),
  220
Quantization, 65, 68

RG (Residential Gateway)
 Features, 178–184
 Impact on channel changing times, 161
 Introduction, 175–178
 Standardization, 184–193

Routers,
  Distribution, 123
  Multicast, 130, 139
RTCP (Real-Time Control Protocol),
  349–350, 447–448
RTP (Real-Time Transport Protocol)
  Benefits, 90
  Feedback mechanism, 447–449
  Overview, 88
  Packet structure, 92
  VoD transport protocols, 349–350, 351
RTSP (Real-Time Streaming Protocol),
  350–358

Sampling, 65
SAP (Session Announcement Protocol),
  171
SARFT (State Administration of Radio,
  Film, and Television), 15
SDH (Synchronous Digital Hierarchy),
  57
SDP (Session Description Protocol), 171
SDTV (Standard definition TV), 12, 367
SDV (Switched Digital Video), 35
Secure Video Processor Alliance,
  277–278
SMPTE (Society of Motion Picture and
  Television Engineers), 79
SNMP (Simple Network Management
  Protocol),
  CableHome implementation, 192
  IPTV network management, 426–428
SOAP (Simple Object Access Protocol),
  321
SONET (Synchronous Optical Network),
  35, 57
SSDP (Simple Service Discovery Protocol),
  321
Statistical multiplexing, 72

TCP (Transmission Control Protocol)
  Comparison to UDP, 104–106
  Troubleshooting, 476–478
  Used to route IPTV content, 97–101,
    102
TDMA (Time division multiple access), 40,
  57
Telco TV, see IPTV
TFTP (Trivial File Transfer Protocol), 44

Transcoding, 120
TS Packets, 85, 89, 203–205
TV-Anytime, 392–393
TV Web browsing, see IPTV application
  types

UDP (User Datagram Protocol)
  Comparison to TCP, 104–106
  Benefits and drawbacks, 101, 103–104
  Relationship to RTP, 89
UDI (Unified Display Interface), 377
Unicast, 124, 126, 152
Universal Plug and Play, see DLNA
UPA-DHS (Universal Powerline
  Alliance-Digital Home Standard),
  305–307
UWB (Ultra-wideband), 327, and also see
  HANA

VBR (Variable Bit Rate), 72
VC-1 (Video Codec 1)
  Access units, 95–96
  Characteristics, 79–80
  Profile levels, 80
VDSL (Very High Speed DSL)
  VDSL 1, 31
  VDSL 2, 31
  VDSL (Long Reach), 31
  VDSL (Short Reach), 31
VESA (Video Electronics Standards
  Association), 375 and also see
  DisplayPort
Video analyzers, 462 and also see QoE
VLANs (Virtual LANs)
  801.1Q, 179, 328, 361
  RG Implementation, 178
  Technical overview, 359–363
VoD (Video on Demand)
  Application types, 335–340
  Evolution, 334–335
  Integration with other IP applications,
    358–364
  Overview 10, 396
  PPV, 332–334
  RAID (Redundant Array of Independent
    Disks), 342–343
  Server clustering, 347–348
  Server testing, 464–465
  Streaming servers, 340–349

WHMN (Whole Home Media Networking)
  Middleware standards, 314–328
  Phone and coaxial cable solutions,
    307–314
  Powerline solutions, 298–307
  QoS implementation, 328
  Structured cabling solutions,
    288–294
  Technology requirements, 286
  Wireless solutions, 294–298

WiMAX
  Fixed, 48–50
  Mobile, 50–51
WiMAX Forum, 48–49
WirelessHD Consortium, 15
WLAN (Wireless LAN), *see* IEEE 802.11n

X.509, *see* digital certificates
XML (Extensible Markup Language),
    169–170, 269–270, 279, 346